Molecular Biology *of* Cancer

Molecular Biology
of Cancer

MECHANISMS, TARGETS, AND THERAPEUTICS

FOURTH EDITION

LAUREN PECORINO
UNIVERSITY OF GREENWICH

Great Clarendon Street, Oxford, OX2 6DP,
United Kingdom

Oxford University Press is a department of the University of Oxford.
It furthers the University's objective of excellence in research, scholarship,
and education by publishing worldwide. Oxford is a registered trade mark of
Oxford University Press in the UK and in certain other countries

First edition 2005
Second edition 2008
Third edition 2012

Impression: 1

Published in the United States of America by Oxford University Press
198 Madison Avenue, New York, NY 10016, United States of America

British Library Cataloguing in Publication Data
Data available

Library of Congress Control Number: 2015958815

ISBN 978–0–19–871734–8

Printed in Great Britain by
Bell & Bain Ltd., Glasgow

This book is dedicated to my mentors:

Raffaela and Joseph Pecorino

Professor Frank Erk

Professor Sidney Strickland

Professor Jeremy Brockes

In memory of:

Marie Favia

Xianzhe Li

Mildred Maiello

Kerry O'Neill

■ PREFACE

Molecular Biology of Cancer: Mechanisms, Targets, and Therapeutics is intended for both undergraduate and graduate-level students (including medical students) and employees in the pharmaceutical industry interested in learning about how a normal cell becomes transformed into a cancer cell and how this knowledge has been applied to new diagnostics and therapeutics. In a normal cell, signaling pathways detect and respond to changes in the environment and regulate cellular activities. Cells contain many receptors on their membrane that allow a signal from outside the cell (e.g. growth factors) to be transmitted to the inside of the cell. Signaling pathways are composed of molecules that interact with other molecules, whereby one triggers the next in a sequence, in a way similar to the actions of team members in a relay race. The relay of information may cause a change in cell behavior or in gene expression, and results in a cellular response (e.g. cell growth). Interference in these signal transduction pathways has grave consequences (e.g. unregulated cell growth) and may lead to the transformation of a normal cell into a cancer cell. The identification of the malfunctions of specific pathways involved in carcinogenesis provides scientists with molecular targets that can be used to generate new cancer therapeutics. I have chosen to present the biology of cancer together with a promise for its application towards designing new cancer drugs. Therefore, for most chapters in the text, the first half discusses the cell and molecular biology of a specific hallmark of cancer, and the last half of the chapter discusses *therapeutic strategies*. To help form a link between particular molecular targets discussed in the first half of the chapter and the therapeutic strategies discussed in the last half of the chapter, a target symbol (◎) is shown in the margin. I hope that this presentation stimulates interest and motivates learning of the subject matter.

Several areas of cancer biology have seen enormous growth over recent years and have been assigned their own chapter in this fourth edition. These include separate chapters for Metastasis, Angiogenesis, Immunology and Immunotherapy, Infectious Agents and Inflammation, and Technology and Drug, and Diagnostics Development. New topics have been added. To name a few, these include immune checkpoints, studying gene function by CRISPR-Cas9, mutational signatures (kataegis) caused by APOBECs, newly proposed mechanisms for the role of obesity in cancer, non-coding RNAs, and the role of exosomes in intercellular communication. Most exciting is the numerous additions of newly approved therapeutics to many chapters.

Personally, I believe that the use of diagrams and illustrations is an extremely powerful tool of learning. A picture paints a thousand words . . . and more. I strongly suggest that the reader studies and enjoys the figures, artistically created by Joseph Pecorino. Major points and new cancer therapeutics are illustrated in red, and the target symbol (◎) is used to identify molecular targets. Detailed descriptions of the figures are found in the body of the text. This edition uses full color and includes data presented in primary research papers.

Several features that are used throughout the text to facilitate learning and interest are described below:

Pause and think

These features are often presented in the margins of the text and are designed to engage the reader in thought and to present additional perspectives of core concepts. Many times questions are posed; sometimes they are answered, and other times they encourage the rereading of particular sections of text.

How do we know that?

These features examine experimental evidence from the scientific literature, and ask the reader to analyze raw data or understand the details of an experimental protocol.

Special interest boxes

Shaded boxes are used to highlight special topics of interest such as the box entitled "Skin cancer" in Chapter 2. They are also used to provide additional explanation of more complex subjects such as "A little lesson about ROS . . . " in Chapter 2 and "A little lesson about the MAP kinase family . . . " in Chapter 4.

Lifestyle tips

These are suggestions about lifestyle choices and habits to minimize cancer risk, based on our current knowledge.

Pioneers in the field of . . .

Scientists around the world have made contributions to the concepts presented in this text. This feature highlights a small selection of scientists who were pioneers in a specialized topic and their major contributions that have had a major impact in our understanding of cancer biology. This feature is meant to give a human touch to the text but, of course, is not comprehensive by any means. It may also be used as a tool for professional use to follow a continuing interest in the research literature.

Analysis of . . .

Specific molecular techniques used to analyze particular biological and cellular events are described. It is important that science and medical professionals ask themselves "How do we know that?". Each of the major concepts underlying our current state of knowledge is the result of numerous experiments that generate data, suggesting possible explanations and mechanisms of cellular events. The information retrieved is governed by the techniques that are used for analysis.

Chapter highlights—refresh your memory

Summary points are listed in order to consolidate major concepts and provide a brief overview of the chapter. These may be particularly useful for revising for examinations.

Self-tests and activities

Several features are included to strengthen your understanding of particular concepts presented in the chapter. Self-tests presented within the text ask you to immediately reinforce material just presented and often refer to a figure. This causes a break from reading and engages you, the student, in "active" learning. Activities that are aimed at strengthening your understanding of particular concepts and encouraging additional self-centered learning are presented towards the end of a chapter. Some require web-based research, while others are more reflective. Multiple choice questions can be found on the companion web site.

Further reading is a list of general references found at the end of each chapter. These references consist mostly of reviews and support the contents of the chapter. They are not referenced in the main body of the text.

Selected special topics mainly lists specific primary research papers that *are* referenced in the main body of the text and may be pursued for further interest. Several relevant web sites are also included.

Appendix 1 is a summary diagram that links key molecular pathways to the cell cycle.

Glossary

Over 170 entries are defined in a clear and concise manner in order to provide students with a handy reference point for finding explanations of unfamiliar words.

It is my hope that the readers of this text will learn something new, become interested in something molecular, and ultimately, somehow, contribute to the field of cancer biology. This field is evolving at a tremendous rate, and so. by the time of printing. some information contained within these pages may need to be updated! This does not concern me, because my aim is to present a *process* of how the pieces of science are put together and how we may attempt to apply our knowledge to cancer therapies. Many new drugs will fail, but a select few will not. These select few will make marked improvements in the quality of life for many.

New to this edition

- Use of full color and the addition of primary data from the literature.

- Updates of immunology and immunotherapies.

- New topics covered include immune checkpoints, studying gene function by CRISPR-Cas9, mutational signatures (kataegis) caused by APOBECs, non-coding RNAs, newly proposed mechanisms for the role of obesity in cancer, and the role of exosomes in intercellular communication, to name a few.

- Additional "How do we know that?" features.

■ ACKNOWLEDGMENTS

First, I would like to express my deepest gratitude to Jonathan Crowe, previously Commissioning Editor and now Editor in Chief (Natural and Social Sciences), at Oxford University Press (OUP). I am indebted to him for his faith that I could turn a one-page proposal into a complete textbook for the first edition and for his continued support during the writing of the following editions. He nurtured the synthesis of the book with special care and provided a wealth of helpful suggestions and advice. With love, I thank my father, Joseph Pecorino, for his never-ending encouragement, and I acknowledge his artistic talent used to translate dozens of my stick drawings into precise illustrations for the book during our visits across the Atlantic Ocean over the years. Stephen Crumly kindly reproduced the illustrations using his fine skills in computer graphics under tight deadlines. Jessica White (Publishing Editor, OUP) has been instrumental in editing and handling of this fourth edition. Thanks to Jennifer Rogers and Sian Jenkins, Production Editors, and the production staff, including Joyce Cheung, Copy-Editor, and Sarah Broadley for additional assistance at Oxford University Press. I also thank Karen Roberts for her careful proofreading of the book.

Kind appreciation is expressed for the precise and critical comments given by my official reviewers for the first edition: Tony Bradshaw, Oxford Brookes University, UK; Moira Galway, St Francis Xavier University, Canada; Maria Jackson, University of Glasgow, UK; Helen James, University of East Anglia (UEA), UK; and Ian Judson, Cancer Research UK, London, UK. The value added to the text by these scientists cannot be underestimated. Their comments have had tremendous impact and provided a foundation for the following editions. Special thanks is expressed for the critical comments given by my official reviewers for the second and third edition—Michael Carty, NUI Galway, Irish Republic; from the UK: Joanna Wilson, Glasgow, Jonathan Bard, Edinburgh, Phillipa Darbre, Reading, Stephanie McKeown, Ulster, Penka Nikolova, KCL, Elana Klenova, Essex; from the USA: Annemarie Bettica, Manhattanville College, Nancy Bachman, Oneonta, and James Olesen, Ball State University; Roger Barraclough, University of Liverpool; Helen Coley, University of Surrey; Meg Duroux, Aalborg University; Paula M. Checchi, University of California, Davis; Jill Johnston, The University of Sydney; Tapas K. Sengupta, Indian Institute of Science Education and Research (IISER), Kolkata; Paola Marignani, Dalhousie University; Andrew Sharrocks, University of Manchester; and Alicia F. Paulson, University of South Dakota. Kind appreciation is expressed to my most recent set of reviewers: Sarah Allinson, Lancaster University, Byram W. Bridle, University of Guelph,

Dawn Coverley, University of York, Christine Edmead, University of Bath, and Helen James, University of East Anglia.

I thank Rita Canipari, Ken Douglas, Dario Tuccinardi, Ricky Rickles, Anne Schuind, Dylan Edwards, UEA, Nicole Bournias-Vardiabasis, University of Californa at San Bernardino, and Young-Joon Surh, Seoul National University, and former students Sarah Thurston, Stephen O'Grady, and Kenny Prat for unofficial critical comments and suggestions to previous editions. Sincere thanks to Tatiana Chirstides, University of Greenwich, David Lyden, Weil Cornell, and Tim Fenton, University College London for critical reading of chapters in this fourth edition.

Many improvements in the fourth edition are a result of casual feedback from many people from different places—so many thanks to all of you and apologies to those I have not named. I thank my colleagues at the University of Greenwich for suggestions and communicating cancer news. Special thanks also to Samer El-Daher, the Head of Department, and Martin Snowden, Pro-Vice-Chancellor of the Faculty of Engineering and Science at the University of Greenwich, for their support of this project. I am thankful for suggestions from my Cancer Biology and Therapeutics class 2014 and 2015, especially Robyn Holden for ideas for figures.

I acknowledge the financial support of the Faculty of Engineering and Science, University of Greenwich, to attend the American Association of Cancer Research (AACR) Annual Meetings over the last few years. The information gained and contacts made were important resources for this edition. Appreciation for their kind gestures of support of this textbook is given to Samuel Adunyah, Meharry Medical College; David Axelrod, Rutgers; Michael Caligiuri, Ohio State University; Jules Harris, University of Arizona; and Candace Ritchie, Merck, who I met at these conferences.

Many fellow scientists have made suggestions or other contributions to the previous editions of this book, including: Jeremy Griggs, several members of the Kuriyan Laboratory, David Lyden, Weil Cornell Medical College, Gerd Pfeifer, Mariann Rand-Weaver, and Jerry Shay. Special thanks are expressed to Andrea Cossarizza, Sarah Cowan, Xiuhuai Liu, M.-A. Shibata, and Kelly Dobben-Annis for helping me to obtain electronic figures.

The Royal Society of Medicine Library, London, provided an ideal scientific sanctuary for the writing of this edition, and I express my thanks to the staff.

I am especially grateful to the support that came from my family, especially Raffaela Pecorino and the late and greatly missed Teresa Rapillo and from friends. I am very grateful for the tremendous support given by Marcus Gibson.

I admire and acknowledge the work of all those scientists whose research efforts have contributed to the field of cancer research.

OUTLINE CONTENTS

DETAILED CONTENTS

Chapter 1

Introduction

Introduction

The aim of this text is to provide a foundation in the molecular biology of cancer and to demonstrate the conceptual process that is being pursued in order to design more specific cancer drugs. Common threads are woven throughout the different chapters so that the terminology becomes familiar and the mechanisms of cellular processes become clear. The text also provides guidance for everyday decisions that may lead to a decrease in cancer risk. The translation of the knowledge of molecular pathways into clinically important therapies (linked throughout the text by the target symbol, ◎) will be communicated and will breathe excitement into learning. Academically, you will gain a foundation in the cell and molecular biology of cancer. More importantly, you will develop an intellectual framework upon which you can add new discoveries that will interest you throughout your lifetime. My goal in writing this book is to inspire. It would be most gratifying for me if, by reading this book, you, the reader, will be compelled to contribute to the cancer research field directly. Knowledge is powerful.

Cancer statistics are shocking. One in three people is likely to suffer from cancer during their lifetime. Siegel *et al.* (2015) estimated that 589,430 Americans would die from cancer in 2015, and the mortality rate (number of cancer deaths per year per 100,000 people) was 199 for men in the UK in 2012 (Mortality in the UK, September 2014, Cancer Research, UK). The worldwide **incidence** (number of new cases) was estimated to be about 14.1 million cases in 2012 (http://globocan.iarc.fr/Pages/fact_sheets_cancer.aspx). These numbers are cold, stark, and impersonal. Hidden behind them are tears, fears, pain, and loss. No one is excluded from the risk. There is a need to understand the disease and to translate our knowledge into effective therapies. In order to understand the process of **carcinogenesis**, whereby a normal cell is transformed into a cancer cell, we must know the intricacies of cell function and the molecular pathways that underlie it. We must consider the cell in the context of the entire body. We have a lot to learn! However, knowledge of the molecular details in important cellular and biochemical pathways can be applied to a new wave of cancer therapies. What better reward for these efforts?

1.1 What is cancer?

Cancer is a group of diseases characterized by unregulated cell growth and the invasion and spread of cells from the site of origin, or primary site, to other sites in the body. Several points within this definition need to be emphasized. First, cancer is considered to be a group of diseases. Over 100 types of cancer have been classified. The tissue of origin gives the distinguishing characteristics of the cancer. Approximately 85% of cancers occur in epithelial cells and are classified as carcinomas. Cancers derived from mesoderm cells (e.g. bone, muscle) are called sarcomas, and cancers of glandular tissue (e.g. breast) are called adenocarcinomas. Cancers of different origins have distinct features. For example, skin cancer has many characteristics that differ from lung cancer. The major factor that causes cancer in each target tissue is different: ultraviolet (UV) radiation from the sun can easily target the skin, while inhalation of cigarette smoke can target the lungs. In addition, as will be examined in detail later, there are differences in the molecular mechanisms involved in carcinogenesis within each cell type and the pattern of spread of cells from the primary site. Treatment must be applied differently. Surgical removal of a cancerous growth is more amenable for the skin than the lungs. This initial view presents layers of complexity which may seem insurmountable to dissect in order to improve the conventional therapeutic approaches. However, even though the underlying cellular and molecular routes may be different, the end result is the same. In the year 2000, upon fine analysis, Hanahan and Weinberg defined six hallmarks of most, if not all, cancers. They proposed that acquiring the capability for autonomous growth signals, evasion of growth inhibitory signals, evasion of apoptotic cell death, unlimited replicative potential, angiogenesis (the formation of new blood vessels), and invasion and metastasis are essential for carcinogenesis. More recently, they have modified their concept to include two enabling characteristics, genome instability and tumor-promoting inflammation, that are crucial for acquiring the six hallmarks of cancer, and they highlight two emerging hallmarks, reprogramming energy metabolism and avoiding immune destruction (Hanahan and Weinberg, 2011). The last two processes are considered as emerging hallmarks because their relationship to the established six hallmarks mentioned previously requires further research, although there is evidence of their importance in carcinogenesis. Established hallmarks and emerging hallmarks, along with enabling characteristics (Figure 1.1; also see Box "The hallmarks of cancer"), will be examined in detail in this text, and each is a potential target pathway for the design of new therapeutics.

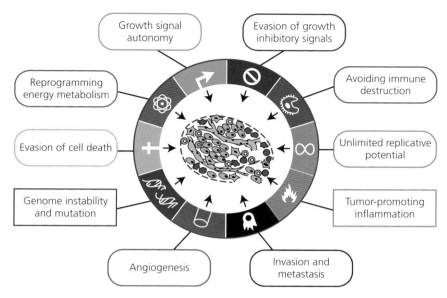

Figure 1.1 The hallmarks of cancer. Reprinted and modified from Hanahan, D. and Weinberg, R.A. (2011) Hallmarks of cancer: the next generation. *Cell* **100**, p. 646, copyright (2011), with permission from Elsevier Science.

The hallmarks of cancer

See Figure 1.1.

- Growth signal autonomy:
  - Normal cells need external signals from growth factors to divide
  - Cancer cells are not dependent on normal growth factor signaling
  - Acquired **mutations** short-circuit growth factor pathways leading to unregulated growth.
- Evasion of growth inhibitory signals:
  - Normal cells respond to inhibitory signals to maintain homeostasis (most cells of the body are not actively dividing)
  - Cancer cells do not respond to growth inhibitory signals
  - Acquired mutations or gene silencing interfere with the inhibitory pathways.
- Avoiding immune destruction (emerging hallmark):
  - There is evidence to support the theory of immune surveillance that states the immune system can recognize and eliminate cancer cells
  - Successful cancer cells may be those that do not stimulate an immune response or can interfere with the immune response so as to avoid immune destruction.
- Unlimited replicative potential:
  - Normal cells have an autonomous counting device to define a finite number of cell doublings after which they become senescent. This cellular counting device ➜

→ is the shortening of chromosomal ends, **telomeres**, that occurs during every round of DNA replication
  - Cancer cells maintain the length of their telomeres
  - Altered regulation of telomere maintenance results in unlimited replicative potential.
- Tumor-promoting inflammation (an enabling characteristic):
  - Virtually all **tumors** contain inflammatory immune cells
  - Inflammation is an immune response that can facilitate the ability of acquiring the core hallmarks of cancer. For example, inflammatory cells can provide growth factors and enzymes that promote angiogenesis and invasion
  - In addition, inflammatory cells can release oxygen species that are mutagenic.
- Invasion and metastasis:
  - Normal cells maintain their location in the body and generally do not migrate
  - The movement of cancer cells to other parts of the body is a major cause of cancer deaths
  - Alterations of the genome may affect the activity and/or levels of enzymes involved in invasion or molecules involved in cell–cell or cellular–extracellular adhesion.
- Angiogenesis (formation of new blood vessels):
  - Normal cells depend on blood vessels to supply oxygen and nutrients, but the vascular architecture is more or less constant in the adult
  - Cancer cells induce angiogenesis, the growth of new blood vessels, needed for tumor survival and expansion
  - Altering the balance between angiogenic inducers and inhibitors can activate the angiogenic switch.
- Genome instability and mutation (an enabling characteristic):
  - Acquiring the core hallmarks of cancer usually depends on genomic alterations
  - Faulty DNA repair pathways can contribute to genomic instability.
- Evasion of cell death:
  - Normal cells are removed by **apoptosis**, often in response to DNA damage
  - Cancer cells evade apoptotic signals.
- Reprogramming energy metabolism (emerging hallmark):
  - Uncontrolled cell division demands increases in fuel and biosynthetic precursors that is obtained by adjusting energy metabolism
  - Unlike normal cells, cancer cells carry out glycolysis, even in the presence of oxygen. Glycolysis intermediates can be used in biosynthetic pathways.

PAUSE AND THINK

Why are malignant tumors life-threatening? They are physical obstructions, and, as they invade other organs, they compromise function. They also compete fiercely with healthy tissues for nutrients and oxygen.

Cancer is characterized by unregulated cell growth and the invasion and spread of cells from their site of origin. This leads to the distinction between a **benign** tumor and a **malignant** tumor. A benign tumor is not evidence of cancer. Benign tumors do not spread throughout the body (that is, they do not metastasize), although some can be life-threatening because of their location (e.g. a benign brain tumor that may be difficult to remove). Malignant tumors, on the other hand, do not remain encapsulated, show features of **invasion**, and metastasize.

> **Cancer cells can be distinguished from normal cells in cell culture conditions**
>
> Normally, cells grow as a single layer, known as a monolayer, in a Petri dish due to a property called contact inhibition; contact with neighboring cells inhibits growth.
>
> Transformed cells (cells that have become cancer cells) acquire the following **phenotypes**:
>
> - They fail to exhibit contact inhibition and instead grow as piles of cells or "foci" against a monolayer of normal cells
> - They can grow in conditions of low serum
> - They adopt a round **morphology** rather than a flat and extended one
> - They are able to grow without attaching to a substrate (e.g. the surface of a Petri dish), exhibiting "anchorage independence".

1.2 Evidence suggests that cancer is a disease of the genome at the cellular level

Interestingly, most agents that cause cancer (carcinogens) are agents that cause alterations to the DNA sequence called mutations (mutagens). Thus, similarly to all genetic diseases, cancer results from alterations in DNA. A large amount of evidence indicates that the DNA of tumor cells contains many alterations ranging from subtle point mutations (changes in a single base pair) to large chromosomal aberrations, such as deletions and chromosomal translocations. The accumulation of mutations in cells over time represents a multi-step process that underlies carcinogenesis. The requirement for an accumulation of mutations over time explains why there is an increased risk of cancer with age and why cancer has become more prevalent over the centuries as human lifespan has increased. There have been more cases of cancer in recent years because we are living longer. World life expectancy more than doubled over the past two centuries, from roughly 25 years to about 65 for men and 70 for women, and some populations, such as Japanese women, currently enjoy life expectancies of almost 85 years (data within Oeppen and Vaupel, 2002). The longer we live, the more time there is for our DNA to accumulate mutations which may lead to cancer. There is evidence, however, that, in a small percentage of tumors, a single catastrophic event in a cell can lead immediately to many mutations and cause cancer (described further in Chapter 2). Interestingly, only 5–10% of the mutations observed are thought to be directly involved in causing cancer, based upon mathematical modeling. This estimate provides the basis for the current optimism in the field of molecular therapies. Almost all of the mutations identified in tumor cells are somatic mutations whereby the DNA of a somatic (body) cell has been damaged. These mutations are not passed on to the next generation of offspring, and therefore cannot be inherited, but they are passed to daughter cells after cell division. Only alterations in the DNA of sperm or egg cells, called germline mutations, are passed on to offspring. Some germline mutations can cause an increased risk of developing cancer but are rarely involved in causing cancer immediately.

Inheritable modifications of the genome and chromatin structure also play a role in carcinogenesis (discussed in Chapter 3). Thus, cancer is considered to be a disease of the genome at the cellular level. But our understanding is evolving, and we now appreciate that cancer is not just a cell-autonomous process but is also dependent on interactions between tumor cells, with the tumor microenvironment, and with long-distance systemic signaling. All cells in a primary tumor arise from a single cell that contains an accumulation of initiating mutations; in other words, the initiation of the development of cancer is clonal. It is generally assumed that only one of the 10^{14} cells in the body needs to be transformed in order to create a tumor. However, cancer cells continue to change their behavior as they progress. The progressive changes of a cell, resulting from an accumulation of additional mutations that confer a growth advantage over its neighbors, proceed in a fashion analogous to Darwinian evolution: chance events give rise to mutations that confer changes in phenotype and allow adaptation to the environment, resulting in subclonal selection and an advantage in fitness. Thus, clonal evolution explains the heterogeneous population of cells within a tumor; primary tumors become composed of subclones. This classifies the mechanism of cancer as a complex Darwinian adaptive system that has been validated by recent studies of the cancer genome (Greaves and Maley, 2012). The accumulation of mutations occurs only after the cell's defense mechanisms (e.g. DNA repair) have been evaded. Any alterations of DNA that are not repaired before the next cell division are passed on to the daughter cells and are perpetuated. The cell relies on several processes to repair damaged DNA. In cases of severe DNA damage, cell suicide is induced in order to protect the whole body from cell transformation. The molecular details of DNA repair and cell suicide and the mutations that compromise them will be described in Chapters 2 and 7. Thus, many mechanisms exist for blocking carcinogenic events, but over-burdening the system increases the probability that a cell carrying a deleterious mutation will escape surveillance.

Growth, apoptosis, and differentiation regulate cell numbers

There are three important processes that contribute to the overall net cell number in an individual. First, cell proliferation (cell division, cell growth) is the most obvious. Cell division results in two daughter cells. Second, the elimination of cells by programmed cell death also affects the net cell number. Last, during the process of differentiation, cells can enter an inactive phase of cell growth, and thus differentiation can affect net cell numbers. DNA mutations that alter the function of normal genes involved in growth, apoptosis, or differentiation can affect the balance of cell numbers in the

(a)   (b)   (c)

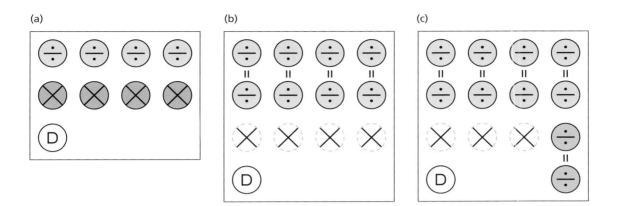

body and lead to unregulated growth. Examine the simplistic model shown in Figure 1.2. If four of the nine cells shown in Figure 1.2(a) divide, and four are programmed to die by apoptosis, and one differentiates (so the cell neither dies nor divides), the cell number will remain the same (Figure 1.2(b); remaining cells shown in color). However, if apoptosis is blocked in one cell and that cell divides instead (red), the total number of cells will increase to 11 (Figure 1.2(c)). Similarly, if differentiation of a stem or precursor cell is blocked and that cell divides, as is the case in some leukemias, the number of cells will also increase. The role of differentiation and stem cells in cancer will be discussed in Chapter 8. Thus, an alteration in the processes of growth, apoptosis, or differentiation can alter cell numbers. Normal genes that can be activated by mutation to be oncogenic are called proto-oncogenes. Proto-oncogenes play functional roles in normal cells. The term reminds us that all normal cells have genes that have the potential to become oncogenic.

Figure 1.2 Growth, apoptosis, and differentiation affect cell number (see text for explanation).

PAUSE AND THINK

So, is a mutation in the hemoglobin gene likely to cause cancer? No, because the function of hemoglobin does not affect cell growth, differentiation, or death and does not lead to unregulated growth of blood cells. The hemoglobin gene is not a proto-oncogene.

Oncogenes and tumor suppressor genes

Growth is regulated by both positive and negative molecular factors. Thus, to increase growth, enhancement of positive factors or depletion of negative factors is required. Note these factors are products of genes, and all cells, except the egg and sperm cells, contain two alleles of every gene. Inheritance over cell generations follows the Laws of Mendelian genetics and includes the concepts of dominant and recessive alleles; that is, if the genotype is heterozygous, one allele may be dominant over the other and is the one that dictates the phenotype (e.g. remember the brown eye allele is dominant over the blue eye allele to produce brown eyes), and two copies of a recessive allele are needed for the recessive allele to be exhibited in the phenotype (e.g. two alleles for blue eyes produce blue eyes).

There are two major types of mutated genes that contribute to carcinogenesis: oncogenes and tumor suppressor genes (Figure 1.3). A general description of an oncogene is a gene mutated such that its protein product

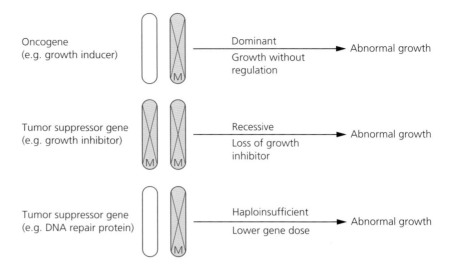

Oncogene
(e.g. growth inducer)

Dominant

Growth without
regulation

→ Abnormal growth

Tumor suppressor gene
(e.g. growth inhibitor)

Recessive

Loss of growth
inhibitor

→ Abnormal growth

Tumor suppressor gene
(e.g. DNA repair protein)

Haploinsufficient

Lower gene dose

→ Abnormal growth

Figure 1.3 Oncogenes and tumor suppressor genes.

is produced in higher quantities or has increased activity and therefore acts in a dominant manner to initiate tumor formation. "Dominant" refers to the characteristic that a mutation in only one allele is sufficient for an effect. For example, one oncogene produces increased quantities of a specific growth factor (e.g. platelet-derived growth factor) which stimulates growth inappropriately. Another example is an oncogene that produces a growth factor **receptor** with increased activity because it has been altered, so that it is always in the "on" state and does not require growth factor to transduce a signal into the cell.

Identification of oncogenes by cell transformation assays

The prototypical experiment used to demonstrate the presence of an oncogene is to test for cell transformation in culture. The DNA of interest is isolated and introduced into a standard cell line called NIH/3T3 (mouse fibroblast cells) by calcium phosphate precipitation or electroporation. If the test DNA contains an oncogene, foci (mentioned earlier) will form and will be easily identifiable against a monolayer of untransformed NIH/3T3 cells.

Tumor suppressor genes code for proteins that play a role in inhibiting both growth and tumor formation. Loss of growth inhibition occurs when mutations cause a loss of function of these genes. Consequently, growth is permitted. Tumor suppressor mutations are mainly **recessive** in nature because one intact allele is usually sufficient to inhibit growth; thus both alleles of the gene must be mutated before the loss of function is actually seen phenotypically. Recessive mutations support Knudson's two-hit hypothesis, the classical model used to explain the mechanism behind tumor suppressor action (see Chapter 6 and Figure 6.2). It states that both alleles need to be mutated (recessive) to trigger carcinogenesis. This model has been used to explain the mechanism behind conditions that predispose individuals to an increased risk of cancer. Patients inherit one mutated tumor suppressor allele and may acquire a second somatic mutation over time. Therefore, these patients have a "head start" towards a cancer pheno-type in the race for accumulation of mutations. Recent evidence suggests there is an alternative mechanism for particular tumor suppressor genes, called haploinsufficiency, whereby only one mutated allele can lead to the cancer phenotype. As the term suggests, one normal allele produces half ("haplo-") of the quantity of protein product produced by normal cells, and this is not enough to suppress tumor formation in these cases. This has been demonstrated for genes that regulate DNA repair and the DNA-damage response, such that reduced activity leads to genetic instability. Gene dosage may also affect the spectrum of tumors observed; haploinsuf-ficiency may cause cancer in some cell types, and recessive mutations may cause cancer in other cell types (Fodde and Smits, 2002). Other exceptions to Knudson's two-hit hypothesis will be discussed in Chapter 6.

Studies of adult stem cells have made recent contributions to our understanding of carcinogenesis. Stem cells are undifferentiated cells that have the ability to self-renew and produce differentiated progeny. Normal stem cells may be a main starting point for carcinogenesis in some cancers, as both cancer cells and stem cells utilize, and rely on, self-renewal molecular programs. Also, cancer is more likely to develop in cells that are actively proliferating, as there is a greater chance for muta-tions to accumulate; normal stem cells continue to proliferate over long periods of time. These concepts will be discussed further in Chapter 8.

The concepts described in this section suggest that cancer is a disease of the genome at the cellular level.

 HOW DO WE KNOW THAT?

Types of evidence

Like all science, cancer biology depends on evidence. Gilbert's textbook, *Developmental Biology*, classifies evidence into three types: *correlative evidence*, *loss-of-function evidence*, and *gain-of-function evidence*.

Correlative evidence ("show it" evidence) documents obser-vations between two events and weakly intimates that one may cause the other. For example, a gene from a tumor sample has a mutation, compared with the same gene isolated from healthy tissue. This type of evidence provides a good starting point but is not particularly strong evidence, and may even be coincidental.

Loss-of-function evidence ("block it" evidence) uses differ-ent techniques to inhibit the function of a gene, gene product, or other factor of interest. Antibodies that block protein func-tion and knock-out mice are common experimental techniques used to investigate loss of function. Appropriate controls must be in place to ensure that only the target is affected.

Gain-of-function evidence ("move it" evidence) is the strongest type of evidence and is obtained when your factor of interest is moved to a new location and triggers a causative event at a time or place where it normally does not occur. This is strong evidence. Recombinant DNA plasmids may be con-structed whereby the coding region of a gene of interest is ➜

➔ placed under the control of a promoter that directs expression of the gene in a different tissue or at a different time. DNA transfection of cells in culture and the production of transgenic animals with such recombinant plasmids are important experimental techniques used to demonstrate gain of function.

As you read the scientific literature, try to classify the evidence presented into "show it," "block it," or "move it" types in order to develop your ability to critically analyze the data (Adams, 2003). Overall, the field of cancer biology relies on the sum of many types of experimental techniques that must be critically evaluated.

1.3 Influential factors in human carcinogenesis

Some researchers suggest that more than half of cancers can be prevented by applying our current knowledge (Colditz *et al.* 2012). Environment, reproductive life, diet and exercise, and smoking are four factors that play an important role in carcinogenesis. Exposure to carcinogens, hormonal modifications influenced by childbirth, birth control and hormone replacement therapies, and exposure to viruses underlie these lifestyle factors. Changes in behavior clearly affect cancer incidence, as we will see in the following sections. Epidemiology, the study of disease in the population, has been instrumental in elucidating the contributions of these factors towards different cancers. Although molecular details will be discussed in later chapters, a brief introduction of each factor is given in the following sections.

Environment

Observations by a British surgeon in 1775 resulted in the first correlation between an environmental agent and specific cancers. Percival Pott concluded that the high incidence of nasal and scrotal cancer in chimney sweeps was due to chronic exposure to soot. Not only where you work, but also the choice of where you relax can contribute to your risk of cancer. Unprotected exposure to the sun exposes your skin to UVB radiation which can directly alter your DNA by forming pyrimidine dimers and cause mutations. Sun blocks that have UV-absorbing ingredients have been developed to protect your skin from UV radiation and are a good defense if you do decide to relax in the sun.

Reproductive life

Another early observation was that nuns are more likely to develop breast cancer than other women. We now know that having children reduces breast cancer risk for women, compared with not having children. The age of a woman at the time of giving birth for the first time and the age of a woman at the initiation and termination of her menstrual cycles also influence cancer risk. Hormonal contraception and fertility treatments also affect cancer risk because they alter a women's ovulation schedule (active ingredients prevent and promote ovulation, respectively). Breast cancer incidence rates in women in the USA have been relatively stable since 2004, after decreasing 7% from 2002 to 2003, primarily due to a reduction in use of hormone replacement therapy. Sexual promiscuity can also contribute to increased risk of cancer. Sexually transmitted human papillomaviruses can be found in all cervical cancers worldwide. It is not surprising, therefore, that nuns have a low incidence of cervical cancer. Barrier methods of contraception and vaccination can help protect against this infectious pathogen. Kaposi's sarcoma is a predominant cancer in Kenyan males, which correlates with the AIDS epidemic in Africa. The causative virus of Kaposi's sarcoma, human herpesvirus type 8, generally requires immunosuppressive conditions created by HIV infection.

It is interesting to note that timing of diagnosis of a specific cancer can vary among different geographical locations. Song *et al.* (2014) reported that the age of diagnosis of breast cancer in China is almost 10 years earlier than in the USA and the European Union, suggesting genetics and lifestyle factors underlie the difference. Such data can inform recommendations for when Chinese women should begin breast screening.

Diet and exercise

The incidence of a specific cancer varies greatly between different populations in different geographical locations. Observation of immigration patterns has revealed that local cancer rates strongly influence cancer risk, with diet being one of the most influential factors. Figure 1.4 shows a comparison of cancer incidence between US males and Japanese males. Stomach cancer is a predominant cancer in the Japanese population and a minor cancer in the population of the USA. Interestingly, the

Figure 1.4 Leading new cases of cancers differ among different populations. The percentage of new cases of a specific cancer is reflected in the degree of red shading. Data from Globocan 2012, IARC (Ferlay *et al.*, 2015).

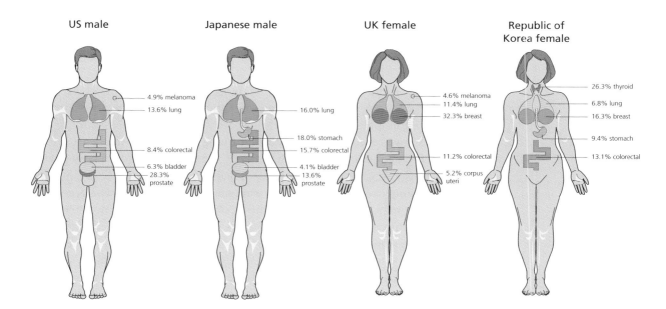

risk of stomach cancer in Japanese people who have migrated to the USA decreases only if they adopt the American diet, but not if they retain a Japanese diet. The Mediterranean diet, which is rich in fresh fruit and vegetables, has been promoted to be beneficial in reducing cancer risks. Recently, studies of the molecular interactions of individual dietary constituents (e.g. polyphenols, carotenoids, and allium compounds) with cellular signaling pathways have begun, and some will be examined in Chapter 11. Modern life has removed physical activity from our daily routines, both in children and adults. The risk of breast cancer can be reduced by at least 25% by more physical activity in women from the onset of menarche throughout the premenopausal years (Colditz *et al.* 2012).

Alcohol

Alcohol was classified as a carcinogen by the International Agency for Research on Cancer in 2007. There is convincing evidence that chronic alcohol drinking increases the risk of cancer of the mouth, esophagus, and breast, and probable evidence for increased risk of liver cancer. Chronic alcohol drinking accounts for 389,000 cases of cancer worldwide. Alcohol and smoking have a synergistic effect, meaning that the risk of cancer from both alcohol and smoking combined is greater than the sum of each risk alone (think of 2 + 2 = 5). Current guidelines recommend a maximum daily alcohol intake of 28 g (approximately a quarter of a bottle of wine) in men and half this amount in women. Molecular mechanisms involved in alcohol-mediated carcinogenesis are discussed in Chapter 11.

Smoking

The clearest example of lifestyle factors underlying a specific cancer is the discovery that smoking causes lung cancer (it is also implicated in pancreatic, bladder, kidney, mouth, stomach, and liver cancer). Since 1985, lung cancer has remained the main cancer worldwide. Smoking accounts for 40% of all cancer deaths: 1.18 million deaths. At least 81 carcinogens have been identified in cigarette smoke. Smoking became particularly fashionable in Europe and the USA during World War I and World War II, and resulted in an epidemic of lung carcinoma. After vast public education campaigns and a subsequent reduction of smoking, lung cancer death rates have fallen dramatically in the USA (Figure 1.5). Unfortunately, lung cancer rates are still rising in other parts of the world, such as China. It may be a good idea for these countries to consider implementing restrictions on smoking in public places and imposing tobacco taxes. It is clear that some future cancer deaths can be avoided by changes in lifestyle factors; tobacco, obesity, and physical inactivity are causes that could be eliminated. Effective change needs to be supported by social commitment. Skepticism that we can affect cancer incidence must be replaced by action, and action must be instilled in individuals earlier, rather than later, in life.

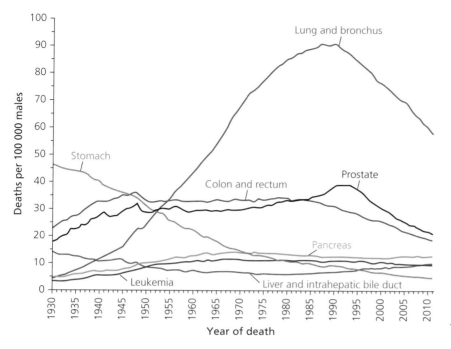

Figure 1.5 Annual age-adjusted cancer death rates among males for selected cancers, USA, 1930–2011. From Siegel, R. *et al.* (2015) Cancer statistics, 2015. *CA Cancer J. Clin.* **65**: 5–29, with permission from John Wiley and Sons.

Additional influences

In addition to lifestyle factors, there are risk factors inherent in our own physiology. By-products of our metabolism and errors that occur during DNA replication contribute to carcinogenesis. Aerobic metabolism produces by-products of oxygen radicals that are mutagenic. Several inherited metabolic diseases also produce mutagenic by-products. For example, tyrosinemia type I patients have a defect in the enzyme fumarylacetoacetate hydrolase which is involved in tyrosine breakdown. As a result of this block, the by-products fumarylacetoacetate and maleylacetate accumulate and increase the risk of cancer, owing to their ability to covalently modify DNA and cause mutations. During DNA replication and repair, polymerases can introduce mutations directly in DNA because of their associated error rates. There is a perpetual inherent risk of mutation during the lifespan of a cell by the nature of cell processes.

1.4 Principles of conventional cancer therapies

The earliest therapeutic strategy used against cancer was to remove surgically as much of the cancer as possible. Obviously this is relatively easy in some types of cancer and impossible in other types. It is not a precise procedure at the cellular level and does not address the question of cells that have spread from the primary site (metastasized cells). Therefore,

PAUSE AND THINK

Let us "create" an example purely to illustrate the concept of the MTD. Two aspirins may be the minimum for an effective dose against a headache, and 30 may be the dose that can be tolerated before harmful side effects are observed. However, if harmful side effects were seen after three aspirins, the therapeutic index would decrease and the drug would be much less favorable.

chemotherapy and radiotherapy have been used to inhibit or eradicate metastasized cells. The objectives of cancer therapies are to prevent proliferation (**cytostatic effect**) and to kill the cancer cells (**cytotoxic effect**). The aim with all drugs is to achieve an effective result with the minimum side effects. This is indicated by the **therapeutic index**. This is the value of the difference between the minimum effective dose and the maximum tolerated dose (MTD) (Figure 1.6). The larger the value, the safer the drug. Many conventional cancer treatments are administered at MTDs.

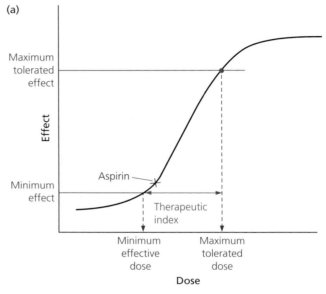

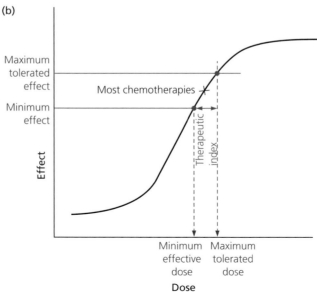

Figure 1.6 The therapeutic index is the value of the difference between the minimum effective dose and the maximum tolerated dose (MTD). (a) The value for the therapeutic index of aspirin is higher than that of (b) the therapeutic index of most chemotherapies.

Chemotherapy

Conventional chemotherapy uses chemicals that target DNA, RNA, and protein to disrupt the cell cycle in rapidly dividing cancer cells and thus has broad specificity. The ultimate goal of cytotoxic chemotherapy is to cause severe DNA damage and to trigger apoptosis in the rapidly dividing cancer cells. The side effects of chemotherapy, which we are all too aware of, such as alopecia (loss of hair), ulcers, and anemia, are due to the fact that hair follicles, stomach epithelia, and hematopoietic cells are also rapidly dividing, and therefore they too are greatly affected by these drugs. Often, prescribed MTDs induce toxicity in sensitive tissues and require a pause in drug administration so that normal cells can recover. We must appreciate that conventional chemotherapies (e.g. cisplatin and methotrexate, discussed in Chapter 2) have had results in treating cancers and continue to extend lives, but at the same time we must desire drugs with better efficiencies and less severe and debilitating side effects. We must strive to develop drugs that rise to these expectations.

1.5 Clinical trials

Testing of new drugs in humans must progress through staged clinical trials (Table 1.1). Phase I trials examine dose responses for assessing drug safety, using a small number (20–80) of healthy volunteers or patients. Many parameters of the metabolism of the drug in humans (e.g. how long does the drug remain in the body?) are obtained at this time. About 70% of drugs tested in Phase I will progress to Phase II studies. Phase II trials are designed to examine efficacy in a larger group of people (100–300). Phase III trials should not be initiated prior to knowing the effective drug dosage. Phase III trials are large-scale studies (1000–3000 people) to confirm drug effectiveness, to monitor side effects, and also to compare the efficacy of the new drug with conventional treatments. Only terminally ill patients may be recruited for clinical trials, by law, in many countries. This has implications for the outcome of testing particular drugs and will be discussed later in the text. About 30% of drugs tested successfully complete Phase III studies. Drugs are also tested against control populations. These people either

Table 1.1 Clinical trials

	Purpose	Number of patients
Phase I	Safety	20–100
Phase II	Efficacy	Up to several hundred
Phase III	Efficacy often tested against conventional treatments	Several hundred to several thousand

receive no treatment or receive a placebo, or an inactive substance. In order to reduce the risk of bias, trials can be randomized. That is, patients are randomly assigned to either a treatment group or a control group, guaranteeing that the two groups are similar. In addition, the trial may be conducted as a single-blind study, whereby patients do not know which group they are in, or as a double-blind study, whereby neither patients nor investigators know who has received the treatment or placebo until after a code is broken that identifies the people in the two groups. Unlike conventional chemotherapies, targeted therapy may not require MTDs. Experience is teaching us that the design of trials for molecularly targeted drugs needs to be well thought out. It is important to consider the stage and type of cancer to be treated, patient populations with the correct molecular profiles (that is, does the patient carry mutations in genes of interest?), assessment of the compound in inhibiting its molecular target, and assessment of the relationship between molecular inhibition and clinical response.

1.6 The role of molecular targets in cancer therapies

The major flaw in the rationale of most conventional therapies is the lack of selectivity against tumor cells versus normal cells. As a result, the side effects of most therapies are very harsh, as mentioned earlier. There is a need to learn about the differences between normal and transformed cells at the molecular level in order to identify cancer-specific molecular targets. In this way, we can design drugs that will be specific for the cancer cells, have increased efficacy, and cause fewer side effects.

Molecules of fame

As we examine the molecular pathways that underlie carcinogenesis, we must keep in mind that the pathways do not act in isolation but are interconnected (see Appendix 1). Despite the many hundreds of molecules involved in carcinogenesis, there are several families of "star players" in the story of carcinogenesis. Many of these "star players" act as nodes that receive signals from many pathways and can exert several effects in response to a specific signal.

The family of protein kinases, encoded by one of the largest families of genes in eukaryotes, must be included in any introduction to cancer biology. Protein kinases phosphorylate (add a phosphate group to) a hydroxyl group on specific amino acids in proteins. Tyrosine kinases phosphorylate tyrosine residues, while serine/threonine kinases phosphorylate serine and threonine residues. Phosphorylation results in a conformational change and is an important mechanism for regulating the activity of a protein. Kinases can be found at the cell surface as transmembrane receptors, inside the cell as intracellular transducers, or inside the nucleus. Kinases play a

PAUSE AND THINK

Is tumor shrinkage a suitable assessment criterion for cytostatic drugs? No, because direct cell death is not anticipated with this class of drugs. Cytotoxic drugs are expected to kill tumor cells.

critical role in major cell functions, including cell cycle progression, signal transduction, and transcription, and are important molecular targets for the design of cancer drugs. Phosphorylation is also regulated by phosphatases—enzymes that remove phosphate groups. Mutational analysis of all known human protein tyrosine phosphatases suggests that several of these act as tumor suppressors in some types of cancer (see Chapter 6).

The Ras family is another "star" set of proteins, encoded by genes which are mutated in over 50% of certain cancers (e.g. colon cancer). Ras is an intracellular transducer protein that acts subsequently to binding of a growth factor to its receptor and is involved in transmitting the signal from the receptor through the cell. As G proteins, they reside on the intracellular side of the plasma membrane and are activated by the exchange of GDP for GTP.

Tumor protein p53 (TP53; p53) and its related family members hold pivotal positions in guarding the integrity of the genome by coordinating responses of the cell (e.g. cell cycle arrest, DNA repair, apoptosis) to different types of stress (e.g. DNA damage, hypoxia). The *p53* gene is a tumor suppressor gene that has a key role in inhibiting carcinogenesis. It is mutated in more than half of all cancers, and over a thousand different mutations have been identified. It acts as a transcription factor and induces the expression of genes required to carry out its functions.

The retinoblastoma gene (*Rb*) is also a tumor suppressor gene that plays a central role in regulating the cell cycle. It is commonly mutated in several cancers. The retinoblastoma protein normally functions as an inhibitor of cell proliferation by binding to, and suppressing, an essential transcription factor of cell cycle progression. Its activity is regulated by phosphorylation by cyclin D and the cyclin-dependent kinases (4/6).

The introduction of cancer genomics

The completion of the Human Genome Project, whereby every nucleotide of a human genome has been sequenced and mapped, has paved the way for cancer genomics. Learning about the details of the genome of a cancer cell and how it differs from a normal cell will provide us with the fine distinctions needed to design more powerful and specific drugs (Stratton, 2011). Both alterations that are common for specific cancers and alterations that are different among individuals with the same specific cancer are important to understand. One of the most significant findings from examining the cancer genome has been that the genomic profile of an individual's tumor is unique and "personal." This understanding is leading us into the era of personalized medicine where a patient's tumor genome will inform doctors of the best treatment for that individual.

Since the Human Genome Project, new sequencing technologies, such as next-generation sequencing (NGS), have developed and greatly advanced the speed at which genomes can be sequenced, and this has encouraged new approaches for investigations. The International Cancer Genome

Consortium (ICGC) has been organized to coordinate large genomic studies across the globe to characterize at least 50 classes of cancer, including the sequencing of several hundreds of tumors for each type, and to make the data available to the entire research community. The information gathered from these efforts promises to yield important insights into the understanding of cancer and will lead to better molecular classifications that will inform drug discovery strategies. Results from specific projects under the ICGC, such as the whole-genome sequencing of four cases of chronic lymphocytic leukemia, have already been published (Puente et al., 2011). Furthermore, the human genome has been subdivided into functional categories such as the "kinome" by the mapping of the complete set of 518 protein kinase genes in the genome. As aberrant regulation and mutation of these genes are involved in carcinogenesis, this will be an important tool for the design of new molecular therapies. The International SNP Map Working Group is analyzing single nucleotide polymorphisms (SNPs) to identify mutations within the genome that may be linked with cancer. Genome-wide association studies (GWAS) investigate the entire genome for common genetic variants in different individuals to see if there are any associations between SNPs and a trait/disease such as cancer. A recent demonstration of a genome-wide association study of SNPs in pancreatic cancer revealed new loci associated with pancreatic cancer risk, and some new loci fall within genes involved in the development of the pancreas and diabetes predisposition (Wolpin et al., 2014).

Studies of functional genomics are important for providing insights into cancer biology. Gene targeting by homologous recombination in mice generated crucially important knock-out mice for functional analysis for over a quarter of a century. But recent discoveries have led to new approaches to help elucidate the function of genes in cancer. Knowledge about RNAs in gene regulation has led to methods such as RNA interference that uses small interfering RNAs (siRNAs) to target the degradation of complementary single-stranded target RNAs, thus knocking down specific protein expression. Even more impactful is the CRISPER system (discussed in Section 14.3) that efficiently uses sequence-specific nucleases to edit the genome. These nucleases are recruited by small, easily generated guide RNA (gRNA) that uses Watson–Crick base pairing to identify a genomic target. The system also allows many mutations to be introduced simultaneously and thus will be a valuable tool for understanding the mechanisms of carcinogenesis.

Analysis of gene function by small interfering RNAs (siRNAs)

Gene function is often determined by abolishing the expression of a gene product and observing the resulting phenotype. RNA interference is a cellular mechanism for regulating **gene expression** in most eukaryotes. Short RNA duplexes (approximately 21 nucleotides long with two nucleotide 3′ overhangs), called small interfering RNAs (siRNAs), ➡

→ mediate the expression of genes by causing the degradation of homologous single-stranded target RNAs. Experimentally, we can use siRNAs to target endogenous genes in mammalian cells. Using the known sequence of a segment of target mRNA, sense and antisense RNAs are designed, synthesized, and annealed to produce siRNA duplexes. The siRNAs are delivered to cells by classical gene transfer methods (e.g. electroporation). Specific antibodies against the targeted protein are often used to ensure that target protein levels have been diminished.

Good news

The nature of the "now" generation of drugs is small molecules and antibodies targeted against selective gene products. Soon it is likely that methods to prevent expression of specific target genes (e.g. antisense RNA, siRNAs) will also enter the clinic. Drugs that boost our immune system to fight cancer are here and are the most promising. Several therapies based on these designs that can treat specific cancers will be described later in the text. The good news is that, in the USA, the cancer death rates have decreased in both men and women from 1998 to 2011 (Siegel *et al.*, 2015), and similar progress can be seen in other countries in the world. Progress is beginning to show.

▨ CHAPTER HIGHLIGHTS—REFRESH YOUR MEMORY

- Cancer is a common disease that will affect one out of three people over their lifetime, worldwide.

- Cancer is a group of diseases characterized by unregulated cell growth and the invasion and spread of mutated cells throughout the body.

- There are ten hallmarks of cancer. They are:
    - Growth signal autonomy
    - Evasion of growth inhibitory signals
    - Avoiding immune destruction
    - Unlimited replicative potential
    - Tumor-promoting inflammation
    - Invasion and metastasis
    - Angiogenesis
    - Genomic instability and mutation
    - Evasion of cell death
    - Reprogramming energy metabolism.
- Most carcinogens are mutagens.

- Most often, carcinogenesis is a multi-step process that requires the accumulation of several mutations. However, there is evidence that, in a small percentage of tumors, a single catastrophic event in a cell can lead immediately to many mutations and cause cancer.

- Cancer is a genomic disease at the cellular level.

- Genes that are involved in growth, differentiation, or cell death when deregulated can give rise to the cancer phenotype.

- A gene containing a dominant mutation that results in inappropriate activation of growth is an oncogene.

- Tumor suppressor genes are usually inactivated by mutations in both alleles (recessive), and this results in inactivation of growth inhibition.

- Haploinsufficiency, whereby only one allele of a tumor suppressor gene is inactivated, also contributes to carcinogenesis.

- Changes in lifestyle factors can affect cancer risk.

- Many conventional therapies are broad-acting drugs administered at MTDs, resulting in severe side effects.

- Protein kinases, enzymes that phosphorylate proteins, are important molecules in carcinogenesis.

- Cancer genomics is being used to define molecular targets for tumor-specific effects.

- Some cancers can already be treated by specific molecular approaches.

ACTIVITY

1. Become familiar with the Globocan 2012 web site, and see how epidemiological data can be formatted in different ways (go to http://globocan.iarc.fr/). On the top navigation bar, select **Online Analysis**, then select **Cancer by population** under Pie Charts. Examine the data for different countries. How do the cancer profiles differ between countries on different continents? Select to run a cancer map for lung cancer. What are your conclusions?

FURTHER READING

Alison, M.R. (2007) *The Cancer Handbook*, 2nd edn. John Wiley and Sons, Inc., New York.

Gilbert, S.F. (2013) *Developmental Biology*, 10th edn. Sinauer Associates, Inc., Sunderland, MA.

Jemal, A., Bray, F., Center, M.M., Ferlay, J., Ward, E., and Forman, D. (2011) Global cancer statistics, 2011. *CA Cancer J. Cli.* **61**: 69–90.

Peto, J. (2001) Cancer epidemiology in the last century and the next decade. *Nature* **411**: 390–395.

Stratton, M.R., Campbell, P.J., and Futreal, P.A. (2009) The cancer genome. *Nature* **458**: 719–724.

WEB SITES

American Cancer Society.*Cancer Facts* and *Statistics* http://www.cancer.org/Research/CancerFactsStatistics/index

GLOBOCAN 2012 http://globocan.iarc.fr/

Cancer Research UK.*Latest UK Cancer Incidence(2008)*and *Mortality (2008) Summary* (2011) http://info.cancerresearchuk.org/prod_consump/groups/cr_common/@nre/@sta/documents/generalcontent/cr_072108.pdf

The U.S. Food and Drug Administration. Running Clinical Trials http://www.fda.gov/ScienceResearch/SpecialTopics/RunningClinicalTrials/default.htm

The International Cancer Genome Consortium http://www.icgc.org/

■ **SELECTED SPECIAL TOPICS**

Adams, D.S. (2003) Teaching critical thinking in a developmental biology course at an American liberal arts college. *Int. J. Dev. Biol.* **47**: 145–151.

Colditz, G.A, Wolin, K.Y., and Gehlert, S. (2012) Applying what we know to accelerate cancer prevention. *Sci. Transl. Med.* **4**: 65–71.

Ferlay, J., Soerjomataram, I., Dikshit, R., Eser, S., Mathers, C., Rebelo, M., *et al.* (2015) Cancer incidence and mortality worldwide: sources, methods and major patterns in GLOBOCAN 2012. *Int. J. Cancer* **136**: E359–E386.

Fodde, R. and Smits, R. (2002) A matter of dosage. *Science* **298**: 761–763.

Greaves, M. and Maley, C.C. (2012) Clonal evolution in cancer. *Nature* **481**: 306–313.

Hanahan, D. and Weinberg, R.A. (2011) Hallmarks of cancer: the next generation. *Cell* **144**: 646–674.

Oeppen, J. and Vaupel, J.W. (2002) Broken limits to life expectancy. *Science* **296**: 1029–1030.

Puente, X.S., Pinyol, M., Quesada, V., Conde, L., Ordóñez, G.R., Villamor, N., *et al.* (2011) Whole-genome sequencing identifies recurrent mutations in chronic lymphocytic leukemia. *Nature* **427**: 101–105.

Siegel, R., Miller, K., and Jemal, A. (2015) Cancer statistics, 2015. *CA Cancer J. Clin.* **65**: 5–29.

Song, Q.K., Li, J., Huang, R., Fan, J.H., Zheng, R.S., Zhang, B.N., *et al.* (2014) Age of diagnosis of breast cancer in China: almost 10 years earlier than in the United States and the European Union. *Asian Pac. J. Cancer Prev.* **15**: 10021–10025.

Stratton, M.R. (2011) Exploring the genomes of cancer cells: progress and promise. *Science* **331**: 1553–1558.

Wolpin, B.M., Rizzato, C., Kraft, P., Kooperberg, C., Petersen, G.M., Wang, Z., *et al.* (2014) Genome wide association study identifies multiple susceptibility loci for pancreatic cancer. *Nat. Genet.* **46**: 994–1002.

Chapter 2

DNA structure and stability: mutations versus repair

Introduction

Genetic information, coded within DNA, requires stability. DNA directs the production of proteins, needed for the structure and function of cells over a lifetime, through an adaptor molecule, RNA. Unlike RNA and protein, which have a limited existence before they are degraded and/or recycled, DNA must maintain its integrity over that lifetime. However, our genes are subject to a myriad of attacks by both environmental agents and endogenous processes that result in mutation and scission. Changes to the DNA sequence may have severe consequences for the cell and its progeny. Cancer is a disease that involves alterations to gene structure and gene expression at the cellular level. The role of the accumulation of mutations gradually over time is well established for the process of carcinogenesis, but recent evidence from advances in sequencing technology shakes this conventional view. Small localized areas of hypermutation, called "**kataegis**" (Greek for "thunderstorm"), have been identified in cancer genomes (Nik-Zainal *et al.*, 2012). Evidence also suggests that, for a small number of cancers, one-off cell catastrophic events may also generate mutations that underlie carcinogenesis (Stephens *et al.*, 2011). A one-off cell crisis that shatters chromosomes and results in tens to hundreds of genomic rearrangements is called **chromothripsis** ("chromo" meaning **chromosome** and "thripsis" meaning breaking into small pieces). In this chapter, we will review the structure of a gene and describe the mutations that occur during carcinogenesis.

When considering the process of carcinogenesis, we must be aware that cells are equipped with defense mechanisms against mutations, such as the detection and repair of DNA damage. Detection and repair of DNA damage are particularly crucial in the time before a cell divides, as errors existing during replication will be passed on to daughter cells. Pausing the cell cycle is sometimes coupled to the repair of DNA damage. Apoptosis, a more hard-line defense, can be triggered as a last resort; thus, cell suicide is the ultimate price to be paid to prevent perpetuation of DNA damage and to protect the individual from carcinogenesis (see Chapter 7). In this chapter, we will also examine how, on one hand, mutations in DNA occur as a consequence of exposure to carcinogens and, on the other hand, examine the DNA repair systems that are in place to maintain the integrity of the genome and suppress tumorigenesis.

2.1 Gene structure—two parts of a gene: the regulatory region and the coding region

We have about 20,000 protein-coding genes! They are encoded in our DNA, an impressively simplistic double-helical molecule made up of two chains of nucleotides. A nucleotide is made up of a sugar, phosphate, and a nitrogenous base (adenine, guanine, cytosine, or thymine), and it is the sequence of the bases that holds the instructional information of our genes. The central dogma of molecular biology states that DNA is transcribed into RNA and RNA is translated into protein (translation). Gene expression refers to the transcription of a gene. For the purpose of simplicity, keep in mind that there are two distinct functional parts to a gene (Figure 2.1). The 5′ end of a gene contains nucleotide sequences that make up the promoter region, and this region is involved in regulating the expression of the gene. These 5′ nucleotide sequences interact with proteins that affect the activity of RNA polymerase and determine when and where a gene is expressed. (Note, however, there are exceptions; for many genes, some regulatory regions can be located elsewhere, such as downstream (the direction along the DNA molecule towards the 3′ end of the gene) or within introns.) The TATA box (TATAAAA), located near the start site of transcription, is one of the most important regulatory elements for most genes. Binding of the TATA box-binding protein (TBP) to the TATA box is crucial for the initiation of transcription. A short sequence of DNA within a promoter that is recognized by a specific protein and

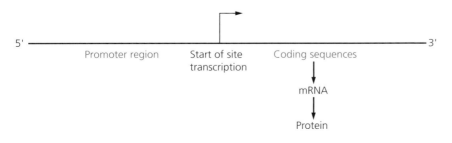

Figure 2.1 A simplistic representation: two functional parts of a gene.

contributes to the regulation of the gene is called a response element (RE). Common response elements identify genes under a common type of regulation. For example, the sequence CCATATTAGG is referred to as the serum response element (SRE) and is found in genes that are responsive to serum. Also, it is not surprising that the response element for a protein that is essential for the regulation of the cell cycle, the transcription factor E2F, is found in the promoters of the cyclin E and cyclin A genes, the products of which are major players in the cell cycle. Enhancer elements are additional regulatory DNA sequences that are position- and orientation-independent relative to a promoter and are important for tissue-specific and stage-specific expression. And recently clusters of enhancer elements, called super-enhancers, have been identified (recently examined by Pott and Lieb, 2015). Downstream of the promoter are the nucleotides that will be transcribed into RNA, and those coding for exons will be translated into protein. These downstream nucleotide sequences represent the coding region of the gene.

2.2 Mutations

As stated previously, most carcinogens are mutagens. These agents induce mutations either by modifying DNA (e.g. forming DNA adducts) or by causing chromosomal damage (e.g. DNA strand breaks). Several types of mutations are illustrated in Figure 2.2: transitions, transversions, insertions, deletions, and chromosomal translocations. Transitions and transversions are two types of base substitutions. A transition is the substitution of one purine for another purine, and a transversion is the substitution of a purine for a pyrimidine or vice versa. Base substitutions during replication may occur for several reasons. First, DNA polymerase is not always 100% accurate. The enzyme may make an error and insert a wrong nucleotide during DNA synthesis. Also, modifications of bases owing to oxidation or covalent additions and alterations of chromatin

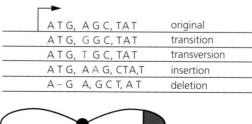

Figure 2.2 Types of mutations.

structure can cause misreading of the DNA template by DNA polymerase. Remember that the genetic code is a triplet code read in a sequential, but non-overlapping, manner. An insertion or deletion of a base can alter the reading frame (marked by a "," in Figure 2.2) and thus can also be referred to as a frameshift mutation. In most cases, this leads to a non-functional or truncated protein product. A chromosomal translocation is the exchange of one part of one chromosome for another part of a different chromosome and results in changes of the base sequence of DNA. Gene amplification, whereby the copy number of a gene increases from the two copies present in the normal diploid genome to sometimes hundreds of copies, may also occur in cancer cells (not shown).

In theory, mutations may occur anywhere across a particular gene, but the location will determine whether some of these mutations give rise to a growth advantage and contribute to carcinogenesis. For example, a mutation may alter the conformation of a cyclin protein and result in unregulated progression of the cell cycle, whereas another mutation may have no effect on the amino acid sequence and therefore no effect on protein conformation or function. Remember the concepts of "wobble" and "degeneracy" in the genetic code whereby the third nucleotide of a codon may be changed and the new codon may code for the same amino acid. "Driver mutations" are located in cancer genes by definition and confer a growth advantage on cells, whereas "passenger mutations" do not confer a growth advantage and "are there for the ride." Some calculations predict that 5–7 driver mutations are required in most cancers, whereas other calculations predict as many as 20 driver mutations are required in some cancers. Some mutations may give rise to a cell that has an increased rate of mutation as a result of, for example, defective DNA repair, and these cells are said to have a "mutator phenotype."

The consequence of a mutation in a gene is determined by its location with respect to the two functional parts of a gene. Mutations occurring in the promoter region may alter the regulation of the gene and affect the levels or temporal/spatial expression of the gene product. The consequence of such mutations may be over- or under-expression of the protein product or the appearance of the protein product at the wrong time or in the wrong place (i.e. the wrong cell type), respectively. Alternatively, mutations occurring in the coding region of genes may affect the structure and thus alter the function of the gene product or cause a truncation (e.g. the introduction of a stop codon) that abolishes the protein's function completely.

As we will see in later chapters, there are many examples of these types of mutations in genes regulating growth, differentiation, and apoptosis that are involved in carcinogenesis. For most cancers, different types of mutations accumulate over time within an individual cell and cause the cell to be transformed into a cancer cell. But, for some cancers, particularly bone cancer (up to 25%) and colorectal cancer, recent evidence suggests that a single event of chromothripsis involving one or a few chromosomes leads to multiple gene rearrangements (in some cases tens to hundreds) that can disrupt tumor suppressor genes and/or produce oncogenic fusion genes. In simple terms, a chromosome shatters, and the attempt to repair the damage results in many incorrect junctions (Figure 2.3). Chromothripsis suggests that several tumor-promoting events may occur at the same time, rather than progressively. The pattern of localized clustering of rearrangements is highly unlikely to have occurred by a gradual process.

 HOW DO WE KNOW THAT?

Chromothripsis

Sequencing strategies had shown, for example, that a kidney cancer cell line, called TK10, contained 55 rearrangements involving chromosome 5. (Note: TK10 is a hyperdiploid cell line and contains more than 46 chromosomes.) In order to investigate the genomic rearrangements of a single chromosome in a cell line that showed evidence of chromothripsis, Stephens and colleagues (2011) designed **f**luorescence **in** **s**itu **h**ybridization (FISH) probes to five widely separated regions of chromosome 5 and analyzed TK10 cells (see "Fluorescence

in situ hybridization (FISH)" on p. 121 for details about FISH hybridization). The probes were each labeled with a different color dye and hybridized to the TK10 cells. As can be seen in Figure 2.3, there were several wild-type copies of chromosome 5 showing the expected arrangement of the chromosomal regions of interest (white, yellow, red, purple, green) and two copies of a chromosome derived from chromosome 5 in which all five FISH probes were closely juxtaposed and rearranged (purple, red, yellow, green, white). This is strong evidence for chromothripsis.

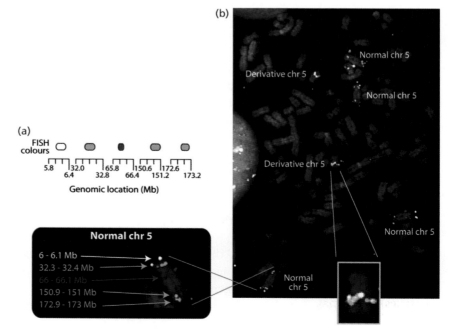

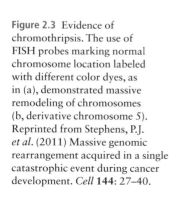

Figure 2.3 Evidence of chromothripsis. The use of FISH probes marking normal chromosome location labeled with different color dyes, as in (a), demonstrated massive remodeling of chromosomes (b, derivative chromosome 5). Reprinted from Stephens, P.J. *et al.* (2011) Massive genomic rearrangement acquired in a single catastrophic event during cancer development. *Cell* **144**: 27–40.

Although the cause of chromothripsis is still unknown, several suggestions have been proposed: ionizing radiation that leads to chromosome breaks; telomere dysfunction which may lead to end–end chromosome fusions (both suggested by Stephens *et al.*, 2011); or aborted apoptosis such that cells which have initiated DNA fragmentation survive (suggested by Tubio and Estivill, 2011). Defective or delayed DNA replication and defective DNA damage responses have also been considered (Forment *et al.*, 2012). This area promises to be an area of future research.

2.3 Carcinogenic agents

The backbone of cancer biology has been the identification of carcinogens responsible for cancer-causing mutations, and the identification of specific mutations as causative factors of carcinogenesis along with the elucidation of the pathways they affect. Several classes of carcinogens will now be described, including radiation, chemicals, infectious pathogens, and particular endogenous reactions.

Radiation as a carcinogen

Radiation is energy. There are two forms of radiation: energy traveling in waves or as a stream of atomic particles. Energy waves include gamma (γ) rays, high-energy electromagnetic radiation that is similar to X-rays. Atomic particles include alpha (α) and beta (β) particles that are emitted by radioactive atoms. (Alpha particles comprise two protons and two neutrons, while beta particles comprise electrons.)

Electromagnetic radiation is naturally occurring radiation which possesses a broad range of energies. Electromagnetic radiation moves as waves of energy, which have peaks and troughs (in a manner analogous to waves at sea). The distance between successive peaks (or troughs) is termed the wavelength. High-energy electromagnetic radiation, such as cosmic radiation, has a short wavelength, while low-energy radiation, such as radio waves, has a long wavelength. The electromagnetic spectrum spans electromagnetic radiation of varying wavelengths, as shown in Figure 2.4. The electromagnetic spectrum extends from long-wavelength radiation (not shown) to extremely short-wavelength radiation, such as X-rays and gamma radiation. The visible spectrum spans those wavelengths that we can detect with our eyes as visible light. Ultraviolet (UV) radiation is emitted from the sun and has a higher energy (and so a shorter wavelength) than visible light.

Several types of radiation (including both energy waves and atomic particles) can damage DNA and act as carcinogens. The amount of energy released by a particular radiation source affects the mechanism and extent

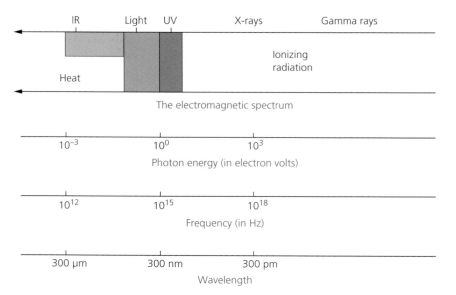

Figure 2.4 The electromagnetic spectrum and corresponding characteristics (infrared, IR; ultraviolet, UV).

of damage to DNA. The amount of energy released by a particular radiation source and absorbed by the body tissue is measured in grays (Gy). One gray is the release to the body tissue of 1 joule (J) of energy per 1 kg of tissue. The real issue is not how much radiation is absorbed by the body tissue, however, but how much damage is done when the radiation has been absorbed. The amount of damage caused depends on the rate at which a particular radiation source releases energy. If a radiation source releases energy at a high rate, then it causes more damage than a source that releases energy more slowly.

Linear energy transfer (LET) is used to help describe the rate at which energy is released. Specifically, it describes the amount of energy released by a radiation source as it travels a fixed distance. High-LET radiation emits more energy than low-LET radiation over the same distance. Therefore, high-LET radiation (such as alpha particles) causes more biological damage than low-LET radiation (such as X-rays). The quantity and type of DNA damage caused by a particular radiation source depend on whether it is high- or low-LET radiation. Double-stranded DNA breaks are more commonly caused by high-LET radiation and lead to chromosomal translocations and deletions.

The amount of biological damage caused by a particular source of radiation is measured in sieverts (Sv). (The numerical value of these units is determined by multiplying the gray units by a factor relating to the LET value of a particular type of radiation.)

Two classes of radiation, ionizing radiation and UV radiation, have been demonstrated to act as carcinogens and damage DNA. Let us examine both types of radiation next.

Ionizing radiation

Ionizing radiation includes both alpha and beta particles (atomic particles) and gamma rays (energy waves). When high-energy radiation, such as gamma rays, strikes molecules in its path, electrons may be displaced from atoms within the molecule. The loss of one or more electrons converts the molecule from being electrically neutral to carrying an electrical charge. The charged molecule is called an ion, and hence the radiation causing the formation of an ion is called ionizing radiation.

Ionizing radiation can damage DNA directly by causing ionization of the atoms comprising DNA, or indirectly by the interaction with water molecules (a process known as radiolysis), to generate dangerous intermediates called reactive oxygen species (ROS) (see Box "A little lesson about ROS . . . "). These ROS may react with DNA, or with other biomolecules, to cause damage within the cell.

A little lesson about ROS ...

Some radiation exerts its biological effect by the generation of damaging intermediates through the interaction of radiation with water, or radiolysis. As our body comprises 55–60% water, radiation is most likely to strike water than any other matter. The striking of water by radiation causes it to lose an electron and become highly reactive. This sets off a chain reaction in which water is converted to oxygen, O_2, through a three-step process. Radiation interacts with a single molecule of water, and thus it cannot split directly into the diatomic gases H_2 and O_2. Equation (1) is **not** possible because this equation is not balanced.

$$H_2O \rightarrow H_2 + O_2 \qquad (1)$$

The balanced equation (Equation 2), does not apply to radiation, as radiation interacts with only a single molecule of water, and not the two molecules required in Equation (2):

$$2H_2O \rightarrow 2H_2 + O_2 \qquad (2)$$

Instead, radiolysis results in the sequential generation of three dangerous reactive oxygen species (ROS), as an electron (e^-) is lost at each step. The three ROS, formed in sequence, are the hydroxyl radical ($\cdot OH$), hydrogen peroxide (H_2O_2), and the superoxide radical (O_2^-):

$$H_2O \xrightarrow{\ e^-\ } \underset{\text{hydroxyl radical}}{\cdot OH} \xrightarrow{\ e^-\ } \underset{\text{hydrogen peroxide}}{H_2O_2} \xrightarrow{\ e^-\ } \underset{\text{superoxide radical}}{O_2^-} \xrightarrow{\ e^-\ } O_2$$

The hydroxyl radical is an extremely reactive molecule; in fact, it is one of the most reactive (and therefore dangerous) molecules known! It immediately removes electrons from any molecule in its path, turning that molecule into a free radical and so propagating a chain reaction. (A free radical is a highly unstable, reactive molecule that possesses an unpaired electron. Both the hydroxyl radical and the superoxide radical shown earlier are free radicals.)

Neither hydrogen peroxide nor the superoxide radical are as reactive as the hydroxyl radical. Hydrogen peroxide is actually more dangerous to DNA than the hydroxyl radical, ➔

➜ however. The slower reactivity of hydrogen peroxide (compared with the hydroxyl radical) gives the hydrogen peroxide molecule time to travel into the nucleus of a cell where it is free to interact with, and wreak havoc upon, DNA.

Oxidation of DNA (the removal of electrons by species such as the free radicals mentioned here) is one of the main causes of mutation and explains why free radicals are such potent carcinogens. Oxidation can produce several types of DNA damage, including oxidized bases. Among the variety of oxidized nitrogenous bases observed, 8-oxoguanine is the most abundant. DNA polymerase mispairs 8-oxoguanine with adenine during DNA replication, leading to a G→T transversion mutation. The presence of iron can exacerbate the consequences of hydrogen peroxide production. If it encounters iron and receives an electron from it, hydrogen peroxide can be reconverted into the hydroxyl radical that may attack DNA. The Fenton reaction (Equation 3), illustrates this:

$$H_2O_2 + Fe^{2+} \rightarrow OH^- + \cdot OH + Fe^{3+} \tag{3}$$

$$O_2^- \cdot + Fe^{3+} \rightarrow O_2 + Fe^{2+} \tag{4}$$

The superoxide radical is the third intermediate before the formation of oxygen. It is not very reactive but acts more as a catalyst for the generation of the other two intermediates mentioned because it helps regenerate iron (Equation 4) in the form needed for the earlier mentioned Fenton reaction. Thus, the ROS intermediates affect one another.

(Lane, 2002)

People are exposed to varying amounts of ionizing radiation. Exposure to gamma rays from cosmic radiation depends on the altitude at which you live or travel. The average exposure for high-altitude flights is about $0.005–0.01$ mSv h^{-1}. A chest X-ray required for medical diagnosis of some conditions exposes patients to 0.1 mSv. The contribution of the accumulation of these varying daily exposures towards cancer risk is relatively unknown.

Studies of the victims of the atomic bombing in Japan continue to contribute to our knowledge of ionizing radiation as a carcinogen. Evidence suggests that the most important damage associated with ionizing radiation-induced carcinogenesis is double-stranded DNA breaks. A long-term and comprehensive study of 80,000 atomic bomb survivors called The Life Span Study (latest report Preston *et al.*, 2007) has revealed three important points: (1) leukemia is the most frequent ionizing radiation-induced cancer; (2) age is an important risk factor, whereby those exposed as children are most affected; and (3) the risks of solid cancer increase with dose in a linear fashion. People exposed at 30 years of age have a risk of solid cancer that is elevated by 35% per Gy for men and 58% per Gy for women at the age of 70. These studies have helped guide recommendations for radiation limits. The US National Council on Radiation Protection and Measurements (NCRP) and the International Commission on Radiological Protection recommends an annual radiation dose limit of 1 mSv (100 mrem) for members of the public.

Ultraviolet radiation

UV radiation from the sun is also carcinogenic and is a principal cause of skin cancer. Of the three types of UV light—UVA (wavelength 320–380 nm), UVB (wavelength 290–320 nm), and UVC (wavelength 200–290 nm)—UVB is the most effective carcinogen. The conjugated double bonds in the rings of the nitrogenous bases of DNA absorb UV radiation. UVB directly and uniquely causes characteristic UV photoproducts: cyclobutane pyrimidine dimers and pyrimidine–pyrimidone photoproducts (Figure 2.5a, b). Cyclobutane pyrimidine dimers are the most prevalent, formed at least 20–40 times more frequently than other UV photoproducts. The formation of a pyrimidone (6–4) photoproduct mimics an abasic site (a nucleotide minus a base) and is more efficiently repaired than cyclobutane pyrimidine dimers. The formation of a pyrimidine dimer causes a bend in the DNA helix, and, as a result, DNA polymerase cannot read the DNA template. Under these conditions, DNA polymerase preferentially incorporates an "A" residue. Consequently, TT dimers are often restored, but TC and CC dimers result in transitions (TC→TT and CC→TT) (Figure 2.5c). Results from a mammalian cell system showed that

PAUSE AND THINK

Marie Curie, who worked with radioactivity all of her adult life, died of leukemia at the age of 67. The multi-stage process of carcinogenesis is manifested by the fairly long incubation periods needed to develop cancer after exposure to carcinogens.

LIFESTYLE TIP

In 2009, The International Agency for Research on Cancer (IARC) raised the classification of UV-emitting tanning devices to "carcinogenic to humans"—the highest risk category—based on evidence of increased risk of melanoma for regular users (El Ghissassi *et al.*, 2009). Even short-term tanning salon exposure (ten treatments in 2 weeks) results in the formation of cyclobutane pyrimidine dimers. A natural look is less taxing for your DNA!

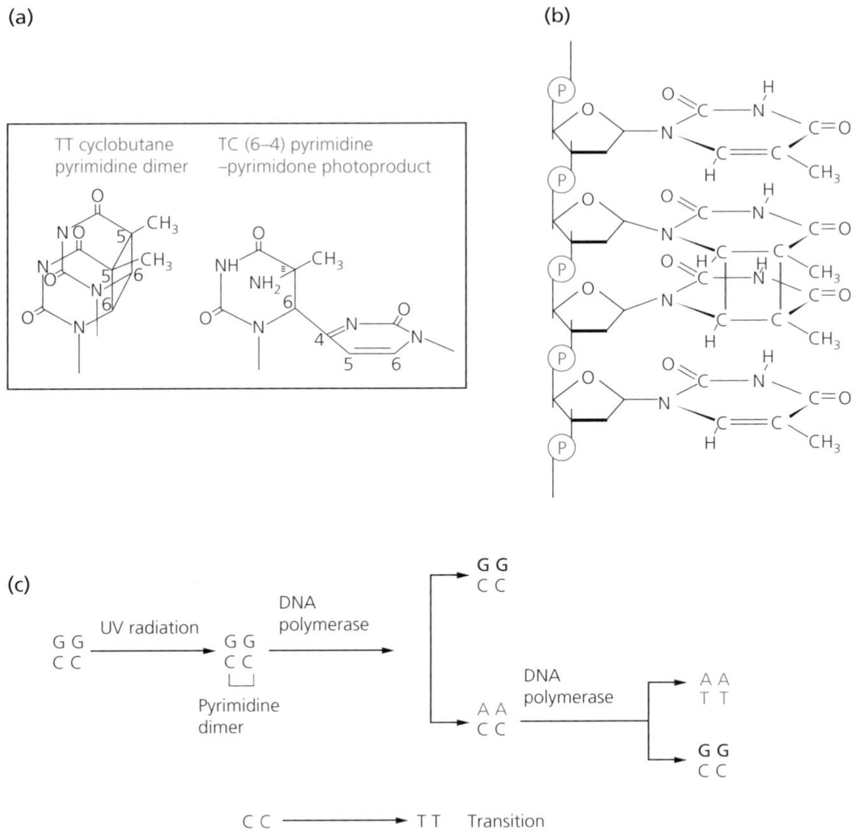

(a)

(b)

(c)

Figure 2.5 (a) UV photoproducts. (b) A pyrimidine dimer in the context of a polynucleotide chain. (c) Steps involved in UV-induced transitions.

cyclobutane pyrimidine dimers are responsible for at least 80% of UVB-induced mutations. The precise class of mutations resulting from pyrimidine dimers is a unique molecular signature of skin cancer (see Box "Skin cancer")—they are not found in any other types of cancer.

UVA indirectly damages DNA via free radical-mediated damage. Water is fragmented by UVA, generating electron-seeking ROS (such as the hydroxyl radical, as mentioned earlier) that cause DNA damage (e.g. oxidation of bases). G→T transversions are characteristic of UVA damage.

Skin cancer

UV light is specifically carcinogenic to the skin because it does not penetrate the body any deeper than the skin. The skin is made up of squamous cells, basal cells, and melanocytes, and skin cancers are classified by the cell type they affect: squamous cell carcinoma (SCC), basal cell carcinoma (BCC), and melanoma, respectively. The depth of transmission of each type of UV light is dependent on the wavelength, the longer the wavelength the deeper the penetration: UVC radiation is almost completely absorbed by the ozone layer and rarely reaches the skin, but humans may be exposed from commercial sources such as germicidal lamps; UVB penetrates into the basal level of the epidermis; and UVA penetrates into the more acellular dermis level. Sunscreens work on the basis of including UV-absorbing organic chemicals (e.g. cinnamates), inorganic zinc-containing pigments, or titanium oxides in their ingredients to minimize UV absorption by the skin. (Note that melanin formation, known to most people as tanning, is a natural defense mechanism against UV absorption.) Additional ingredients in sunscreens must be used with care, as we have learned that some compounds may be photosensitized carcinogens, chemicals that can be activated by UV to become carcinogenic. Ironically, some early sunscreens included bergamot oil which contains 5-methoxy psoralen, a photosensitized carcinogen! Some drugs, such as fluoroquinolone antibiotics, are also photosensitized carcinogens, which explains the reasons for the precautions from doctors to stay out of the sun during their administration.

A cellular mechanism for the elimination of UV-damaged skin cells is to initiate apoptosis. This phenomenon is familiar to us as the peeling of the skin after a sunburn. The tumor suppressor p53 protein (introduced in Chapter 1 and discussed in detail in Chapter 6) is an important regulator of apoptosis. Mutation of the *p53* gene is important for the initiation of SCC and BCC, but not melanoma. The characteristic mutations (CC→TT transitions) caused only by UV and no other carcinogen were identified in the *p53* gene. Mutations in the *p53* gene, which disrupt normal *p53* function and provide cells with a growth advantage, may induce the formation of tumor cells. The pattern of mutation is not random but rather tends to be localized to nine places, called hotspots. This suggests that *p53* mutations are causal for skin cancer. Further investigation of why there are so few hotspots within the context of hundreds of sites with adjacent pyrimidine dimers in the *p53* gene yielded an explanation. The hotspots in *p53* are not repaired efficiently. Removal of cyclobutane pyrimidine dimers is particularly slow at these sites. The resulting loss of *p53* function causes a block in apoptosis and consequently allows the proliferation of mutated *p53* cells. Thus, UV radiation not only induces *p53* mutations, but also selects for the clonal expansion of the *p53* mutated cells, by inducing apoptosis in normal cells with wild-type *p53*. →

→ Different pathways seem to be central for melanoma. The elucidation of one of these pathways was one of the first successes of the Cancer Genome Project. It identified mutations in the *BRAF* gene in 66% of malignant melanomas (Davies *et al.*, 2002). BRAF is a serine/threonine kinase that functions in the signal transduction pathway downstream of a melanocyte-stimulating hormone and may explain why there is a high frequency of *BRAF* mutations in melanoma relative to other cancers. Surprisingly, the major mutation identified (T→A) in the kinase domain is not characteristic of UV-induced mutations (CC→TT). See Chapter 4 for a newly approved drug designed to target BRAF.

Chemical carcinogens

Many chemicals in our environment and in our diet play a role in human carcinogenesis. The common mechanism of action of chemical carcinogens is that an electrophilic (electron-deficient) form reacts with nucleophilic sites (sites that can donate electrons) in the purine and pyrimidine rings of nucleic acids. Some chemical carcinogens can act directly on DNA, but others become active only after they are metabolized in the body, forming what are called ultimate carcinogens, the molecules that execute the damage. A family of enzymes called the cytochrome P450 enzymes is involved in the metabolism of chemicals in the liver and is important in the activation of carcinogens to ultimate carcinogens. Genetic polymorphisms and variable expression account for differences in responses to chemical carcinogens among individuals. For example, the expression of one of the P450 enzymes called CYP1A1 (aryl hydrocarbon hydroxylase) can vary 50-fold in human lung tissue and may be responsible for the delivery of varying doses of ultimate carcinogens among smokers (Alexandrov *et al.*, 2002).

Carcinogens can be segregated into ten groups:

(i)   Polycyclic aromatic hydrocarbons

(ii)   Aromatic amines

(iii) Azo dyes

(iv)   Nitrosamines and nitrosamides

(v)   Hydrazo and azoxy compounds

(vi)   Carbamates

(vii) Halogenated compounds

(viii) Natural products

(ix)   Inorganic carcinogens

(x)   Miscellaneous compounds (alkylating agents, aldehydes, phenolics).

Four major classes of carcinogens are described in the following sections: polycyclic aromatic hydrocarbons (PAHs), aromatic amines, nitrosamines, and alkylating agents. These carcinogens exert their effects by

Figure 2.6 (a) Examples of polycyclic aromatic amines. (b) Metabolic activation of BP.

adding functional groups covalently to DNA. Chemically modified bases, called DNA adducts, mask the identity of the base or distort the DNA helix, causing replication errors. The resulting mutations initiate cancer.

Polycyclic aromatic hydrocarbons (PAHs)

The first demonstration that chemicals could be used to induce cancer in animals was carried out in 1915. Coal tar, containing carcinogenic PAHs, induced skin carcinomas on the ears of rabbits. Carcinogenic PAHs are derived from phenanthrene (Figure 2.6a). Additional rings and/or methyl groups in the bay region of the three aromatic rings can convert inactive phenanthrene into an active carcinogen. Benzo[a]pyrene (BP), the most well-known carcinogen in cigarette smoke, and 7,12-dimethyl benz[a] anthracene (DMBA), one of the most potent carcinogens, are examples of PAHs. PAHs must be metabolized further in order to give the ultimate carcinogen that will form adducts with purine bases of DNA. The P450 enzyme, CYP1A1, is the predominant enzyme that metabolizes BP to the highly reactive mutagenic BP diol epoxides (Figure 2.6b). BP results mainly in G→T transversions.

(a)

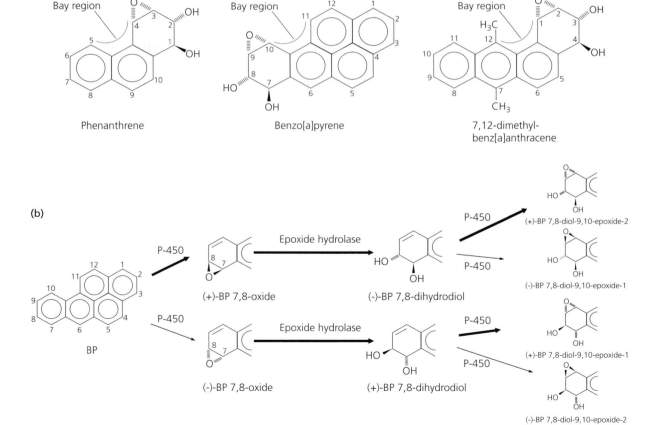

(b)

Aromatic amines

Heterocyclic amines (HCAs) are carcinogens produced by cooking meat, formed from heating amino acids and proteins. About 20 HCAs have been identified. Three examples, Phe-P-1, IQ, and Mel Q, are shown in Figure 2.7. It is important to be aware of these, as they illustrate an example of carcinogens to which we may be exposed daily and which are produced in our own kitchens.

A PIONEER IN THE FIELD . . . of molecular carcinogenesis: Gerd Pfeifer

Gerd Pfeifer has made important contributions to determining the molecular mechanisms of cancer. Investigations into skin and lung cancer provided strong evidence that UV radiation and carcinogens in cigarette smoke are causative agents for each cancer, respectively. Pfeifer and his colleagues demonstrated that the mutational hotspots of the *p53* gene observed in skin cancer cells are due to low-efficiency repair of DNA at these sites, as discussed in Box "Skin cancer." By mapping DNA adducts of the *p53* gene that are formed after exposure to benzo[a]pyrene diol epoxide (a potent cigarette carcinogen), Pfeifer and colleagues showed that the locations of these adducts matched the distribution of *p53* gene mutations in lung tumors from smokers. This seminal work, reported in *Science* in 1996, provided a direct causal link between a defined carcinogen and lung cancer.

Nitrosamines and nitrosamides

Many nitrosamines and nitrosamides are found in tobacco or are formed when preservative nitrites react with amines in fish and meats during smoking. The structure of alkylnitrosoureas, examples of nitrosamines, is shown in Figure 2.8(a). Their principal carcinogenic product is alkylated

PAUSE AND THINK

What is the structural difference between IQ and Mel Q?

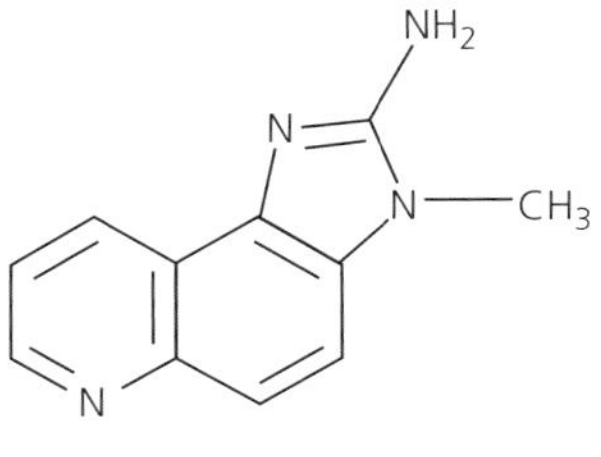

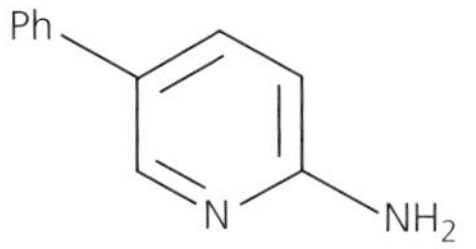

2-Amino-5 phenylpyridine
(Phe-P-1)

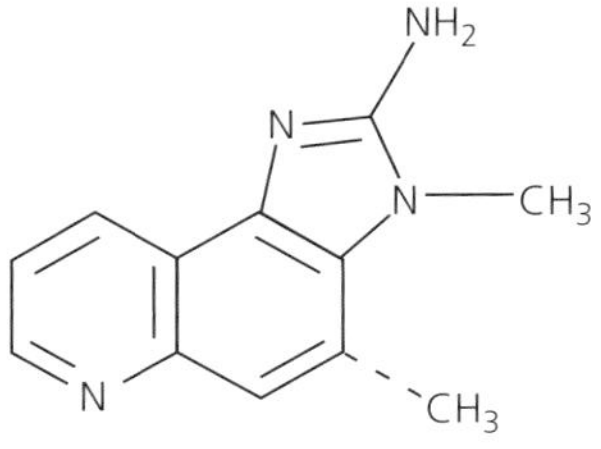

2-Amino-3 methylimidazo
[4,5-f] quinoline (IQ)

2-Amino-3,4 dimethylimidazo
[4,5-f] quinoline (Mel Q)

Figure 2.7 Heterocyclic amines.

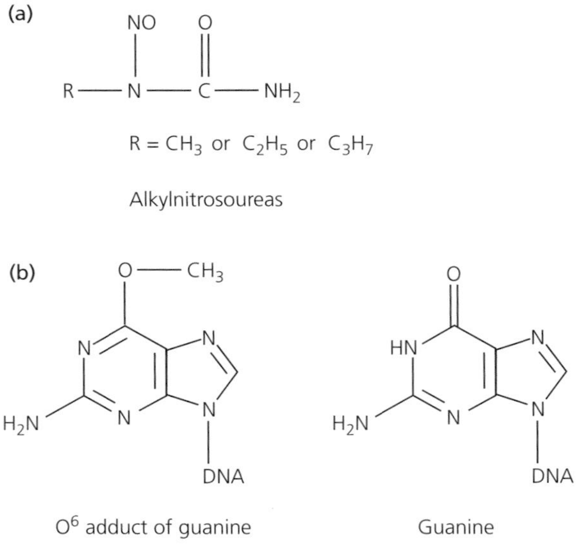

(a)

$R = CH_3$ or C_2H_5 or C_3H_7

Alkylnitrosoureas

(b)

O^6 adduct of guanine   Guanine

Figure 2.8 (a) An example of nitrosamines: alkylnitrosoureas. (b) A potential carcinogenic product of nitrosamines: O^6 adduct of guanine. Guanine is shown for comparison.

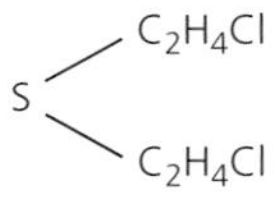

Figure 2.9 Structure of mustard gas.

LIFESTYLE TIP

It has been suggested that changes in the way we prepare food can reduce the amounts of HCAs produced. Oven-roasting, marinating, and coating food with breadcrumbs before frying are modifications that may reduce the formation of HCAs.

O^6 guanine derivatives, as shown in Figure 2.8(b) (guanine is depicted next to it for comparison).

Alkylating agents

Mustard gas (sulfur mustard; Figure 2.9) is the most well-known example of an alkylating agent because of its use and consequences observed during World War I. It is a bi-functional (having two reactive groups) carcinogen that is able to form intra- and inter-chain cross-links on DNA directly.

Fibrous minerals: asbestos and erionite

Asbestos and erionite are naturally occurring fibrous minerals that act as chemical/physical carcinogens and mutagens. In 2009, the International Agency for Research on Cancer (IARC) concluded, from a study of women heavily exposed to asbestos in the workplace, that asbestos induced mutagenicity, chromosomal aberrations, aneuploidy, polyploidy, and epigenetic alterations in the disease process. Asbestos is a group of fibrous silicate minerals that was used extensively in building materials because of its insulating properties but is now prohibited in several countries (but is still used in the developing world) because of its association with several diseases of the lung, including lung cancer and mesothelioma. Erionite is a fibrous zeolite mineral formed from volcanic rock. Although the exact mechanisms of carcinogenesis remain unclear, they may include generation of ROS, physical interaction with the mitotic machinery, and induction of a chronic inflammatory response (see Chapter 13), in addition to genetic and epigenetic alterations. Genetics may predispose some people to the carcinogenic effects of fibrous materials.

HOW DO WE KNOW THAT?

Family pedigrees

Malignant mesothelioma is a rare cancer that is linked to fibrous minerals. It is rare in the USA and the UK; however, there is an epidemic of malignant mesothelioma in several small villages of Turkey that has been linked to exposure to erionite. Fifty percent of all deaths in these regions are due to malignant mesothelioma! Pedigree studies of families that live in these villages uncovered a link between genetics and fiber carcinogenesis. Chemical and physical analysis, including scanning electron microscopy, mass spectrometry, and X-ray diffraction, showed that there was no difference between the type of erionite in villages that exhibited the epidemic and those that did not. However, malignant mesothelioma was prevalent in some families and absent in others, and marriages between these two groups led to some offspring developing the disease. Thus, the data demonstrate that a genetic predisposition can influence mineral fiber carcinogenesis.

Dogan *et al.* (2006) and family pedigree data within: available online at http://cancerres.aacrjournals.org/cgi/reprint/66/10/5063.

Infectious pathogens as carcinogens

Early in the 20th century, viruses were shown to cause tumors in animals. As we will see in Chapter 4, they have been invaluable tools for investigating the molecular events of cell transformation. Viruses that are oncogenic can be classified as DNA tumor viruses or RNA tumor viruses (also called retroviruses), depending on the nucleic acid that defines their genome. The mechanisms of carcinogenesis for these two classes of virus differ. DNA tumor viruses encode viral proteins that block tumor suppressor genes, often by protein–protein interactions (discussed in Chapter 6). Many retroviruses cause cancers in animals by encoding mutated forms of normal genes (i.e. oncogenes) that have a dominant effect in host cells (discussed in Chapter 4). Mechanisms of replication between DNA and RNA viruses also differ. Some DNA viruses, such as human papillomavirus and Epstein–Barr virus, replicate strictly as episomes within host cells. Retroviruses replicate by integration of the viral genome into the host DNA and utilize the host's translational machinery to produce viral proteins. Integration may lead to deregulated gene expression.

Direct causation of cancer by specific viruses and bacteria has been demonstrated and will be discussed in detail in Chapter 13. However, a few named examples are given below. The IARC has classified 12 types of human papillomavirus (types 16, 18, 31, 33, 35, 39, 45, 51, 52, 56, 58, 59) as human carcinogens and causative agents of cervical cancer. In addition, Kaposi's sarcoma-associated herpesvirus (KSHV) causes Kaposi's sarcoma; hepatitis B virus is associated with liver cancer, and Epstein–Barr virus (EBV) with nasopharyngeal carcinoma. The human T-cell lymphotropic virus type 1 (HTLV-1) is the only retrovirus known to cause cancer in humans. It causes acute T-cell leukemia (ATL). *Helicobacter pylori*, a Gram-negative spiral bacterium, establishes chronic infection and ulcers in the stomach and alters host cell function, which is associated with carcinogenesis. The IARC has classified *H. pylori* as a human carcinogen and one of the causative agents of gastric cancer. The typhoid pathogen, *Salmonella enterica* serovar Typhi (*S. typhi*), establishes chronic infection in the gallbladder and has been linked to hepatobiliary and gallbladder carcinoma. The molecular events behind the mechanism of bacteria-induced transformation are the subject of current studies. The promotion of host cell proliferation, the generation of oxygen free radicals and subsequent DNA damage, and the activation of oncogenes are areas of investigation.

Endogenous carcinogenic reactions

In addition to carcinogens, endogenous cellular reactions generate mutations. Oxidative respiration and lipid peroxidation, two processes of normal cell metabolism, produce ROS that can react with DNA and lipids to produce oxidized products (e.g. 8-oxoguanine) also seen by exposure to radiation (see Box "A little lesson about ROS ... "). During respiration, the initiating radical, superoxide anion ($O_2^-\cdot$), is produced upon reduction of NADH and formation of ubisemiquinone during oxidative phosphorylation. Therefore, breathing generates ROS intermediates similar to those generated by radiation! However, the distribution of these intermediates differs between the two sources: radiation produces extremely reactive hydroxyl radicals immediately and randomly within a cell, while respiration produces the less reactive superoxide radical immediately and only at specific locations within the cell.

Endogenous chemical reactions (e.g. hydrolysis of the glycosidic bond between a base and deoxyribose, producing an abasic site) also contribute to the formation of mutations. Deamination of cytosine to form uracil is the most common. A family of cytidine deaminases, called APOBECs (apolipoprotein B mRNA editing enzyme, catalytic polypeptide-like), produces a mutational signature that is present in many cancer types. APOBECs play a role in immune responses that protect against retroviruses, such as HIV, but these enzymes also deaminate cytosines in the host genome. Cytosine deamination by these enzymes occurs preferentially at TCA or TCT sequences and results in the conversion of cytosine to uracil. Uracils are removed from the DNA by uracil-DNA glycosylase enzymes, creating abasic sites, and DNA polymerases then insert an A across from the abasic site, which results in the C→T transition.

Clusters of TC→TT transitions on the same strand of DNA and within localized regions from APOBEC activity is one mechanism for the formation of kataegis. Some members of the APOBEC family are overexpressed in several human cancer types. Errors during DNA replication and DNA recombination contribute to the formation of mutations, although the DNA polymerases used possess proofreading ability to help minimize the number of mutations caused in this way. The proofreading function is dependent on the 3′–5′ exonuclease activity of the polymerase. If an incorrect nucleotide is added to the growing 3′ end of the newly synthesized strand, the DNA double helix exhibits melting; that is, the strands remain separated at this point. Melting causes the polymerase to pause, and the strand is transferred to the exonuclease site. Here, the incorrect nucleotide is removed; the strand is transferred back to the original polymerase binding site, and DNA synthesis reoccurs. Overall, it is estimated that 10^4–10^6 mutations occur in a single human cell per day. By and large, under normal circumstances, this immense error burden is successfully dealt with by the highly efficient cellular DNA repair mechanisms.

2.4 DNA repair and predispositions to cancer

DNA repair is an important line of defense against mutations caused by carcinogens and by endogenous mechanisms. If DNA lesions are not repaired before a cell replicates, they may contribute to carcinogenesis. Repair of the various types of DNA damage is accomplished by several different DNA repair mechanisms. Five types of DNA repair systems are described in the following sections: one-step repair, nucleotide excision repair, base excision repair, mismatch repair, and recombinational repair. Defects in most of these pathways result in a predisposition to cancer.

One-step repair

One-step repair involves the direct reversal of DNA damage. The repair enzyme alkyltransferase directly removes an alkyl group from the O^6 atom of guanine after exposure of DNA to alkylating carcinogens such as N-methylnitrosourea. In this case, a methyl group is transferred to a cysteine residue on the alkyltransferase, and the alkyltransferase becomes inactive.

Nucleotide excision repair (NER)

NER is specific for helix-distorting lesions such as pyrimidine dimers and bulky DNA adducts induced by environmental agents (UVB and PAHs, respectively). This damage interferes with transcription and replication, as described earlier. Two subpathways exist: global genome NER surveys the genome for helix distortion, and transcription-coupled repair identifies damage that interferes with transcription. The lesion, along with some (24–32) adjacent nucleotides, is excised out by endonucleases, and DNA polymerase δ/ε is used to fill in the gap using the opposite strand as a template. Proliferating cell nuclear factor is part of the polymerase holoenzyme and physically forms a ring that encircles and binds the damaged region. Xeroderma pigmentosum (XP) is an inherited disorder characterized by a defect in NER. Affected individuals are hypersensitive to the sun and have a 1000-fold increased risk of skin cancer. Seven XP gene products (XPA–XPG) have been identified out of the 25 proteins involved in NER.

Base excision repair (BER)

BER targets chemically altered bases (e.g. 8-oxoguanine) induced mostly by endogenous mechanisms; in the absence of such repair, the damage will cause a point mutation. The 8-oxoguanine lesion can functionally mimic T and form a stable 8-oxoguanine:A base pair, often bypassing detection by replicative DNA polymerases. Failure to remove 8-oxoguanine before replication results in a G→T transversion mutation. The first step of BER is carried out by a family of DNA damage-specific glycosylases (such as OGG1 and MUTYH) which scan millions of base pairs per second for 8-oxoguanine lesions. These glycosylases then flip the lesion outside of the helix and cleave the base from the DNA backbone, creating an abasic site. Subsequently, an endonuclease cleaves the DNA strand at the abasic site, and DNA polymerase β replaces the nucleotide and ligase fills the gap. Poly (ADP-ribose) polymerase (PARP) interacts with single strand break intermediates formed during BER and synthesizes a poly (ADP-ribose) chain that signals to other DNA repair proteins and also leads to modification of histones and relaxed chromatin structure for increased DNA accessibility. Mutations in the *OGG1* gene that codes for the principal glycosylase responsible for the repair of 8-oxoguanine:C base pairs have not been identified in tumors to date. No inherited defects in BER had been identified in humans until recently—mutations in the *MUTYH* (formerly *hMYH*) gene that encodes a DNA glycosylase responsible for the removal of mismatched adenines paired with 8-oxoguanine may be the principal cause of multiple colorectal adenoma syndrome (David *et al.*, 2007).

Mismatch repair

Mismatch repair corrects replication errors that have escaped editing by polymerases. It includes repair of insertions and deletions produced as a result of slippage during the replication of repetitive sequences, as well as nucleotide mismatches. The molecular events can be described in brief as follows:

- recognition of the mismatch is carried out by proteins hMSH2/6 and hMSH2/3
- hMLH1/hPMS2 and hMHL1/hPMS1 are recruited
- the newly synthesized strand is identified (flagged by the replication machinery)
- endonucleases and exonucleases remove the nucleotides around, and including, the mismatch
- DNA polymerases resynthesize a newly replicated strand.

Hereditary non-polyposis colorectal cancer (HNPCC) is one of the most common cancer syndromes in humans. Half of all patients with HNPCC carry a germline mutation in *hMLH1* or *hMSH2*. Loss of function of the protein products encoded by these genes is responsible for complete loss of mismatch repair. Thus, cells are vulnerable to mutations.

Recombinational repair

Homologous recombination and non-homologous end-joining are two types of recombinational repair that mend double-strand DNA breaks. Homologous recombination depends on the presence of sister chromatids formed during DNA synthesis as a template for recombining severed ends. Many members of the same protein family make up a complex that performs what has been nicknamed DNA gymnastics. The molecular events shown in Figure 2.10 are described in brief:

(a) A double-strand break activates the ataxia telangiectasia mutated (ATM) kinase.

(b) The RAD50/MRE11/NBS1 complex (a substrate of ATM) uses its 5'–3' exonuclease activity (depicted by scissors in Figure 2.10) to create single-stranded 3' ends.

(c) BRCA1/2 aids in the nuclear transport of RAD51 (shown as gray circles).

(d) RAD52 facilitates RAD51 binding to these exposed ends to form a nucleoprotein filament.

(e) RAD51 can exchange a homologous sequence from a single strand within a double-stranded molecule (shown in red; e.g. a sister chromatid), with a single-stranded sequence.

(f) The sequences from the double-stranded molecule are then used as a template sequence for repair.

(g) Resolvases restore the junctions formed as a result of homologous recombination, called Holliday junctions.

(h) Two copies of intact DNA molecules are produced with rarely any errors.

Ataxia telangiectasia is an inherited syndrome whereby patients have a mutation in the ataxia telangiectasia mutated (ATM) kinase. Patients are sensitive to X-rays and have an increased risk of lymphoma. Germline mutations in the *BRCA1* and *BRCA2* genes give rise to an increased risk of breast and ovarian cancers. Those who develop cancer show that they acquired a second mutation later in life and illustrate that *BRCA1* and *BRCA2* act as tumor suppressor genes. Loss of homologous recombination leads to genomic instability and carcinogenesis. Note that loss of function of these genes also occurs in sporadic tumors.

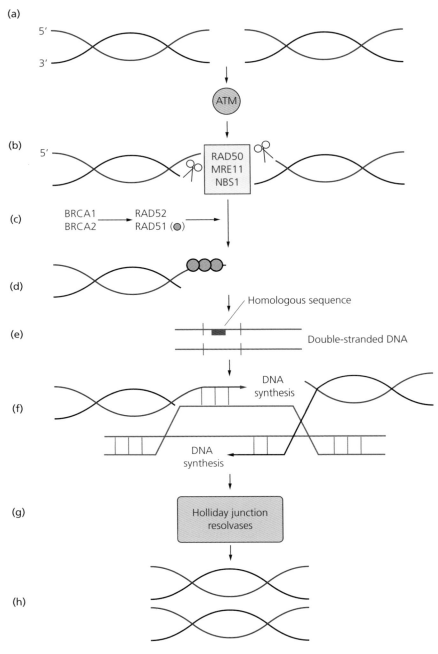

(a)

5′
3′

ATM

(b)

5′

RAD50
MRE11
NBS1

(c)

BRCA1 ⟶ RAD52
BRCA2   RAD51 (●)

(d)

Homologous sequence

(e)

Double-stranded DNA

(f)

DNA
synthesis

DNA
synthesis

(g)

Holliday junction
resolvases

(h)

Homologous recombination complete

Figure 2.10 Recombinational repair.

Suffice it to say that the other type of recombinational repair, end-joining, links non-homologous ends and is therefore error-prone, and can possibly result in chromosomal translocations.

One of the main molecular players involved in carcinogenesis is p53 and should be mentioned here. p53, "the guardian of the genome," is a protein that plays an important role in the molecular events that protect the integrity of DNA; it is central in the orchestration of DNA repair. The details of this important tumor suppressor protein will be discussed in Chapter 6.

PAUSE AND THINK

How do you think DNA repair proteins locate DNA damage among the genome's 3 billion nucleotide base pairs? One conventional proposal is that proteins constantly scan the DNA for damage. But is this really feasible? A more recent proposal focuses on the ability of DNA to act as a "live wire" and carry an electric current, as discussed in a feature in *Science* magazine (December 2014). The bases of DNA consist of aromatic rings that have overlapping electron clouds that create a continuous path for electrons. Breaks in the charge transport system in a DNA molecule caused by abnormalities in the DNA structure (mismatches and lesions) could act as a sensor for DNA damage and a way to monitor the integrity of DNA. It will be interesting to follow the data that support this proposal (Grodick *et al.*, 2015).

 Therapeutic strategies

2.5 Conventional therapies: chemotherapy and radiation therapy

Conventional therapies continue to extend and save lives. It is important to understand their rationale before moving to more molecular approaches discussed later in the text. Several conventional therapies aim to induce extensive DNA damage in order to trigger apoptosis, and paradoxically include agents classified as carcinogens. Other conventional therapies inhibit DNA metabolism in order to block DNA synthesis in the rapidly dividing cancer cells. DNA synthesis is essential to produce a new set of chromosomes for the daughter cells produced by cell division. Still other drugs interfere with the mechanics of cell division. Both chemotherapies and radiotherapy will be discussed.

Chemotherapy

A brief description and examples of the three main types of classical chemotherapy are given in the following sections.

Alkylating agents and platinum-based drugs

Alkylating agents and platinum-based drugs work by a similar mode of action. Alkylating agents have the ability to form DNA adducts by covalent bonds via an alkyl group. They may act during all phases of the cell cycle. Chlorambucil (Figure 2.11a) is one example of a member of the nitrogen mustard family of drugs. Its usual target is the N7 position

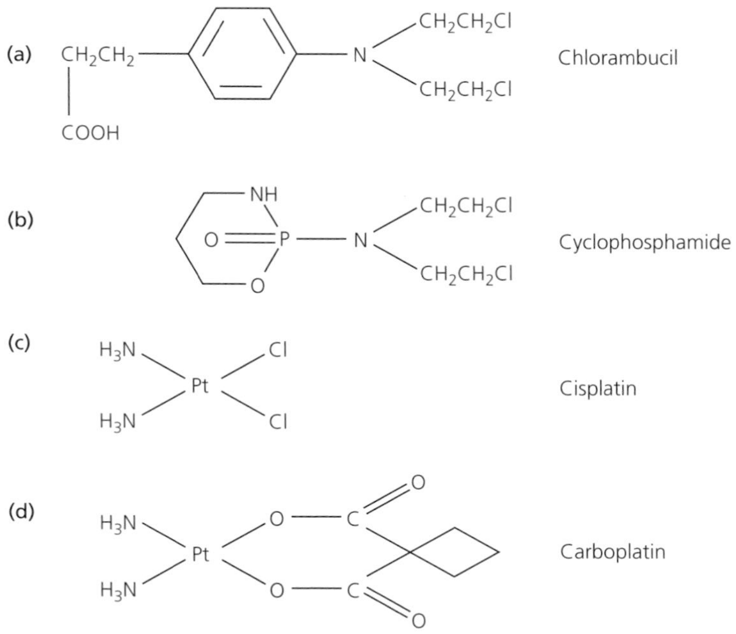

Figure 2.11 Examples of alkylating agents and platinum-based drugs.

PAUSE AND THINK

Do you recall a similar mechanism of action for any carcinogens?

of guanine residues. Bi-functional alkylating agents (compounds with two reactive groups) form intra-strand and inter-strand cross-links in DNA that alter the conformation of the double helix or prevent separation of the DNA strands and interfere with DNA replication. They are much more potent than monofunctional analogs, indicating that cross-linking is the basis of their function, as monofunctional analogs cannot cross-link.

Some drugs require metabolic activation within the body. The alkylating agent cyclophosphamide (Figure 2.11b) is one example. Oxidases in the liver produce an aldehyde form that decomposes to yield phosphoramide mustard, the biologically active molecule.

The platinum-based drugs, such as cisplatin [cis $Pt(II)(NH_3)_2Cl_2$] and carboplatin (Figure 2.11c and 2.11d, respectively), form covalent bonds via the platinum atom. Cisplatin is a water-soluble molecule that contains a Pt atom bound to four functional groups. The Pt–N bond has a covalent character and is essentially irreversible, whereas that with Cl is more labile. Cl is replaced with water in the plasma and cytosol, before the molecule binds to the N7 position of guanine and adenine in its DNA target. The GG, AG, and GXG (where X can be any base) adducts comprise over 90% of the total. The resulting DNA damage triggers apoptosis. Although cisplatin had a major impact on some cancers, such as ovarian cancer, it was associated with irreversible kidney damage. Later, carboplatin was identified as a less toxic platinum analog.

Antimetabolites

Antimetabolites are compounds that are structurally similar to endogenous molecules (e.g. nitrogenous bases of DNA), and therefore can mimic their role and inhibit nucleic acid synthesis. Two examples, fluorodeoxyuridylate (F-dUMP) and methotrexate, are shown alongside similar endogenous molecules, deoxyuridylate (dUMP) and dihydrofolate, respectively, in Figure 2.12. 5-fluorouracil (5-FU) is a derivative of uracil and is converted into F-dUMP. F-dUMP competes with the natural substrate dUMP for the catalytic site of thymidylate synthase, the enzyme that produces deoxythymidylate (dTMP) (Figure 2.13). F-dUMP forms a covalent complex with the enzyme and acts as a suicide inhibitor, generating an intermediate that

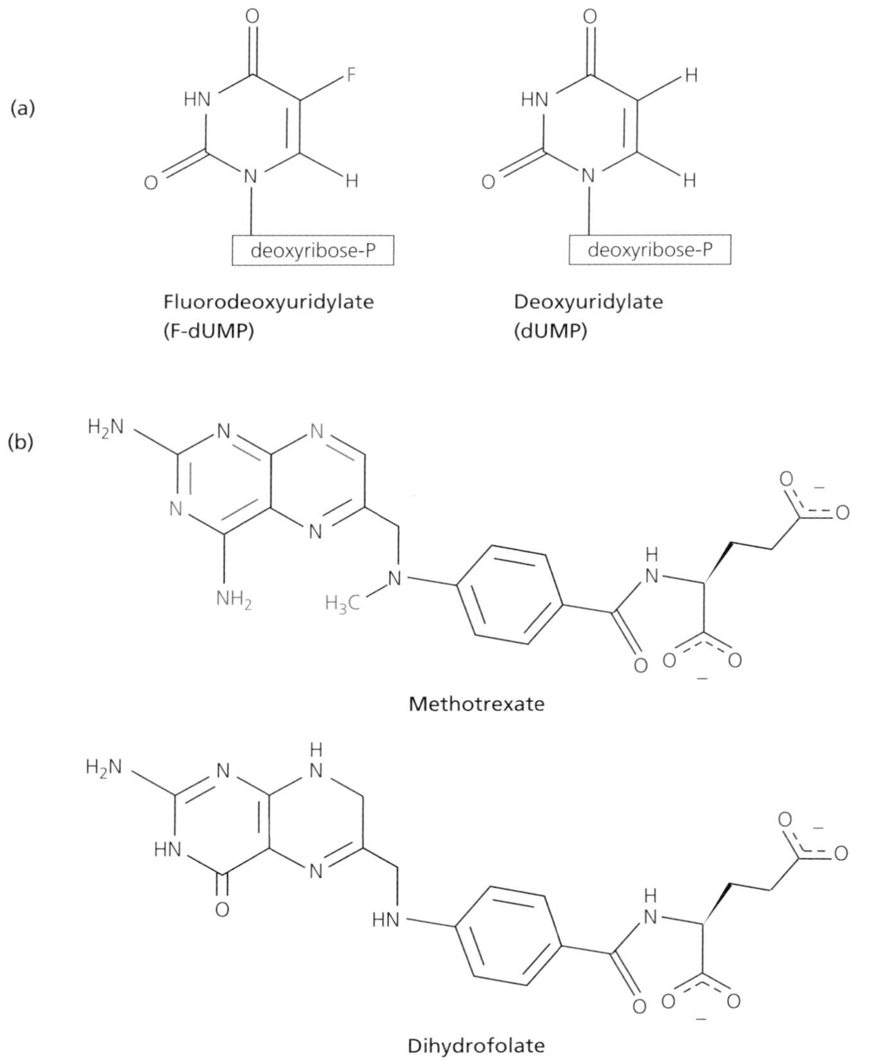

Figure 2.12 Antimetabolites: (a) fluorodeoxyuridylate (F-dUMP) and (b) methotrexate. Structural differences between the antimetabolite and endogenous molecule are shown in red.

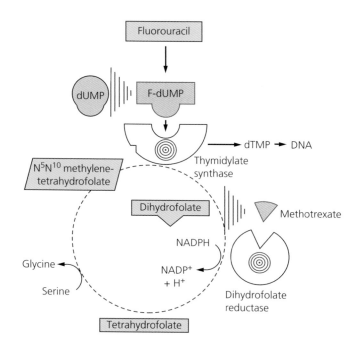

Figure 2.13 Action of antimetabolites fluorodeoxyuridylate (F-dUMP) and methotrexate (both shaded red). The enzyme thymidylate synthetase uses N^5N^{10} methylene-tetrahydrofolate as a methyl donor and catalyzes the methylation of dUMP to form dTMP. The cancer drug fluorouracil is converted into the antimetabolite F-dUMP (red rectangular shape), which competes (/////) with dUMP and targets thymidylate synthetase (target symbol, ◎, shown). Methotrexate (red triangle) is an antimetabolite that competes (/////) with dihydrofolate, and methotrexate targets the enzyme dihydrofolate reductase (target symbol, ◎, shown).

inactivates the thymidylate synthase through covalent modification. As a result, the dTMP and dTTP pools are depleted; dUMP and dUTP accumulate, and DNA synthesis in rapidly dividing cells is severely compromised. Another important antimetabolite, methotrexate, targets an accessory enzyme of the same reaction. As an analog of dihydrofolate, methotrexate is a competitive inhibitor of dihydrofolate reductase, the enzyme used to regenerate tetrahydrofolate. Tetrahydrofolate produces N^5N^{10} methylene-tetrahydrofolate that is required in the thymidylate synthase reaction (Figure 2.13; see Chapter 11 for further discussion of tetrahydrofolate).

Organic drugs

Doxorubicin is a microbial anthracycline antibiotic that inhibits topoisomerase II, an enzyme that releases torsional stress during DNA replication, by trapping single-strand and double-strand DNA intermediates. It can also form DNA adducts and intercalate into DNA. Doxorubicin diffuses across cell membranes and accumulates in most cell types. Cardiac damage is its most severe side effect, but new compounds (e.g. ICRF-187) that can block the cardiac toxicity are being investigated. These drugs are primarily used to treat solid tumors (e.g. of the breast or lung).

The plant alkaloids vincristine and vinblastine (from the Madagascar periwinkle plant) bind to tubulin and prevent microtubule assembly, in contrast to the drug paclitaxel (taxol) which binds to the β-tubulin subunit in polymers and stabilizes the microtubules against depolymerization. Thus, two opposing strategies can be used to disrupt the mitotic spindle.

Radiation therapy

Radiation therapy, either alone or in combination with other therapies, is received by approximately 60% of cancer patients in the USA. Ionizing radiation is usually delivered to the tumor by electron linear accelerators. Radiation reacts with water inside cells to generate ROS that damage DNA. Apoptosis will be induced in cells that contain large amounts of DNA damage. The supply of oxygen affects the potency of ionizing radiation and is thought to be caused by the generation of ROS. Oxygen can assist in making radiation-induced damage permanent. More double-strand breaks occur in cells irradiated in the presence of oxygen than in cells irradiated in the absence of oxygen. Therefore, the number of zones of hypoxia within a solid tumor influences the outcome of radiation treatment. Targeting of the tumor has been made more precise by modern techniques such as magnetic resonance imaging (MRI) and computed tomography (CT) which produce three-dimensional images of the tumor within the body.

Heterogeneous cell sensitivity and drug resistance: obstacles to these treatments

A major obstacle in achieving long-term effects with chemotherapy (and molecularly targeted therapy) is drug resistance. Tumor resistance can be a characteristic that is present before treatment (intrinsic) or acquired during treatment (extrinsic). Pharmacokinetic effects (absorption, distribution, metabolism, and elimination (ADME)) influence the amount of a systemically administered drug that reaches the tumor. The cells of a tumor are heterogeneous with respect to drug exposure and inherent characteristics such as mutational profiles. Cancer cells, as part of a large tumor mass, will receive different doses of treatment, depending on the location of individual cells within the mass. Cells deep within the tumor, and therefore furthest from the blood supply, will receive lower doses than cells on the surface of the tumor. Cells within the same tumor may have acquired different mutations, some that lead to drug resistance. Cells within the tumor that are classified as cancer stem cells (discussed in Chapter 8) are intrinsically resistant to the therapies.

The effects of a drug can also be affected by or within a cancer cell. Anticancer drugs impose a strong force for the selection of cells that can acquire drug resistance. There are several mechanisms that a cancer cell may utilize to become resistant to chemotherapy (Figure 2.14). Cells may become resistant by increasing the efflux of the drug, decreasing the intake of the drug, increasing the number of target molecules within the cell, or altering drug metabolism or DNA repair processes. Increasing the efflux of a drug is regulated at the cell surface. There is a family of ATP-dependent transporters that are involved in the movement of nutrients and other molecules across membranes. The multi-drug resistance gene (*MDR1*) codes for one

PAUSE AND THINK

We are familiar with the concept that overuse of antibiotics can lead to the creation of resistant bacteria. Their rapid division and high frequency of mutation allow for the selection of bacterial strains that can evolve to survive in the presence of the drug. Selection of a drug-resistant clone of human tumor cells that are genetically unstable can evolve in a similar manner.

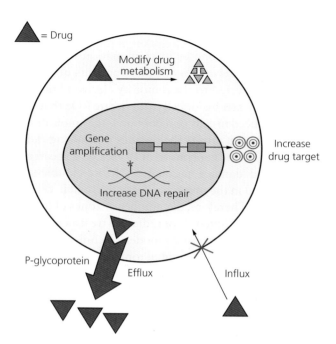

Figure 2.14 Mechanisms of drug resistance. Based on a diagram printed in *Annu. Rev. Med.* 53, Copyright (2002) by Annual Reviews (http://www.annualreviews. org/).

member of this family called P-glycoprotein (P-gp) or the multi-drug transporter. This protein, normally a chloride ion efflux pump, can bind a variety of chemotherapeutic drugs, including doxorubicin, vinblastine, and taxol. Upon binding, ATP is hydrolyzed and causes a conformational change of P-gp. As a result, the drug is released extracellularly. The transporter can be recycled by a second hydrolysis of ATP and continue to increase the efflux of the drug. Some drugs utilize specific transporters to enter cells. Mutations in these receptors may render them non-functional and decrease influx of the drug. Resistance to methotrexate commonly occurs by mutation of the folate transporter. An increase in the number of drug target molecules by gene amplification is another means of developing resistance against methotrexate. The *DHFR* gene is amplified in some cancer cells. An increase in the efficiency of DNA repair, such as increased alkyltransferase activity, can give rise to resistance from alkylating agents such as doxorubicin. Levels of this enzyme are found to be highly variable in different tumors.

2.6 Strategies that target DNA repair pathways

Synthetic lethal strategies

The name "synthetic lethal strategies" sounds like it was used in a James Bond film. It involves interactions whereby inhibiting the function of one gene is cytotoxic only in the presence of an additional mutation. Let us look at the development of poly-(ADP-ribose) polymerase (PARP) inhibitors for use in tumors with *BRCA1/2* mutations as an example of a synthetic lethal strategy. PARPs are enzymes that are key in base excision repair, a pathway that repairs DNA single-strand breaks. These enzymes synthesize poly (ADP-ribose) polymers that rapidly bind to DNA strand breaks to amplify the DNA damage signal and recruit DNA repair proteins. Inhibition of PARP causes impaired base excision repair and the accumulation of single-strand breaks that can lead to double-strand breaks, normally repaired by homologous recombination pathways. The synthetic lethal strategy takes advantage of the characteristic DNA repair defect caused by mutations in tumors in some cases of inherited breast cancer. Some of these patients carry a germline mutation in either tumor suppressor gene, *BRCA1* or *BRCA2*, and acquire a somatic mutation in the other *BRCA1/2* allele later in life. This results in a tumor that is deficient in homologous recombination and double-strand repair; the non-tumor cells in the patient do not carry the homozygous *BRCA1/2* mutated genotype and retain the homologous recombination DNA repair pathway. Tumors cells that lack BRCA1/2 function and are treated with PARP inhibitors have impaired homologous recombination and base excision repair pathways, and these characteristics are lethal; healthy cells treated with PARP inhibitors have a functional homologous recombination pathway and are viable (Figure 2.15). The success of the oral small molecule PARP inhibitor called olaparib (Lynparza™, AstraZeneca) in clinical trials led to its approval by the Food and Drug Administration (FDA) and European Commission in 2014 for patients with ovarian cancer with *BRCA1* or *BRCA2* mutations. The approval of olaparib with its companion diagnostic for treatment in BRCA-positive ovarian cancer is a great stride for personalized cancer therapy. Additional PARP inhibitors are being tested in clinical trials for other cancer types. Prostate cancer cells with *BRCA1* and *BRCA2* mutations or a specific gene fusion (TMPRSS2:ERG) shows

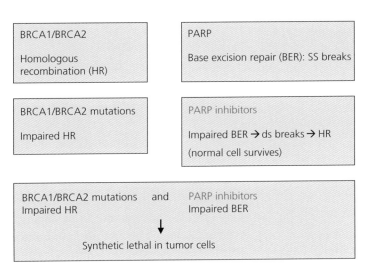

Figure 2.15 The rationale of synthetic lethal interactions: PARP inhibitors.

PAUSE AND THINK

Look at other molecules involved in recombination repair. Do you think that cancer patients with tumors that harbor mutations in genes, such as *ATM* or *MRE11*, would also benefit from PARP inhibitors?

synthetic lethality with PARP inhibitors, and clinical trials of PARP inhibitors are being carried out in prostate cancer patient cohorts that carry these mutations (Tangutoori *et al.*, 2015).

Combination therapies

Inhibition of DNA repair pathways promises to be an important strategy to maximize a tumor's sensitivity to chemotherapy and radiotherapy, as they rely on damaging DNA to exert their effect. Many clinical trials are testing combinations of PARP inhibitors with conventional DNA-damaging chemotherapies, such as platinum-based drugs and alkylating agents, and also with radiotherapy. The therapeutic index of drug cocktails is important to monitor so as not to increase toxicity in healthy cells.

▨ CHAPTER HIGHLIGHTS—REFRESH YOUR MEMORY

- In simplistic terms, a gene consists of a regulatory region and a coding region. Mutations in the former may alter gene expression, while mutations in the latter may affect the gene product.

- Most carcinogens are mutagens.

- Several types of mutations include: base substitutions (transitions and transversions), frameshift mutations (insertions or deletions), and chromosomal translocations.

- An accumulation of mutations over time is responsible for transforming a normal cell into a cancer cell for most cancers, but a single event of chromothripsis may underlie a small number of cancers.

- Small localized areas of hypermutation, called "kataegis" (Greek for "thunderstorm"), have been identified in cancer genomes and may be due to APOBEC activity.

- Carcinogens include radiation, chemicals, and infectious pathogens.

- Radiation can damage DNA directly or indirectly through the formation of ROS.

- Three intermediate ROS formed from the radiolysis of water are the hydroxyl radical, hydrogen peroxide, and the superoxide radical.

- The hydroxyl radical is one of the most reactive substances.

- Many carcinogens need to be metabolized to form an ultimate carcinogen that covalently binds to DNA.

- Many chemical carcinogens add functional groups covalently to DNA.

- DNA adducts can mask specific base pairs or distort the DNA helix and cause misreading during replication resulting in mutations.

- Both viruses and bacteria have been classified as carcinogens for specific cancers.

- One-step repair, nucleotide excision repair, base excision repair, mismatch repair, and recombinational repair are five systems for repairing damaged DNA.

- Patients with XP have an inherited defect in NER and have a 1000-fold increased risk of skin cancer.

- Many patients with hereditary non-polyposis colorectal cancer (HNPCC) have an inherited defect in mismatch repair.

- The major types of chemotherapies are:

  – alkylating agents—two examples are chlorambucil and cisplatin

  – antimetabolites—two examples are 5-FU and methotrexate

  – organic drugs—two examples are vincristine and vinblastine.

- The development of drug resistance is a major problem for chemotherapy.

- Targeting DNA repair pathways can help sensitize tumors to chemotherapy and has also been used to create synthetic lethal interactions.

■ ACTIVITY

1. Make a list of five carcinogens and the mutations they cause. Describe the method of DNA repair used to correct each type of mutation.

■ FURTHER READING

Alexandrov, L.B., Nik-Zainal, S., Wedge, D.C., Aparicio, S.A.J.R, Behjati, S., Biankin, A.V., *et al.* (2013) Signatures of mutational processes in human cancer. *Nature* **500**: 415–421.

Burns, M.B., Temiz, N.A., and Harris, R.S. (2013) Evidence for APOBEC3B mutagenesis in multiple human cancers. *Nat. Genet.* **45**: 977–983.

Chabner, B.A. and Roberts, T.G. (2005) Chemotherapy and the war on cancer. *Nat. Rev. Cancer* **5**: 65–72.

Evers, B., Helleday, T., and Jonkers, J. (2010) Targeting homologous recombination repair defects in cancer. *Trends Pharmacol. Sci.* **31**: 372–380.

Hecht, S.S. (2003) Tobacco carcinogens, their biomarkers and tobacco-induced cancer. *Nat. Rev. Cancer* **3**: 733–737.

Holohan, C., Van Schaeybroeck, S., Longley, D.B., and Johnston, P.G. (2013) Cancer drug resistance: an evolving paradigm. *Nat. Rev. Cancer* **13**: 714–726.

Huang, S.X.L., Jaurand, M.-C., Kamp, D.W., Whysner, J., and Hei, T.K. (2011) Role of mutagenicity in asbestos fiber-induced carcinogenicity and other diseases. *J. Toxicol. Environ. Health B Crit. Rev.* **14**: 179–245.

Ichihashi, M., Ueda, M., Budiyanto, A., Bito, T., Oka, M., Fukunaga, M., *et al.* (2003) UV-induced skin damage. *Toxicology* **189**: 21–39.

Jackson, S.P. and Bartek, J. (2009) The DNA-damage response in human biology and disease. *Nature* **461**: 1071–1078.

Pfeifer, G.P., You, Y.-H., and Besaratinia, A. (2005) Mutations induced by ultraviolet light. *Mut. Res.* **571**: 19–31.

Stratton, M.R. (2009) The cancer genome. *Nature* **458**: 719–724.

Williams, G.M. and Jeffrey, A.M. (2000) Oxidative DNA damage: endogenous and chemically induced. *Regul. Toxicol. Pharmacol.* **32**: 283–292.

■ SELECTED SPECIAL TOPICS

Alexandrov, K., Cascorbi, I., Rojas, M., Bouvier, G., Kriek, E., and Bartsch, H. (2002) CYP1A1 and GSTM1 genotypes affect benzo[a]pyrene DNA adducts in smokers' lung: comparison with aromatic/hydrophobic adduct formation. *Carcinogenesis* **23**: 1969–1977.

David, S.S., O'Shea, V.L., and Kundu, S. (2007) Base-excision repair of oxidative DNA damage. *Nature* **447**: 941–950.

Davies, H., Bignell, G.R., Cox, C., Stephens, P., Edkins, S., Clegg, S., *et al.* (2002) Mutations of the *BRAF* gene in human cancer. *Nature* **417**: 949–954.

Dogan, A.U., Baris, Y.I., Dogan, M., Emri, S., Steele, I., Elmishad, A.G., *et al.* (2006) Genetic predisposition to fiber carcinogenesis causes a mesothelioma epidemic in Turkey. *Cancer Res.* **66**: 5063–5068.

El Ghissassi, F., Baan, R., Straif, K., Grosse, Y., Secretan, B., Bouvard, V., *et al.* (2009) A review of human carcinogens—Part D: radiation. *Lancet Oncol.* **10**: 751–752.

Fong, P.C., Boss, D.S., Yap, T.A., Tutt, A., Wu, P., Mergui-Roelvink, M., *et al.* (2009) Inhibition of poly (ADP-ribose) polymerase in tumors from *BRCA* mutation carriers. *N. Engl. J. Med.* **361**: 123–134.

Forment, J.V., Kaidi, A., and Jackson, S.P. (2012) Chromothripsis and cancer: causes and consequences of chromosome shattering. *Nat. Rev. Cancer* **12**: 663–670.

Grodick, M.A., Muren, N.B., and Barton, J.K. (2015) DNA charge transport within the cell. *Biochemistry* **54**: 962–973.

Henderson, S., Chakravarthy, A., Su, X., Boshoff, C., and Fenton, T.R. (2014) APOBEC-mediated cytosine deamination links PIK3CA helical domain mutations to human papilloma-driven tumor development. *Cell Rep.* **7**: 1833–1841.

Ishikawa, K., Takenaga, K., Akimoto, M., Koshikawa, N., Yamaguchi, A., Imanishi, H., *et al.* (2008) ROS-generating mitochondrial DNA mutations can regulate tumor cell metastasis. *Science* **320**: 661–664.

Lane, N. (2002) *Oxygen—The Molecule that Made the World*. Oxford University Press, Oxford.

Nik-Zainal, S., Alexandrov, L.B., Wedge, D.C., Van Loo, P., Greenman, C.D., Raine, K., *et al.* (2012) Mutational processes molding the genomes of 21 breast cancers. *Cell* **149**: 979–993.

Pott, S. and Lieb, J.D. (2015) What are super-enhancers? *Nat. Genet.* **47**: 8–12.

Preston, D.L., Ron, E., Tokuoka, S., Funamoto, S., Nishi, N., Soda, M., *et al.* (2007) Solid cancer incidence in atomic bomb survivors:1958–1998. *Radiat. Res.* **168**: 1–64.

Smith, C.J., Perfetti, T.A., Garg, R., and Hansch, C. (2003) IARC carcinogens reported in cigarette mainstream smoke and their calculated log P values. *Food Chem. Toxicol.* **41**: 807–817.

Stephens, P.J., Greenman, C.D., Fu, B., Yang, F., Bignell, G.R., Mudie, L.J., *et al.* (2011) Massive genomic rearrangement acquired in a single catastrophic event during cancer development. *Cell* **144**: 27–40.

Tangutoori, S., Baldwin, P., and Sridhar, S. (2015) PARP inhibitors: a new era of targeted therapy. *Maturitas* **81**: 5–9.

Tubio, J.M.C. and Estivill, X. (2011) Cancer: When catastrophe strikes a cell. *Nature* **470**: 476–477.

Chapter 3

Regulation of gene expression

Introduction

Cancer is a disease of the genome at the cellular level that may be manifested by alterations in gene expression, altering the quantity or timing or localization of onco-gene and tumor suppressor gene products. Mutations in the promoter region of genes can cause these alterations. Recent and unexpected findings have revealed that up to 70% of the genome is transcribed into RNA, but only 2% of the genome codes for proteins! In fact, many disease-associated variations occur in non-coding DNA sequences. RNA molecules that do not code for protein are referred to as **non-coding RNAs** (ncRNAs), and some play an important role in the regulation of gene expression. ncRNAs can be transcribed from either the coding or non-coding DNA strand, from sequences that code for introns (intronic), or from regions between genes (intergenic). Figure 3.1 shows that the majority of cancer-related SNPs are located in non-coding regions of the genome (intergenic or intronic), and only a small number are found in coding regions. Indeed, we have a lot to learn.

Gene expression may be modulated in various ways: through the regulation of transcription, chromatin structure, and post-transcriptional mechanisms. In this chapter, we will review the molecular components involved in gene expression, including transcription factors, chromatin modifications and chromatin-binding proteins, ncRNAs, and telomeres, and how they can contribute to the processes underpinning cancer. As mentioned in Chapter 2, mutations in the promoter region of genes can alter the regulation of gene expression and lead to carcinogenesis. An additional mechanism of regulating gene expression, called **epigenetics** (Greek for "upon" the genome), is also important for carcinogenesis. It involves heritable alterations in gene expression encoded by modifications of the genome and its chromatin components, but not by changes in the DNA sequence. The structure of a

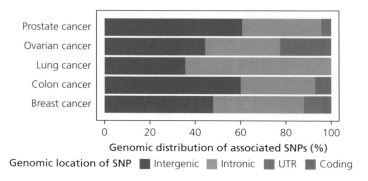

Figure 3.1 Genomic distribution (%) of SNPs in selected cancer types. Adapted by permission from Macmillan Publishers Ltd on behalf of Cancer Research UK: *British Journal of Cancer*, from Cheetham, S.W. *et al.* (2013) Long noncoding RNAs and the genetics of cancer. *Br. J. Cancer* **108**: 2419–2425, copyright (2013).

gene within the context of chromatin is described in order to elucidate how gene and chromatin structure affects gene expression. Throughout this chapter, there is a focus on DNA–protein interactions in transcriptional regulation, chromatin configuration, and telomere extension. In addition, the roles of ncRNAs, including long non-coding RNAs (lncRNAs) and microRNAs (miRNAs) that play a role in the regulation of post-transcriptional gene expression, are presented.

3.1 Transcription factors and transcriptional regulation

Transcription factors are proteins that bind to gene promoters and regulate transcription. About 3000 transcription factors regulate the 20,000 genes or so encoded in the human genome. Transcription factors contain a set of independent protein modules or domains, each having a specific role important for the function of transcription factors. They include DNA-binding domains, transcriptional activation domains, dimerization domains, and ligand-binding domains.

Four common types of DNA-binding domains are the helix–turn–helix motif, the leucine zipper motif, the helix–loop–helix motif, and the zinc finger motif. These domains are characteristic protein conformations that enable a transcription factor to bind DNA. It is the conformation of these protein domains that facilitates binding to DNA. Take the helix–turn–helix

PAUSE AND THINK

To illustrate how a domain works, let us look at an analogy in which an electrical plug represents a domain. There are many different household appliances with vastly different functions, such as a toaster, an iron, and a television. However, each appliance contains a plug that has a function independent of the rest of the appliance. It has a specific conformation that fits into an electrical socket and conducts electricity to power the appliance. A DNA-binding domain is that part of the transcription factor whose function is to recognize specific DNA promoter sequences and bind DNA. There is some variety in the structure of a domain in the same way that there are US, UK, and Continental European plugs.

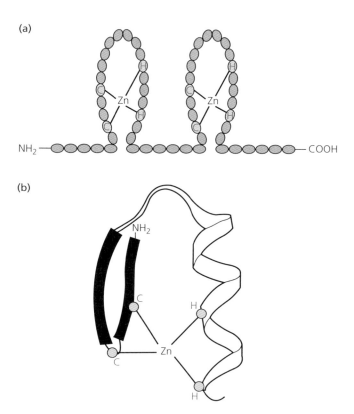

Figure 3.2 The zinc finger DNA-binding domain: (a) primary and (b) secondary structure.

and zinc finger domains as examples. The amino acid side-chains of the alpha helix portions of the helix–turn–helix motif lie in the major groove of the DNA helix and hydrogen bond to specific DNA base pairs. The zinc finger domain (approximately 30 amino acids long; Figure 3.2a) is configured around a zinc atom that links two cysteines and two histidines (shown in red) (or two cysteines and two cysteines). It consists of a simple ββα fold (Figure 3.2b). The side-chains of specific amino acids recognize a specific DNA sequence (about five nucleotide pairs). Transactivation domains function by binding to other components of the transcriptional apparatus in order to induce transcription by RNA polymerase, the main enzyme required for transcription. Some transcription factors work in pairs (forming what we call a "dimer") and require a dimerization domain which facilitates protein–protein interactions between the two molecules. Interactions between transcription factors are a common theme in transcriptional regulation. Some transcription factors only function upon binding of a ligand and therefore require a ligand-binding domain (this is analogous to the space for a coin in a pinball machine). The activity of a transcription factor can be regulated by several means: synthesis in particular cell types only, covalent modification such as phosphorylation, ligand binding, cell localization, and/or, if dimeric, exchange of partner proteins.

HOW DO WE KNOW THAT?

Experimental methods used to examine transcription factor binding (see Figure 3.3)

The interactions of transcription factors and their DNA response elements can be examined by several methods of molecular biology. These protein–DNA interactions are often first detected by gel/band shift assays (also called electrophoretic mobility shift assays or EMSAs). This assay involves the incubation of a protein, usually within a cell or nuclear extract, with a labeled DNA fragment containing the promoter sequences of interest. The products of the incubation reaction are analyzed on a non-denaturing polyacrylamide gel. DNA that has bound protein will be observed as a band that has migrated more slowly (been retarded) than unbound DNA (Figure 3.3a). Competition experiments that use irrelevant DNA fragments or known irrelevant proteins can be used to establish specificity.

Another technique that is used to examine protein–DNA interactions is DNase footprinting. DNase, an enzyme that cleaves DNA, is used to probe a promoter region. Protein bound to a DNA fragment will protect the DNA from DNase cleavage. The DNA is end-labeled and, after enzymatic treatment, is analyzed using gel electrophoresis and autoradiography. Areas of DNA that have been protected by protein binding will result in a clear region referred to as a "footprint" (Figure 3.3b).

A third method of investigating transcription is to construct deletions or point mutations in the promoter fragment, clone them into a reporter plasmid (e.g. luciferase, whose activity can easily be detected), and examine the effects of the mutations upon transfection into cells in culture or using *in vitro* transcription assays (Figure 3.3c). Deletion or mutation of a promoter region that is important for transcription is indicated by a decrease in transcription.

As we will see in the coming chapters, many oncogenic signaling pathways leading to uncontrolled growth, evasion of apoptosis, or aberrant differentiation converge on a single transcription factor that regulates a set of genes to produce a transformed phenotype. Thus, misregulation of a single transcription factor can cause cancer.

Many of the key points of transcriptional regulation can be demonstrated by two examples: the AP-1 transcription factor family and the steroid hormone receptors. The AP-1 transcription factor is important for the processes of growth, differentiation, and death, and therefore plays a role in carcinogenesis. AP-1 binds either to the 12-O-tetradecanoylphorbol-13-acetate (TPA) response element or to the cAMP response element in the promoter region of their target genes.

The AP-1 transcription factor is actually composed of two components and can be produced by dimers of proteins from the Jun and Fos families (Jun, Jun B, Jun D, Fos, Fos B, FRA1, and FRA2) (Figure 3.4). Eighteen possible combinations are possible. Both Jun and Fos members contain a basic leucine zipper dimerization domain. Because the processes of growth, differentiation, and apoptosis need to be carefully regulated, AP-1 is itself activated in response to specific signals such as growth factors, ROS, and radiation. The specific combination of dimers influences the biological response. The antagonism displayed between Jun and Jun B with regard to cell proliferation in some cell types supports this: Jun acts as a positive regulator of proliferation, while Jun B acts as a negative regulator in the presence of Jun.

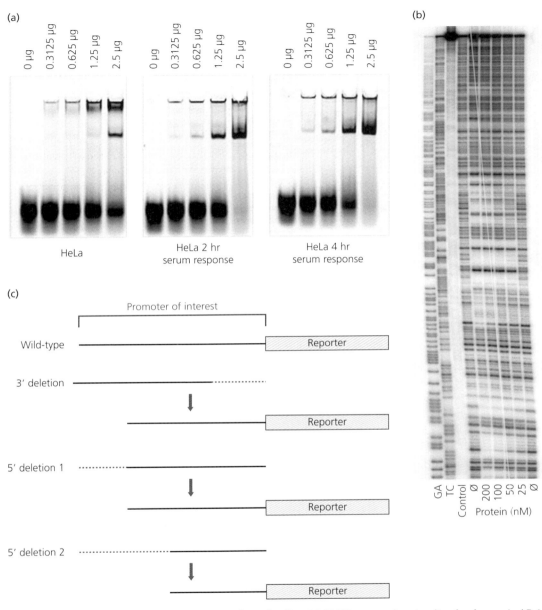

Figure 3.3 Methods used to examine transcription factor binding. (a) EMSA was used to visualize the changes in AP-1 binding following different treatments of HeLa cells: HeLa (control), HeLa 2 h serum response, and HeLa 4 h serum response nuclear extracts were serially diluted and assayed using an infrared dye (IRDyeR700) end-labeled DNA fragment containing the AP-1 response element. Gel electrophoresis and imaging followed. Courtesy of LI-COR Biosciences. (b) A sample DNase footprinting autoradiograph. Courtesy of George P. Munson. (c) A schematic diagram of possible mutant promoter constructs linked to a reporter gene to be used for promoter analysis. The effects of promoter sequence alterations may be indicated by changes in transcriptional activity. The constructs are transfected into cells, and reporter enzyme activity is used as an indicator of transcriptional activity.

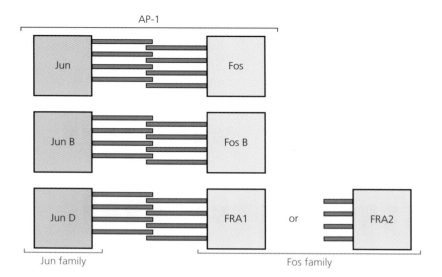

Figure 3.4 Members of the Jun and Fos transcription factor family.

Both the Fos and Jun family of transcription factors that make up the AP-1 complex play a role in tumorigenesis (see references in Milde-Langosch, 2005). As TPA is a tumor promoter and the AP-1 complex binds to the TPA response element, an association of AP-1 with carcinogenesis was implicated early after this property was characterized. The first members of AP-1 identified, c-Jun and c-Fos, were able to transform normal cells in culture to cancer cells and are frequently overexpressed in tumor cells.

Steroid hormones are lipid-soluble signaling molecules that exert their effects by regulating the transcription of sets of genes via specific receptors (Table 3.1). The superfamily of steroid hormone receptors acts as ligand-dependent transcription factors. There are currently 48 members of the nuclear receptor family (see the Nuclear Receptor Signaling Atlas web site http://www.nursa.org/). They contain a zinc finger type of DNA-binding domain, a ligand-binding domain for a specific steroid hormone, and a dimerization domain as they activate transcription as a dimer. Each domain functions independently and in a manner that is specific for a particular steroid hormone receptor. This feature has been utilized as a molecular tool by scientists in so-called domain swap experiments which

Table 3.1 Examples of members of the steroid hormone receptor superfamily

Class: steroid receptors	TR/RAR/PPAR/VDR-like receptors
Androgen receptor (AR)	Peroxisome proliferator activated receptor (PPAR)
Estrogen receptor (ER)	Retinoic acid receptor (RAR)
Glucocorticoid receptor (GR)	Thyroid hormone receptor (TR)
Mineralocorticoid receptor (MR)	Vitamin D receptor (VDR)
Progesterone (PR)	

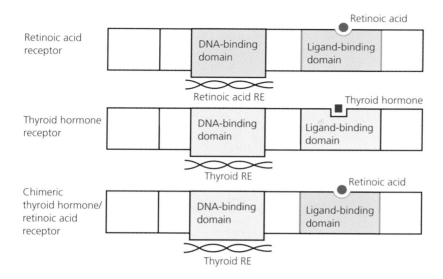

Figure 3.5 A chimeric steroid hormone receptor is shown below its two parental receptors.

produce chimeric receptors. For example, if the ligand-binding domain of the thyroid hormone receptor is swapped with the ligand-binding domain of the retinoic acid receptor, the newly formed chimeric receptor (Figure 3.5) will retain the DNA-binding domain of the thyroid hormone receptor and will activate thyroid hormone-responsive genes. However, these genes will be activated by retinoic acid via the retinoic acid ligand-binding domain, and not by thyroid hormone. Such experiments clearly demonstrate the functional independence of these domains.

Steroid hormones pass through the cell membrane and bind to their particular intracellular receptors in the cytoplasm (note: some members of the steroid hormone receptor superfamily bind to their ligands in the nucleus). Upon binding, the receptors move into the nucleus and activate transcription of their target genes through specific DNA response elements (Figure 3.6).

The retinoic acid receptor (RAR), as a member of the steroid hormone receptor family, acts as a retinoic acid (RA)-dependent transcriptional regulator and is important during differentiation. RA is derived from vitamin A. The RAR is constitutively located in the nucleus and acts as a transcriptional repressor in the absence of RA. It binds to the RA response element (RARE) in target genes as a heterodimer with another member of the family called RXR. Aberrant forms of RARs are characteristic of several leukemias. As we will see throughout the text, members of the steroid hormone receptor superfamily (e.g. estrogen receptor, vitamin D receptor) play an important role in many different types of cancer.

In the examples above, we discuss aberrant transcription factor function due to mutation. On the opposite side of the coin, somatic mutations in the promoter region can lead to altered gene regulation of oncogenes.

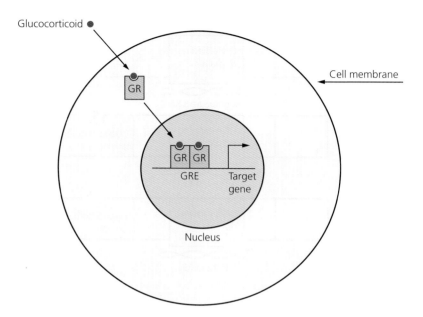

Glucocorticoid

GR

Cell membrane

GR GR

GRE   Target
gene

Nucleus

Figure 3.6 Mechanism of action of the glucocorticoid receptor (GR, red) (GRE, glucocorticoid response element).

Somatic mutations upstream of the *TAL1* oncogene, which codes for a basic helix–loop–helix transcription factor, have been shown to create a super-enhancer that upregulates the expression of this oncogene in T-cell leukemia (Mansour *et al.*, 2014). Super-enhancers are clusters of regulatory DNA elements that recruit transcriptional and chromatin-modifying proteins and may be transcribed into RNA. Many are associated with key genes that control cell state. In this study, the researchers showed that DNA response elements for the MYB transcription factor are created by acquired insertional mutations. MYB then recruits important transcriptional (e.g. GATA3, Pol II) and epigenetic (e.g. a histone lysine acetylase) regulators for strong expression of the *TAL1* oncogene (see Box "How do we know that?"). A similar phenotype is exhibited when a deletion occurs that causes the *TAL1* gene coding sequence to come under the regulation of a ubiquitously expressed promoter of the *STIL* gene.

? HOW DO WE KNOW THAT?

Chromatin immunoprecipitation (ChIP)-sequencing is a technique that is used to identify protein–DNA interactions at precise DNA sequences. The typical protocol involves cross-linking proteins to DNA, sonicating DNA to produce small fragments, adding bead-bound antibodies to capture your protein of interest for immunoprecipitation, unlinking protein from DNA, purifying DNA, and sequencing.

Mansour *et al.* (2014) used a MYB-specific antibody to produce genomic maps of MYB binding in cells from the Jurkat T-cell leukemia cell line and MOLT-3 cells that harbored insertional mutations of interest. Analysis showed precise alignment of MYB binding with other transcriptional and epigenetic regulators such as GATA3 and Pol II (see Figure 3A in Mansour *et al.*, 2014). These data suggest the formation ➜

→ of a super-enhancer. How do we know that this sequence is responsible for the overexpression of the *TAL1* oncogene in these cells? Could you examine DNA sequences from this region in normal cells? How would you test if these sequences are able to drive gene expression? Hint: Look at the third experimental approach under "Experimental methods used to examine transcription factor binding".

Yes, you can recombine the mutant enhancer with a reporter gene, such as luciferase, and compare expression using a similar construct but containing the wild-type promoter.

3.2 Chromatin structure

Human DNA is present in the nucleus of cells in the form of 46 chromosomes. Chromosomes are made of chromatin: a thread of DNA (60%) plus associated RNA (5%) and protein (35%). It is astonishing to think that the actual length of DNA in the nucleus of a cell is over a meter when fully extended. A high level of packaging (Figure 3.7) is required to neatly organize the DNA to fit into the nucleus of a cell and to allow it to assume necessarily organized conformations for transcription and replication. Both of these processes involve unwinding of the double helix and reading of template strands.

The simplest or primary level of organization of chromatin is the wrapping of DNA around a protein "spool" and is referred to as the "beads-on-a-string" array. The beads represent the nucleosome, which contains 147 base pairs (bp) of DNA wrapped 1.7 times around a core of histone proteins. The histone core is an octomer of **histones** containing two copies of histones H2A, H2B, H3, and H4. Each histone contains domains for histone–histone and histone–DNA interactions, and NH_2-terminal lysine-rich and COOH-terminal "tail" domains which can be post-translationally modified (e.g. acetylated, methylated, or phosphorylated). Histone

> **PAUSE AND THINK**
>
> How would you organize long pieces of thread in a sewing box? Most thread is wrapped around a spool for orderly and easy unwinding. Few people would just leave a disorganized bunch of thread for fear of getting tangles and knots.

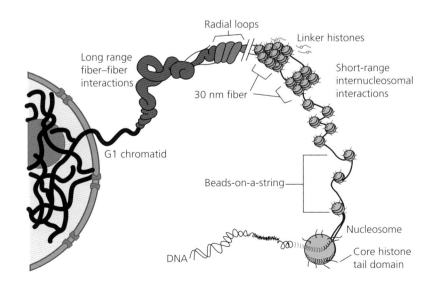

Long range fiber–fiber interactions

Radial loops

Linker histones

Short-range internucleosomal interactions

30 nm fiber

G1 chromatid

Beads-on-a-string

Nucleosome

Core histone tail domain

DNA

Figure 3.7 Multiple levels of chromatin structure. Reprinted, with permission, from *Annu. Rev. Biophys. Biomolec. Struct.* **31**, p. 362, copyright 2002 by Annual Reviews (http://www.annualreviews.org).

H1 is a linker histone and binds to DNA located outside the core. Ten to 60 bp of DNA separate the "beads." The secondary level of organization is the formation of 30 nm fibers, and these can associate to form a tertiary structure of radial loops.

Chromatin has an important role beyond being a structural scaffold. The degree of compaction or relaxation of chromatin determines how readily the DNA in a portion of chromatin can be transcribed: highly compact chromatin cannot be transcribed (it is transcriptionally "silent"), whereas more relaxed chromatin is accessible for transcription. The structure of chromatin—from compacted to relaxed—can change, and it is this feature that enables it to have a regulatory role in transcription. Chromatin conformation is encoded by heritable (that is, they can be passed on from one cell generation to the next) epigenetic modifications, as we will examine in the following section.

3.3 Epigenetic regulation of transcription

Epigenetics refers to heritable information that is encoded by modifications of the genome and chromatin components. These modifications affect the structure and conformation of chromatin and, consequently, transcriptional regulation. Epigenetic alterations in gene expression do not cause a change in the nucleotide sequence of the DNA and therefore are not mutations. Stable epigenetic switches are important during normal cell differentiation. For example, stem cells are maintained in a stem cell state by epigenetic silencing of genes important for differentiation. (Note: it is differential gene expression that makes one cell type different from another.) Two types of epigenetic mechanisms will be discussed in the following sections: histone modifications and DNA methylation. Both can be acquired or inherited, and both affect transcriptional activity by regulating the access of transcription factors to appropriate nucleotide sequences in gene promoters. It is important to keep in mind that all epigenetic processes work together to regulate open and closed states of chromatin, and that regulation of transcription factor binding links genetic and epigenetic mechanisms.

Histone modification

Histone proteins are subject to diverse post-translational modifications such as acetylation, methylation, phosphorylation, and ubiquination. The histone code hypothesis predicts that the pattern of these multiple histone modifications helps to specify the components and activity of the transcription regulatory molecular machinery. Let us focus on acetylation.

The acetylation pattern of histones alters chromatin structure and affects gene expression (Figure 3.8). Acetylation acts as a docking signal

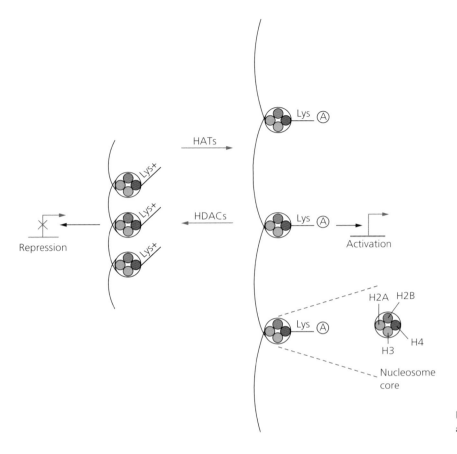

Figure 3.8 Histone acetylation affects gene expression.

for the recruitment or the repulsion of chromatin-modifying factors. Histone acetyltransferases (HATs; add acetyl groups) and histone deacetylases (HDACs; remove acetyl groups) are two families of enzymes that produce the pattern. HATs acetylate specific histone-tail lysines and other non-histone proteins, including transcription factors (e.g. E2F and p53). Acetylation of histones neutralizes the positive charge on lysine residues and relaxes chromatin folding. This correlates with enhanced transcriptional elongation by RNA polymerase II. HDACs remove acetyl groups and restore a positive charge to lysine residues of the histone tails which stabilize chromatin compaction and higher-level packaging. This configuration of chromatin limits the accessibility of transcription factors and results in the repression of transcription.

In addition, transcriptional activators often recruit HATs and other chromatin-remodeling enzymes to the promoter region. The retinoblastoma tumor suppressor protein, mentioned as a "molecule of fame" in Chapter 1, exerts its effects, in part, by recruiting HDACs to specific gene promoters (see Chapter 5). Thus, a signaling network seems to underlie chromatin modeling.

PAUSE AND THINK

In general, HATs activate transcription, and HDACs repress transcription.

DNA methylation

Another epigenetic process that affects transcriptional regulation is DNA methylation. DNA methylation is the addition of a methyl group to position 5 of cytosine. Only 3–4% of all cytosines in DNA are methylated. Methylation only occurs at cytosine nucleotides which are situated 5′ to guanine nucleotides (CpGs). Methylcytosine deaminates spontaneously and results in C→T transitions (Figure 3.9). It is thought that evolution has selected against this dinucleotide as a result of the high rate of mutation, as CpG is under-represented and unequally distributed in the genome. CpG clusters, called CpG islands, are located in the promoter region of 50% of human genes. In general, the CpG islands found in gene promoter regions are not methylated in normal tissues, and transcription may occur. Methylated cytosines are found mainly in repetitive sequences and in the CpG islands found in the promoter region of repressed genes such as X-chromosome inactivated genes, imprinted genes, and some tissue-specific genes. In these cases, methylation is a heritable signal that is associated with a compacted chromatin structure and maintains gene silencing.

   Enzymes, called DNA methyltransferases (DNMTs), mediate the covalent addition of a methyl group from the methyl carrier, S-adenosyl-methionine. Three methyltransferases are known: DNMT1, DNMT3a, and DNMT3b. DNMT1 is involved in the conversion of hemi-methylated DNA to fully methylated DNA during replication. This mechanism allows methylation patterns to be inheritable; if only one strand remained methylated, the signal would be lost in half of its daughter cells after replication. The other two methyltransferases are mainly involved in *de novo* methyltransferase activity (methylation of new sites).

It has been suggested that the mechanism by which methylation results in silencing is by recruiting methyl-binding domain (MBD) proteins, which have been shown to interact with HDACs and chromatin-remodeling enzymes. Therefore, epigenetic regulation of transcription includes cross-talk between methylation, chromatin-remodeling enzymes, and histone modification.

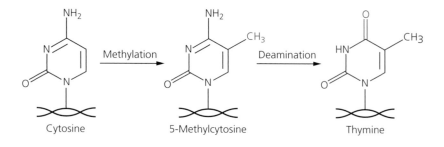

Figure 3.9 Spontaneous deamination of methylcytosine leads to a C→T transition.

3.4 Evidence of a role for epigenetics in carcinogenesis

As we discussed in Chapter 2, cancer mutations are not uniformly distributed along the genome. New evidence suggests that the features of the chromatin (epigenomic features such as the degree of compaction and histone modification) are a major determinant of the mutational landscape of cancer cells. Areas that are relaxed and indicative of active transcription are associated with low mutation density, whereas condensed regions indicative of gene repression are associated with regions of high mutation density. The epigenetic pattern of the tumor is associated with its cell of origin (Polak *et al.*, 2015). There are clinical cases of cancer whereby the primary origin is unknown. The study by Polak *et al.* (2015) demonstrates the cell of origin of a cancer can be determined based on the distribution of mutations along its genome.

Misregulated epigenetic silencing plays an important role in cancer. Abnormal epigenetic programs may inactivate large groups of genes. Epigenetic events can cause other epigenetic events, leading to genome-wide alterations and genomic instability. Recent work suggests that hundreds of epigenetically silenced genes may exist in individual tumors. Epigenetic inactivation and the inactivation of genes by mutation cooperate during carcinogenesis. On one hand, mutations can disable enzymes that are crucial for epigenetic regulation, and, on the other hand, epigenetic silencing can provide the second "hit" (discussed further in Section 6.1, "Definitions of tumor suppressor genes" on p. 128) required for a tumor suppressor gene to be inactivated in a recessive manner (Figure 1.3). A brief examination of some of the accumulating supporting evidence is described next.

Histone modification and cancer

Altered HAT or HDAC activity has been observed in several cancers. Interestingly, one gene, *EP300*, which codes for a HAT has been found to be mutated in epithelial cancers. Several of the mutations predicted a truncated protein, and inactivation of the second allele was observed in five out of six cases, suggesting that it functions as a tumor suppressor. Genomic sequencing of tumor DNA from specific types of lymphoma identified frequent mutations of histone-modifying genes, suggesting that epigenetic alterations play an important role in lymphoma development (Morin *et al.*, 2011).

Acute promyelocytic leukemia is characterized by a chromosomal translocation that produces a fusion protein called PML–RAR. This novel fusion protein retains the DNA-binding domain and ligand-binding domain of the RAR, in addition to PML sequences. PML–RAR recruits HDAC to the promoter region of RA target genes and represses the expression of these genes. The lack of activation of RAR target genes causes the block of differentiation that characterizes the leukemia. Other tumors have also been associated with aberrant recruitment of HDACs.

PAUSE AND THINK

RAR is a member of which family of transcription factors? See Section 3.1, "Transcription factors and transcriptional regulation" on p. 54.

Methylation and cancer

Cancer-specific changes in DNA methylation have been recognized, and many studies have focused on hypermethylation observed in normally unmethylated CpG islands of gene promoters (see Box "Analysis of DNA methylation by sodium bisulfite treatment and methylation-specific PCR"). Gene silencing by methylation may be an important mechanism of carcinogenesis whereby critical genes normally involved in tumor suppression may be switched off. Inactivation of gene expression by methylation of the promoter regions of such genes has been observed in cancer cell lines and human tumors. For example, there is strong evidence for epigenetic regulation of the estrogen receptor in breast and other cancers, as hypermethylation and silencing of the estrogen receptor promoter is common (see Hervouet *et al.*, 2013 for review). Thirty percent of breast cancer patients are estrogen receptor-α-negative. Loss of expression is due to DNA hypermethylation in 41% of these cases. As another example, the breast cancer susceptibility gene, *BRCA1*, is often mutated in a recessive manner in inherited breast cancer. Thus, the loss of function of the gene product suggests that normal BRCA1 acts to suppress breast cancer. Mutation of *BRCA1* is very rarely observed in non-inherited breast cancer. However, interestingly, hypermethylation is associated with the inactivation of *BRCA1* in non-inherited breast cancer, and therefore this may be another way of accomplishing loss of function. These findings support the view that epigenetics may be an additional mechanism for carcinogenesis. Additional examples of some key target genes affected by methylation include the *retinoblastoma* (*Rb*) gene, inhibitor of the cell cycle *p16 INK4a*, *APC* and pro-apoptotic death-associated protein kinase (*DAPK*). Evidence also indicates that abnormal methylation occurs in a subset of pre-malignant cells, and it has been suggested that this may "addict" cells to altered signaling pathways and facilitate subsequent mutational events that provide the cell with a selective advantage and promote tumorigenesis.

Analysis of DNA methylation by sodium bisulfite treatment and methylation-specific PCR

Molecular biology procedures used for standard genetic analysis erase DNA methylation information, and so specialized methods for methylation analysis were developed. Sodium bisulfite treatment of genomic DNA converts unmethylated cytosine residues to uracil by deamination. Note, 5-methylcytosines are not modified under these same conditions. Treated unmethylated DNA is no longer complementary to expected PCR primers so PCR amplification requires specially designed primers. Most commonly, primers are designed to hybridize specifically with the sodium bisulfite modified sequences (i.e. Us instead of Cs). PCR that uses these types of primers is called methylation-specific PCR. Methylation-specific PCR provides information about particular methylation patterns in single gene candidates. This approach is now complemented by current comprehensive genome-scale approaches that generate DNA methylation maps.

Sites of methylated DNA interact with additional epigenetic regulatory proteins such as histone-modifying enzymes (described earlier), methyl cytosine-binding proteins, and DNMTs. These proteins collaborate to define the structure of chromatin. In addition to recruiting HDAC (discussed previously), PML–RAR has also been shown to recruit methyltransferase, resulting in the DNA methylation of a promoter region of a specific gene (Di Croce *et al.*, 2002). The *RARβ2* gene has a RARE in its promoter and is one of the target genes of PML–RAR. It has been demonstrated that PML–RAR forms stable complexes with DNMTs at the *RARβ2* promoter and that the resulting hypermethylation contributes to carcinogenesis. It has been suggested that DNMTs, in addition to mediating methylation, may act as a platform for the assembly of chromatin-modifying factors.

As not all carcinogens are mutagens, it may be possible that some non-genotoxic carcinogens (agents that do not mutate genes) are epigenetic carcinogens. Hypermethylation has been observed in tumor-sensitive mice treated with phenobarbital, a non-genotoxic carcinogen in rodents. Research that examined regions of altered methylation after phenobarbital treatment for the promotion of liver cancer in a mouse model suggests that a disruption to normal methylation patterns is related to tumor formation and that non-genotoxic carcinogens may act via methylation (Phillips *et al.*, 2007). In addition, nutritional deficiencies (methionine, choline) seem to affect the cellular level of *S*-adenosylmethionine, an important methyl group donor. This suggests that perturbation of methylation can be produced through the diet (see Chapter 11). It has also been suggested that epigenetic gene silencing that is characteristic of normal stem cells and progenitor cells may be "locked in" during chronic injury and inflammation (see Chapter 8 for further discussion about cancer stem cells, p. 178) and contribute to carcinogenesis.

Mutation of DNA methyltransferases has been demonstrated in several cancers and leads directly to altered methylation. Whole cancer genome sequencing has identified recurrent somatic mutations in the *DNMT3A* gene in about 25% of acute myeloid leukemia patients. Another role of methylation in transformation may be to promote mutation. Methylated cytosine residues have a tendency to deaminate spontaneously, causing C→T transitions. This may account for the increased mutation rate observed in methylated CpG islands.

Paradoxically, the genome of a cancer cell overall can have 20–60% less methylation than a normal cell. This global hypomethylation (mainly in the coding region of genes, transposable elements, and repetitive DNA sequences) occurs in the cancer cell at the same time as the hypermethylation of specific gene promoters described earlier. Although this phenomenon has not been vigorously studied, several mechanisms have been proposed: activation of oncogenes and genes involved in metastasis, activation of transposable elements leading to genomic rearrangements, and alteration in the sequestering of transcription factors at repetitive repeats. Many different animal models have been used to study hypomethylation and cancer. They include genetic manipulation such as knocking out the *DNMT1* gene or drug-induced demethylation. Results from the varied animal models have given contrasting results; some models show decreases in precancerous lesions, and others show increases in tumors with increased hypomethylation. Overall, misregulation of many epigenetic processes is intimately involved in cancer development.

3.5 Long non-coding RNAs

Long non-coding RNAs (lncRNAs) are endogenous RNAs that are longer than 200 nucleotides in length and that lack an open reading frame (less than 100 amino acids). They are polyadenylated and can produce splicing variants. They are involved in a wide variety of cellular processes, including the processes involved in the hallmarks of cancer. Importantly, they have been shown to be transcribed from gene enhancers and play a role in gene expression. They also play a role in epigenetic modifications and post-transcriptional regulation. Let us look at a few named examples and examine their links with cancer. Upon DNA damage, the crucial tumor suppressor, P53, binds to the promoter region of one lncRNA called lincRNA-p21. This lncRNA plays a role in tumor suppression by acting as a transcriptional repressor of genes known to be downregulated in the P53 response. HOTAIR and XIST are lncRNAs that guide chromatin-modifying factors to specific sites to facilitate epigenetic regulation. Misexpression of these lncRNAs has been linked to breast and female cancers, respectively. Also, lncRNAs from cell cycle gene promoters show periodic expression that is disrupted in cancer cells. And lastly, some cancer-associated SNPs alter the expression of tumor suppressor lncRNAs.

3.6 MicroRNAs (miRNAs) and regulation of mRNA expression

MicroRNAs (miRNAs) are small, non-protein-coding RNAs (18–25 nucleotides in length) that regulate the expression of mRNAs. Each miRNA may be able to repress hundreds of gene targets post-transcriptionally. Therefore, they are powerful regulators of gene expression.

Mature miRNAs require several steps of processing. After they are transcribed by RNA polymerase II from intergenic regions or from regions that code for introns, the primary transcript is processed by ribonucleases

Drosha and DGCR8 in the nucleus. This processing produces pre-miR-NAs, hairpin-shaped intermediates of 70–100 nucleotides. Exportin-5 transports pre-miRNAs into the cytoplasm where they are further processed by ribonuclease Dicer into a double-stranded miRNA. The strands separate, and a mature single-stranded molecule joins an RNA-induced silencing complex (RISC). This effector complex of the miRNA pathway is made up of miRNAs and specific proteins.

The repression of gene targets by miRNAs happens in one of two ways. Under the first scenario, the miRNA hybridizes perfectly to the 3′ untranslated region (UTR) of their target mRNA. The formation of this complex in the RISC leads to mRNA cleavage and subsequent degradation. Alternatively, miRNAs may bind to imperfect complementary sites in the 3′ UTR of their target mRNAs. The formation of this complex in the RISC blocks translation. The net result for both scenarios is a decrease in the amount of the protein encoded by the gene from which the mRNA has been transcribed.

miRNAs are involved in regulating a diverse set of biological processes, including growth, differentiation, and apoptosis. Not surprisingly, some miRNAs may play a role in cancer as oncogenes (miRNAs that are oncogenic are called oncomirs), and others as tumor suppressors. Amplified or upregulated miRNAs that suppress tumor suppressor mRNAs can act as oncomirs. It has been demonstrated that overexpression of a single miRNA, called mir-155, can lead to cancer in mice. However, miRNAs that normally repress the expression of oncogenes act as tumor suppressors and are lost in tumor cells. Endogenous miRNAs have also been implicated in suppressing metastasis (Tavazoie *et al.*, 2007).

Some of the mechanisms of miRNA deregulation are common with traditional oncogenes, such as base pair substitutions, deletions, and aberrant epigenetic modifications (see Section 2.2, "Mutations" on p. 24). However, miRNA expression can also be altered at the level of processing, adding another layer of regulation. The products of oncogenes have been shown to regulate transcription of miRNAs, illustrating that the interplay of oncogenes and miRNAs can be complex. In addition, there can be interactions between lncRNAs and miRNAs. The tumor suppressor PTEN (discussed in detail in Chapter 6) has a pseudogene *PTENP1* that codes for a lncRNA with a conserved 3′ UTR. This region acts as a decoy for miRNAs that inhibit PTEN and can increase the expression of the PTEN tumor suppressor gene.

Selected groups of specific miRNAs are commonly altered in particular cancers, and recent data have shown that miRNA expression profiles of tumors are able to discriminate between different types of cancer. Thus, miRNA profiles will be useful for cancer diagnosis and prognosis (Esquela-Kerscher and Slack, 2006; Yanaihara *et al.*, 2006), whereas mRNA profiles have been proven to be unreliable.

3.7 Telomeres and telomerase

One of the hallmarks of cancer cells is that they acquire a limitless replicative potential (Figure 1.1). Normal cells have an autonomous program that allows for a finite number of replication cycles. This phenomenon is well known in that cells in culture only undergo a certain number of doublings before they stop dividing and enter senescence (permanent growth arrest). Telomeres, repetitive DNA sequences and specialized proteins at the ends of chromosomes, have been shown to function as a molecular counter of the cell's replicative potential (see Verdun and Karlseder, 2007 for review). Telomeres protect the ends of chromosomes from digestion by nuclear enzymes and also prevent induction of mechanisms that repair DNA double-stranded breaks. Telomeres are composed of several thousand repeats of the sequence TTAGGG bound by a set of associated proteins called the shelterin complex, which functions to control telomere length and protect the chromosomal ends. Telomeres shorten by 100–200 bases with each round of DNA replication, owing to the limits of DNA polymerases during DNA replication, referred to as the end-replication problem (see Box "A little lesson about DNA replication . . . "). DNA polymerases proceed only in the 5′–3′ direction and require an RNA primer to initiate DNA synthesis. The RNA primers are removed after replication is complete. As a result, the 3′ end of the parental chromosomal DNA is not replicated, and thus chromosomes progressively erode during each round of replication (Figure 3.10). When the chromosomes reach a threshold length, cells enter a stable and irreversible state of growth arrest called cellular senescence. If cells bypass this stage, because of mutation, and telomeres become critically short, chromosomal instability results and apoptosis (or cell transformation discussed later) is induced.

   Maintaining telomere length in stem cells of renewal tissue (e.g. the basal layer of the epidermis) is important for providing a longer replicative potential. Telomerase, a ribonucleoprotein containing human telomerase reverse transcriptase (hTERT) and a human telomerase RNA (hTR), maintains telomere length in certain cell types such as stem cells. Reverse transcriptases are enzymes that synthesize DNA from RNA—an exception to the central dogma of molecular biology which states that RNA

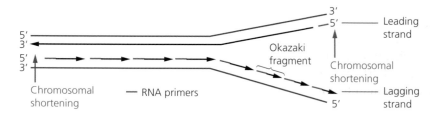

Figure 3.10 Chromosomal shortening after DNA replication.

(a)

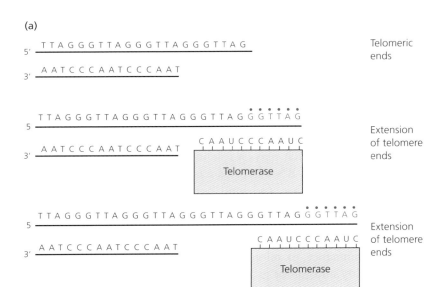

(b)

Figure 3.11 Telomerase and telomeres. a) Telomere extension by telomerase. b) Telomeres shown by yellow fluorescence. Courtesy of Jerry Shay, UT Southwestern Medical Center.

is synthesized from DNA. The hTR contains 11 complementary base pairs to the TTAGGG repeats and acts as a template for the reverse transcriptase to add new repeats (shown in red) to telomeric DNA on the 3′ ends of chromosomes (Figure 3.11).

A little lesson about DNA replication …

DNA replication proceeds in a semi-conservative manner: each of the two parental strands acts as a template for the synthesis of a newly replicated strand (Figure 3.12; new DNA strand synthesis is shown in red). Each of the polynucleotide strands that make up the DNA helix has a sense of direction; that is each has a 5′ end and a 3′ end. The two strands are arranged in an antiparallel manner. As DNA polymerases only work in a 5′–3′ direction, each strand is replicated differently as the DNA helix unwinds. For one strand, the leading strand, replication proceeds in a continuous manner from the 5′ to 3′ end. For the other strand, the lagging strand, replication occurs in a discontinuous manner through the 5′–3′ synthesis of short Okazaki fragments. After removing the RNA primers and filling in the gaps, these fragments are ligated together by the enzyme DNA ligase to form one continuous strand. The requirement of DNA polymerase for an RNA primer and the subsequent removal of this primer causes the strands to shorten at the extreme chromosomal ends during each round of replication.

It has been shown *in vitro* that the telomere ends are not linear but rather complicated structures forming t-loops and may form four-stranded DNA conformations called G quadruplexes. It is important that chromosome ends are distinguishable from DNA double-stranded breaks. If they were not, the DNA repair processes would produce chromosomal fusions and other aberrations in an attempt to repair the damage.

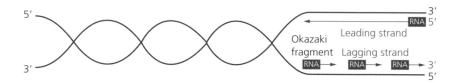

Figure 3.12 Semi-conservative and semi-discontinuous DNA replication.

Telomeres can be transcribed into a lncRNA that contains telomeric repeat-containing RNA (TERRA). That is, TERRA contains UUAGGG repeat sequences. TERRA plays a role in regulating the length of telomeres. It can form DNA–RNA hybrids with telomeric DNA, interact with heterochromatin proteins, and bind both telomerase RNA and TERT. These interactions stabilize telomere heterochromatin, inhibit telomerase activity, and promote telomere shortening (Wang *et al.*, 2015).

Several lines of evidence have linked telomerase activity with cancer. The maintenance of telomeres seems to be important for cancer cell immortality and tumor growth, and approximately 90% of tumors accomplish this by upregulating telomerase. Germline and recurrent somatic mutations in melanoma and other cancers have been identified in the TERT promoter. The mutations create new transcription factor binding sites (Ets/TCF) in the TERT promoter, and this is one mechanism that underlies the increased telomerase activity seen in cancer cells (Heidenreich *et al.*, 2014). Somatic mutations in the TERT promoter included C→T and CC→TT tandem transitions characteristic of UV-induced carcinogenesis known to underlie melanoma. Telomerase activity was clearly a distinguishing feature in one classical study where it was detectable in cultured immortal cell lines (98 of 100) and tumor tissue biopsies (90 of 101), but undetectable in cultured normal somatic cells (22) or benign tissue samples (50). It has been found that telomerase, in addition to two oncogenes, is essential in the protocol to transform normal fibroblasts to cancer cells *in vitro*, thus providing a strong link between telomerase and tumorigenesis. Several oncogenes have been demonstrated to regulate the expression of telomerase. For example, the transcription factor c-myc (an oncogene discussed in later chapters) increases the expression of the *hTERT* gene via specific response elements in the promoter region. As mentioned earlier, if cells bypass the replicative senescence stage because of mutation, telomeres become critically short and chromosomal instability results. This genetic catastrophe may lead to the loss of tumor suppressor mechanisms and evasion of apoptosis. Subsequently, transformed cells emerge. Most transformed cells have upregulated telomerase activity that helps establish cell immortality.

Interestingly, modifications of the telomere hypothesis of senescence described previously have recently been suggested and have strong implications for cancer. The telomere hypothesis would predict that telomeres shorten at a constant rate, yet great heterogeneity of replicative lifespan exists among cells within a clonally derived population (i.e. some cells arrest after a few divisions and some after many divisions). It has been reported that telomere shortening is accelerated by oxidative stress, which suggests that the problem of replicating the ends of chromosomes is not the only determining factor for telomere length and replicative potential. Telomeric DNA is repaired less proficiently, compared with the bulk of the genome, in response to oxidative damage. Unrepaired single-strand breaks accelerate telomere shortening, although the mechanism by which this occurs is unclear. These observations suggest that telomeric DNA may act as a sensor for DNA damage and may explain why there is great heterogeneity in the rate of telomeric shortening among individual cells. Therefore, telomere shortening may act as a tumor suppression mechanism by limiting replicative potential in response to genome damage.

 Therapeutic strategies

3.8 Epigenomic and histonomic drugs

It is a fairly recent view that epigenetic silencing may be as important as mutation as a mechanism for carcinogenesis. Currently, the concept of reversing somatic mutations is difficult to envisage because mutations are not reversible. However, the concept is conceivable for epigenetic changes, as these are modifications that are reversible. A large number of genes known to play important roles in carcinogenesis have been shown to display hypermethylation of their promoter regions. As the increased methylation seen in tumor cells is not observed in normal cells, it provides a tumor-specific target for DNA methylation inhibitors. Similarly, enzymes that alter chromatin structure, such as HDACs that modify histones, provide other molecular targets. As described earlier, several forms of leukemia and lymphoma are associated with transcriptional repression due to recruitment of HDACs. Reversal of epigenetic silencing is an approach that may lead to new therapeutics.

DNA methylation inhibitors

Drugs that block DNA methylation are predicted to show anti-tumor effects, as inactivation of tumor suppressor genes by methylation may be an important mechanism in carcinogenesis. Recall that DNA methylation occurs at position 5 on cytosine. Two 5'-modified analogs of deoxycytidine, 5-azacytidine (5-azaC) and 5-aza-2'-deoxycytidine, have been used to target DNA methyltransferases (Figure 3.13). These drugs are incorporated into DNA and/or RNA. They covalently link with DNA methyltransferases (DNMTs, left blue target in Figure 3.13) and sequester its action such that there is significant demethylation after several rounds of

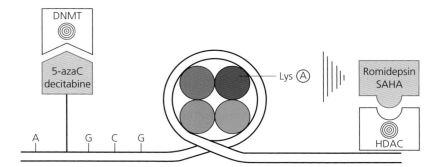

Figure 3.13 Drugs designed to target epigenetic mechanisms (shown in red).

replication. These drugs may result in DNA instability that parallels anti-metabolite chemotherapeutic agents. A potential hindrance is that aberrant methylation and gene repression return after treatment is stopped, dictating that administration of the drug must be prolonged. Both of these drugs showed anti-leukemic activity in clinical trials but were not successful in solid tumors. This may be explained by the fact that leukemia patients often have mutations in genes involved in regulating DNA methylation patterns such as DNMT3A. 5-azaC (azacitidine; Vidaza™) and 5-aza-2'-deoxycytidine (decitabine) have been approved by the US Food and Drug Administration (FDA) for the treatment of leukemia.

Inhibitors of histone deacetylases

Histone-modifying enzymes have been targeted for the development of new cancer drugs (Figure 3.13). Recall that HDACs generally repress gene transcription and that aberrant recruitment is characteristic of some cancers such as leukemias. Reactivation of silenced genes involved in growth, differentiation, or apoptosis provides the rationale for treating such cancers, with inhibitors of HDACs. Several classes of drugs that bind to the catalytic site of HDACs (right blue target in Figure 3.13) and block the binding to their substrates (acetylated lysines of histone proteins) are being tested in clinical trials: short-chain fatty acids such as sodium *n*-butyrate; hydroxamic acids such as SAHA; cyclic peptides such as romidepsin (formerly FK-228); and benzamide derivatives such as entinostat. In general, these drugs are well tolerated, and many can be administered orally. Alteration of gene expression appears to be selective. Interestingly, many HDAC inhibitors induce p21^{WAF1}, a cyclin-dependent kinase inhibitor important for growth arrest. These drugs seem to have little, or no, effect on normal cells. Molecular effectiveness was demonstrated by the detection of acetylated histones, in particular white blood cells and tumor cells, and this was associated with clinical improvement. Romidepsin (FK-228) and SAHA (vorinostat; Zolinza™) have been approved by the FDA for use in the clinic.

PAUSE AND THINK

In addition to the acetylation of histones, methylation of histones is another layer of epigenetic regulation. Can you think of a drug strategy to target this process? Indeed, there are several small molecule inhibitors targeted to histone methylases/demethylases in development. An inhibitor of lysine-specific demethylase 1 (LSD1), ORY-1001, is one such molecule that is likely to enter clinical trials.

3.9 Non-coding RNAs for diagnosis

The lncRNA called PCA3 is strongly overexpressed (66-fold) in prostate tissue, compared to normal prostate tissue, in more than 95% of primary prostate cancer samples. It is not expressed in other healthy tissues. This makes it the most prostate cancer-specific gene known thus far. Translation of this knowledge led to the development of an RNA-based urine test, PROGENSA PCA3 test, for the diagnosis of prostate cancer patients which has been approved by the FDA (Ronnau *et al.*, 2014).

3.10 Telomerase inhibitors

The relatively tumor-specific expression of telomerase and its pivotal role in the ability of a cancer cell to divide indefinitely suggest that it may be a valuable molecular target for new cancer therapies. However, several parameters need to be examined when considering the inhibition of telomerase as a cancer therapy. Effectiveness may depend on initial telomere length, and thus this should be assessed from tumor biopsies prior to treatment. Also the response may be slow, owing to the time needed for the telomeres of cancer cells to shorten enough to trigger senescence or apoptosis, and long-term treatment may be necessary. In general, long-term treatment increases the probability of drug resistance. Several different strategies, targeting either the RNA component or the catalytic protein component, have been explored in pre-clinical studies. As telomerase is dependent on its functional RNA molecule, antisense oligonucleotides and ribozymes have been popular agents used to target hTR. Antisense oligonucleotides are complementary to part of the target RNA and hybridize by Watson–Crick base-pairing. Hybridization can inhibit function directly or trigger degradation by the recruitment of RNases. Hammerhead ribozymes contain antisense sequences for target recognition and an internal endonuclease activity that cleaves the target RNA. Reverse transcriptase inhibitors against the catalytic domain of hTERT and nucleoside analogs have also been investigated. BIBR1532 is a synthetic small molecule inhibitor that directly binds hTERT non-competitively and has been shown to induce telomere-driven senescence. However, problems with solubility and bioavailability have prevented progression to clinical trials. G-quadruplex binding molecules that prevent interaction between the enzyme and substrate have also been developed (e.g. telomestatin). High-throughput screening has identified several natural compounds as telomerase inhibitors such as components of mistletoe and a green tea catechin.

One anti-telomerase drug candidate that has entered Phase II clinical trials is imetelstat (GRN 163L; Geron). Imetelstat, a modified 13-mer oligonucleotide, binds directly to the template region of the RNA component of hTR, which lies in the catalytic site of hTERT. Imetelstat binding to hTR results in direct, competitive inhibition of telomerase enzymatic activity. It is promising that pre-clinical studies of this drug suggest that it may be effective against pediatric brain tumors, a group of cancers with high mortality in children (Castelo-Branco *et al.*, 2011).

▣ CHAPTER HIGHLIGHTS—REFRESH YOUR MEMORY

- Cancer may be caused by alterations in gene expression.
- Seventy percent of the genome is transcribed into RNA, but only 2% of the genome codes for proteins; some ncRNAs are involved in regulating gene expression.
- Transcription factors recognize DNA response elements and are essential for the regulation of gene expression.
- Steroid hormone receptors act as ligand-dependent transcription factors.
- Chromatin has several levels of DNA packaging: the nucleosome, the 30-nm fiber, and radial loops.
- Epigenetic changes also regulate gene expression. These involve modification of nucleotides or chromatin components.
- Histone modification and DNA methylation are two types of epigenetic mechanism.
- HATs add acetyl groups to histones and activate transcription.
- HDACs remove acetyl groups and repress transcription.

- Methylation at CpG islands represses transcription.
- Epigenetic alterations are genome-wide in carcinogenesis.
- lncRNAs play a role in transcriptional regulation, chromatin modifications, and regulation of telomerase.
- miRNA expression is deregulated in cancer.
- Telomeres play a role in the replicative potential of a cell.
- Telomeres shorten with each round of replication, but the rate of shortening may also be influenced by oxidative stress.
- Telomerase is an enzyme that maintains telomere length, and its activity is increased in 90% of tumors.
- Strategies for the design of drugs target DNMTs, HDACs, and telomerase. Some drugs that target epigenetics have been approved.
- Some ncRNAs may be important for diagnosis and prognosis.

▣ ACTIVITY

1. Formulate evidence for your view on the statement that epigenetics is as important as mutation for carcinogenesis. Include an examination of epigenetic diseases that lead to an increased risk of cancer (Feinberg, 2007). Contribute to a class debate on this issue.

2. Look at the evidence supporting the role of oncomirs in cancer. Start with Esquela-Kerscher and Slack (2006) and Yanaihara *et al.* (2006). Critically discuss strategies and challenges for designing drugs that target miRNAs in cancer (Garzon *et al.*, 2010).

▣ FURTHER READING

Baylin, S.B. and Ohm, J.E. (2006) Epigenetic silencing in cancer—a mechanism for early oncogenic pathway addiction? *Nat. Rev. Cancer* **6**: 107–116.

Buseman, C.M., Wright, W.E., and Shay, J.W. (2012) Is telomerase a viable target in cancer? *Mut. Res.* **730**: 90–97.

Chen, H., Li, Y., and Tollefsbol, T.O. (2009) Strategies targeting telomerase inhibition. *Mol. Biotechnol.* **41**: 194–199.

Dawson, M.A. and Kouzarides, T. (2012) Cancer epigenetics: from mechanism to therapy. *Cell* **150**: 12–27.

Ehrlich, M. (2009) DNA hypomethylation in cancer cells. *Epigenomics* **1**: 239–259.

Gutschner, T.C and Diederichs, S. (2012) The hallmarks of cancer: a long non-coding RNA point of view. *RNA Biol.* **9**: 703–719.

Helin, K., and Dhanak, D. (2013) Chromatin proteins and modifications as drug targets. *Nature* **502**: 480–488.

Herman, J.G. and Baylin, S.B. (2003) Gene silencing in cancer in association with promoter hypermethylation. *N. Engl. J. Med.* **349**: 2042–2054.

Houben, J.M., Moonen, H.J., van Schooten, F.J., and Hageman, G.J. (2008) Telomere length assessment: biomarker of chronic oxidative stress? *Free Radic. Biol. Med.* **44**: 235–246.

Jones, P.A. and Baylin, S.B. (2007) The epigenomics of cancer. *Cell* **128**: 683–692.

Laird, P.W. (2003) The power and the promise of DNA methylation markers. *Nature* **3**: 253–266.

Marks, P.A., Richon, V.M., Breslow, R., and Rifkind, R.A. (2001) Histone deacetylase inhibitors as new cancer drugs. *Curr. Opin. Oncol.* **13**: 477–483.

Orom, U.A. and Shiekhattar, R. (2013) Long ncRNAs usher in a new era in the biology of enhancers. *Cell* **154**: 1190–1193.

Rodriguez-Paredes, M. and Esteller, M. (2011) Cancer epigenetics reaches mainstream oncology. *Nat. Med.* **17**: 330–339.

Schreiber, S.L. and Bernstein, B.E. (2002) Signaling network model of chromatin. *Cell* **111**: 771–778.

Shaulian, E. and Karin, M. (2002) AP-1 as a regulator of cell life and death. *Nat. Cell Biol.* **4**: E131–E136.

Thorne, J.L., Campbell, M.J., and Turner, B.M. (2009) Transcription factors, chromatin, and cancer. *Int. J. Biochem. Cell Biol.* **41**: 164–175.

WEB SITE

Nuclear Receptor Signaling Atlas http://www.nursa.org/

SELECTED SPECIAL TOPICS

Castelo-Branco, P., Zhang, C., Lipman, T., Fujitani, M., Hansford, L., Clarke, I., *et al.* (2011) Neural tumor-initiating cells have distinct telomere maintenance and can be safely targeted for telomerase inhibition. *Clin. Cancer Res.* **17**: 111–121.

Cheetham, S.W., Gruhl, F., Mattick, J.S., and Dinger, M.E. (2013) Long noncoding RNAs and the genetics of cancer. *Br. J. Cancer* **108**: 2419–2425.

Di Croce, L., Raker, V.A., Corsaro, M., Fazi, F., Fanelli, M., Faretta, M., *et al.* (2002) Methyltransferase recruitment and DNA hypermethylation of target promoters by an oncogenic transcription factor. *Science* **295**: 1079–1082.

Esquela-Kerscher, A. and Slack, F.J. (2006) Oncomirs—microRNAs with a role in cancer. *Nat. Rev. Cancer* **6**: 259–269.

Feinberg, A.P. (2007) Phenotypic plasticity and the epigenetics of human disease. *Nature* **447**: 433–440.

Garzon, R., Marcucci, G., and Croce, C.M. (2010) Targeting microRNAs in cancer: rationale, strategies and challenges. *Nat. Rev. Drug Discov.* **9**: 775–789.

Gaudet, F., Hodgson, J.G., Eden, A., Jackson-Grusby, L., Dausman, J., Gray, J.W., *et al.* (2003) Induction of tumors in mice by genomic hypomethylation. *Science* **300**: 489–492.

Heidenreich, B., Rachakonda, P.S., Hemminki, K., and Kumar, R. (2014) *TERT* promoter mutations in cancer development. *Curr. Opin. Genet. Dev.* **24**: 30–37.

Hervouet, R., Cartron, P.-F., Jouvenot, M., and Delage-Mourroux, R. (2013) Epigenetic regulation of estrogen signalling in breast cancer. *Epigenetics* **8**: 237–245.

Mansour, M.R., Abraham, B.J., Anders, L., Berezovskaya, A., Gutierrez, A., Durbin, A.D., *et al.* (2014) An oncogenic super-enhancer formed through somatic mutation of a noncoding intergenic element. *Science* **346**: 1373–1377.

Milde-Langosch, K. (2005) The Fos family of transcription factors and their role in tumourigenesis. *Eur. J. Cancer* **41**: 2449–2461.

Morin, R.D., Mendez-Lago, M., Mungall, A.J., Goya, R., Mungall, K.L., Corbett, R.D., *et al.* (2011) Frequent mutation of histone-modifying genes in non-Hodgkin lymphoma. *Nature* **476**: 298–303.

O'Doherty, A.M., Church, S.W., Russell, S.E.H., Nelson, J., and Hickey, I. (2002) Methylation status of oestrogen receptor—a gene promoter sequence in human ovarian epithelial cell lines. *Br. J. Cancer* **86**: 282–284.

Phillips, J.M., Yamamoto, Y., Negishi, M., Maronpot, R.R., and Goodman, J.I. (2007) Orphan nuclear receptor constitutive active/androstane receptor-mediated alterations in DNA methylation during phenobarbital promotion of liver tumorigenesis. *Toxicol. Sci.* **96**: 72–82.

Polak, P., Karlić, R., Koren, A., Thurman, R., Sandstrom, R., Lawrence, M.S., *et al.* (2015) Cell-of-origin chromatin organization shapes the mutational landscape of cancer. *Nature* **518**: 360–364.

Ronnau, C.G.H., Verhaegh, G.W., Luna-Velez, M.V., and Schalken, J.A. (2014) Noncoding RNAs as novel biomarkers in prostate cancer. *BioMed Res. Int.* **2014**: 591703.

Tavazoie, S.F., Alarcon, C., Oskarsson, T., Padua, D., Wang, Q., Bos, P.D., *et al.* (2007) Endogenous human microRNAs that suppress breast cancer metastasis. *Nature* **451**: 147–152.

Verdun, R.E. and Karlseder, J. (2007) Replication and protection of telomeres. *Nature* **447**: 924–931.

Wang, C., Zhao, L., and Lu, S. (2015) Role of TERRA in the regulation of telomere length. *Int. J. Biol. Sci.* **11**: 316–323.

Yanaihara, N., Caplen, N., Bowman, E., Seike, M., Kumamoto, K., Yi, M., *et al.* (2006) Unique microRNA molecular profiles in lung cancer diagnosis and prognosis. *Cancer Cell* **9**: 189–198.

Chapter 4

Growth factor signaling and oncogenes

Introduction

One of the fundamental characteristics of cells is their ability to self-reproduce. The process of cell division (also known as cell proliferation or cell growth) must be carefully regulated, and DNA replication must be precisely coordinated in order to maintain the integrity of the genome for each cell generation. As emphasized earlier in this volume, unregulated growth is a quintessential characteristic of cancer.

Normally, cell division is only initiated in response to a signal from outside of the cell. An extracellular growth factor stimulates cell growth by transmitting a signal into the cell, and ultimately to the nucleus, to regulate gene expression in order to produce proteins that are essential for cell division. There are four types of proteins involved in the transduction of a growth factor signal: growth factors, growth factor receptors, intracellular signal transducers, and nuclear transcription factors which elicit the mitogenic effect through the regulation of gene expression. Examining the normal mechanism of growth will allow a better understanding of the abnormalities that occur during carcinogenesis.

It is important to identify a common thread in many growth factor signal transduction pathways: many growth factor receptors are tyrosine kinases. All kinases catalyze the transfer of the γ phosphate group from ATP/GTP to hydroxyl groups on a specific amino acid in a target protein. Tyrosine kinases phosphorylate tyrosine amino acids in target proteins (Figure 4.1). Serine/threonine kinases phosphorylate serine and threonine residues. The addition of the phosphate group, a bulky charged molecule, may cause a conformational change in the protein, resulting in the activation or inactivation of an enzymatic activity and/or serve as a recognition site for new protein–protein interactions. Examples will be described in the following sections.

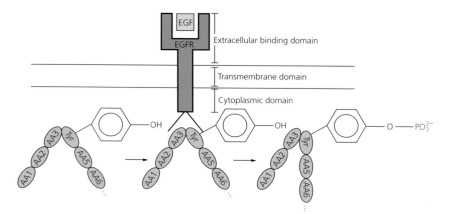

Figure 4.1 Tyrosine kinase receptors phosphorylate tyrosine residues on target proteins. Phosphorylation usually results in a conformational change of the protein target.

4.1 Epidermal growth factor signaling: an important paradigm

Epidermal growth factor (EGF) and its family of receptor tyrosine kinases serve as an important paradigm for how a signal from an extracellular growth factor can be transduced through a cell, regulate gene expression, and trigger cell proliferation. It is a model that we know a great deal about. The EGF receptor (EGFR; also known as ErbB1 or HER1) is a tyrosine kinase receptor and was the first to be discovered. Three additional family members have since been identified: ErbB2 (HER2), ErbB3 (HER3), and ErbB4 (HER4). As members of the receptor tyrosine kinase receptor family, these receptors contain an extracellular ligand-binding domain, a single transmembrane domain, and a cytoplasmic protein tyrosine kinase domain (Figure 4.1), with two exceptions: HER2 does not bind to a known ligand but acts as a co-receptor for the other members of the family, and HER3 has only weak kinase activity. Getting the signal from a growth factor outside the cell to inside the nucleus where gene expression is regulated requires several steps:

- Binding of the growth factor to the receptor
- Receptor dimerization
- Autophosphorylation
- Activation of intracellular transducers (including the "star player" RAS)
- Activation of a cascade of serine/threonine kinases (Raf, MEK, MAPK)
- Regulation of transcription factors for gene expression (to produce proteins needed for cell growth).

Each of the steps involved in the signal transduction pathway of EGF is illustrated in Figure 4.2 and described in the following sections. For ease of

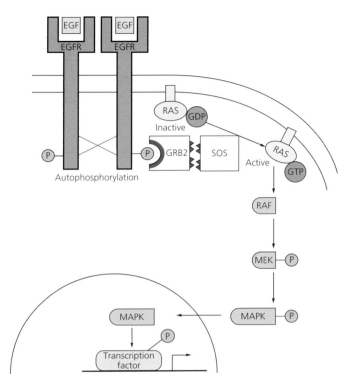

Figure 4.2 The signal transduction pathway of EGF. This pathway is characterized by the sequential steps of growth factor binding, receptor dimerization, autophosphorylation, activation of intracellular transducers, a serine/threonine kinase cascade, and transcription factors, and regulation of gene expression. Details are described in the text.

learning, the complete pathway has been divided into two parts: early steps of the pathway that occur at the cell membrane (Figure 4.5) and later steps of the EGF pathways that occur away from the cell membrane (Figure 4.6). It is essential that you learn this model system, because it will enable you to understand many other signal transduction pathways and it will be the basis for illustrating the mechanisms of carcinogenesis. Most interestingly, the components of this pathway have been targets for the design of new cancer therapeutics, some of which will be described at the end of the chapter.

Growth factor binding

The first step in the EGF signal transduction pathway is the binding of EGF to its receptor, EGFR. Extracellular domains (I and III) of EGFR form a binding pocket for the ligand. Note that, in addition to EGF, there are other ligands that can bind to the EGFR family, including amphiregulin and neuregulins 1–4, but, for simplicity, these will not be discussed in the following sections.

Dimerization

Dimerization is the process of two EGFR monomers interacting to form a dimer. The mechanism for receptor dimerization, as suggested from structural studies, is described here and illustrated in Figure 4.3. The binding

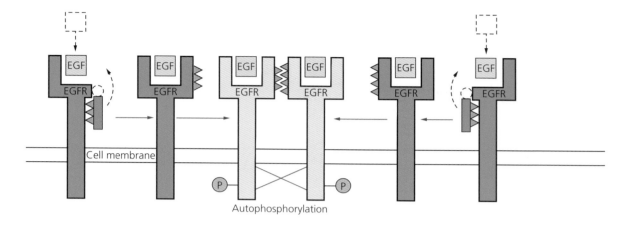

Autophosphorylation

Figure 4.3 EGF receptor dimerization. Growth factor binding causes a conformational change that unmasks a dimerization domain (shown as gray triangles) required for receptor dimerization.

of one EGF molecule to one receptor causes a conformational change that reveals an extracellular receptor dimerization domain (shown in gray). This facilitates the binding to a similar domain in another EGF-bound receptor monomer, resulting in a receptor dimer. It is important to note that EGFR can also form heterodimers with other members of the ErbB family.

Autophosphorylation

The conformational change of the receptor upon ligand binding also disrupts intramolecular (cis) autoinhibitory interactions, leading to kinase activation. The change in shape of the receptor permits access of ATP and substrate to the catalytic kinase domain. The close proximity of two receptors, facilitated by dimerization, enables the kinase domain of one receptor of the dimer to phosphorylate the other receptor of the dimer, and vice versa (see detection of phosphorylated EGFR by antibody staining in Figure 4.4). This intermolecular (between molecules) autophosphorylation on the cytoplasmic domain of the receptors (shown by red

Figure 4.4 Phosphorylation of EGFR after EGF treatment detected by Phospho-EGF Receptor (Tyr992) Antibody (no. 2235; green). A human epithelial carcinoma cell line was analyzed by confocal immunofluorescence microscopy. (a) Untreated cells. (b) EGF-treated cells. DNA was stained with a blue fluorescent dye. Courtesy of Cell Signaling Technology, MA, USA (http://www.cellsignal.com).

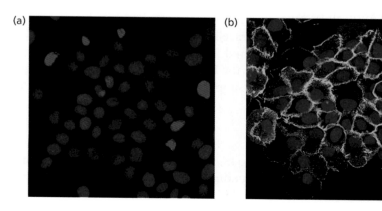

(a)   (b)

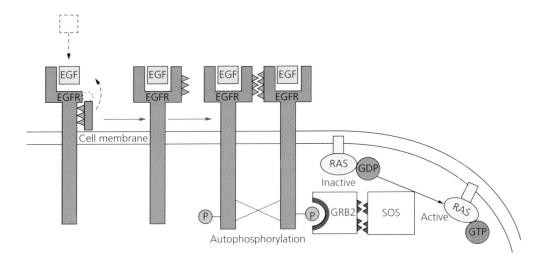

Figure 4.5 The early steps of the EGF pathway. The EGF growth factor binds to its receptor and causes the receptor to change shape. The change in shape allows two receptors to come together and induces kinase activation. The receptors undergo autophosphorylation shown by a "P." The phosphorylated receptor recruits GRB and SOS proteins to the membrane. SOS activates RAS. Activation occurs by the exchange of GDP with GTP. Note the SH2 domain of GRB2 that interacts with "P" and the SH3 domains at the interface of GRB2 and SOS are shaded in red.

crossed lines in Figures 4.2, 4.3, and 4.5) occurs at multiple tyrosine residues and is crucial for the recruitment of cytoplasmic substrate proteins that will pass the signal from the receptor to the signal transducers. At this stage, a signal from outside the cell has been transduced to inside the cell.

Note that activation of the tyrosine kinase receptor needs to be turned off after a particular length of time. Mechanisms for the termination of kinase activity include additional phosphorylation triggering a conformational change that inhibits extracellular ligand binding and kinase activity, dephosphorylation of regulatory phosphorylated tyrosine residues by tyrosine phosphatases, binding of negative regulators (e.g. RALT) to the kinase domain, and receptor endocytosis and degradation. However, some data suggest that the receptor can transmit signals from endosomes and that internalization of the receptor may also play a role in the transport of the receptor to the nucleus and the induction of specific genes.

Translocation of specific proteins to the membrane

Some phosphorylated tyrosine residues resulting from autophosphorylation create high-affinity binding sites for proteins that contain Src homology 2 (SH2) domains.

SH2 domains (approximately 100 amino acids long) and Src homology 3 (SH3) domains (approximately 50 amino acids long) mediate protein–protein interactions in pathways activated by tyrosine kinases. SH2 domains recognize and bind to distinct amino acid sequences (1–6 residues) C-terminal to phosphorylated tyrosine residues, and SH3 domains recognize and bind to proline and hydrophobic amino acid residues on partner proteins. Both SH2 and SH3 domains are frequently found in the

PAUSE AND THINK

Remember that a domain is a part of a protein with a specific configuration that has a specific function analogous to an electrical plug (see Chapter 3, p. 54).

same protein. Proteins mentioned later in the text that contain SH2 and SH3 domains include Grb2, SRC, ABL, and PI3-K.

Grb2, an intracellular protein that contains SH2 and SH3 domains, recognizes the phosphorylated EGFR via its SH2 domains and facilitates the recruitment of specific proteins to the membrane via its SH3 domains. Specifically, the two SH3 domains of Grb2 interact with the SH3 domains of exchange protein SOS (son of sevenless), which facilitates the activation of the pivotal intracellular transducer RAS. Thus, SOS, the activator of RAS, is translocated from the cytoplasm to the membrane in response to growth factor stimulation (Figure 4.5).

RAS activation

The RAS proteins are "star players" in regulating cell growth because of their position in the signal transduction pathway; they act as a pivotal point for the integration of a growth factor signal initiating from the membrane with a number of crucial signaling pathways that carry the signal through the cytoplasm and into the nucleus. N-, H-, and K-RAS are the three members of the family. They are GTP-binding proteins such that, when they are bound to GDP, they are inactive, and, when they are bound to GTP, they are active. Guanine nucleotide exchange factors, such as SOS, mediate the exchange of GDP for GTP by catalyzing the release of GDP from the guanine nucleotide binding pocket on RAS. Because GTP is tenfold more abundant in the cytoplasm, it then binds to RAS. GTP binding causes a conformational change and results in RAS activation, allowing interaction with downstream signal transducers (Figure 4.5).

GTPase activating proteins (GAPs) catalyze the hydrolysis of GTP to GDP to terminate the signal, although RAS proteins possess some intrinsic GTPase activity which allows for self-regulation.

RAS proteins undergo a series of post-translational modifications that direct their trafficking in the cell. Farnesylation, the addition of the C15 farnesyl isoprenoid lipid to the C-terminal CAAX motif (where C represents cysteine, A represents an aliphatic amino acid, and X represents any amino acid), is one modification that is required for localizing RAS to the cell membrane. It is interesting to note that it has been demonstrated that endogenous RAS is capable of activating downstream signaling pathways from subcellular membrane compartments (e.g. endoplasmic reticulum, Golgi) and that subcellular location may dictate function (Fehrenbacher *et al.*, 2009). This adds a new dimension to our knowledge, as it was previously thought that localization to the plasma membrane was essential for RAS activity. Furthermore, it has been demonstrated that RAS can transform cells from their subcellular compartments, as well as from the membrane. These observations must be considered in the rationale for designing new cancer therapies.

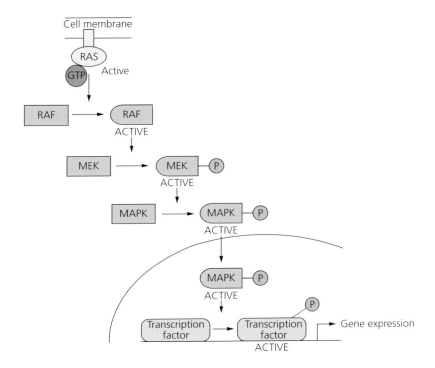

Figure 4.6 The later steps of the EGF pathway. Activated RAS activates signal transducer RAF. RAF phosphorylates and activates MEK. MEK phosphorylates and activates MAPK. This illustrates a serine/threonine kinase cascade. Activated MAPK enters the nucleus, and phosphorylates and activates transcription factors (TF) that turn on specific sets of genes needed for cell growth.

Raf activation

RAS–GTP binds to, and contributes to, the activation of the serine/threonine kinase Raf, one of its main effectors. The recruitment of Raf to the cell membrane is necessary for its activation by RAS–GTP. Activation involves the relief of a Raf auto-inhibitory mechanism. Activated Raf is a signal transducer that carries the signal away from the membrane (Figure 4.6). As a kinase, it phosphorylates mitogen-activated protein kinase kinase (MAPKK; MEK).

> **A little lesson about the MAP kinase family: MAP kinase kinase kinases, MAP kinase kinases, and MAP kinases . . .**
>
> The nomenclature may seem confusing at first, but it is because of the fact that a series of phosphorylation steps is necessary for signal transduction: a kinase (MAPKKK) phosphorylates another kinase (MAPKK), which itself phosphorylates yet another kinase (MAPK; originally called ERK). A unique feature of **MAP kinase** activation is that it requires both tyrosine and threonine phosphorylation. MAPKK is a dual-specificity kinase that phosphorylates both tyrosine and threonine residues. Raf is a MAPKKK, and MEK is a MAPKK.

The MAP kinase cascade

The activated MAPKKs (MEK), dual tyrosine and serine/threonine kinases, go on to phosphorylate the mitogen-activated protein kinases

(MAPKs) (also known as extracellular signal-regulated kinases; ERKs). The MAPKs are a family of serine/threonine kinases that provide the cytoplasmic link between the activated RAS on the plasma membrane and regulation of gene expression, as activated MAPKs can enter the nucleus. The activity of many transcription factors is regulated by phosphorylation, and thus MAPKs can affect the activity of transcription factors via phosphorylation (Figure 4.6).

Although this discussion focuses on MAPK, there are actually three distinct, but parallel, MAP kinase pathways: MAPK, JNK, and p38. As noted in this section, MAPK is activated by growth factors. JNK and p38 are activated by various environmental stress signals such as UV and ionizing radiation. The JNK and p38 pathways usually trigger apoptosis. All three MAPK pathways act as a common mechanism that serves multiple signaling pathways and results in various cellular responses.

Regulation of transcription factors

The AP-1 transcription factor is an important target of the MAPK cascade. As a transcription factor, it binds to DNA and regulates the expression of genes involved in growth, differentiation, and death. One mechanism whereby AP-1 induces cell cycle progression, is by binding to, and activating, the cyclin D gene, a critical regulator of the cell cycle. AP-1 is not a single protein but rather is made up of the products of two gene families, *jun* and *fos*. These proteins contain basic leucine zipper domains that facilitate their binding as dimers to either the cAMP response element (CRE) or the 12-O-tetradecanoylphorbol-13-acetate (TPA) response element in a target gene. AP-1 activity is induced by two mechanisms. First, direct phosphorylation of members of the Fos family by MAPK affects their DNA-binding activity. Second, MAPK phosphorylation and subsequent activation of other transcription factors increase the expression of both *fos* and *jun* genes. As a result, AP-1 activity increases, and subsequent transcriptional regulation proceeds.

The Myc family of transcription factors (Myc, Max, Mad, Mxi) can dimerize in different ways and lead to distinct biological effects of growth, differentiation, and death. Several seem to be targets of MAPKs. Myc is a short-lived protein that promotes proliferation by regulating the expression of specific target genes. Gene targets of Myc include *N-Ras* and *p53*, but the identification of additional targets is the subject of ongoing research. Myc requires the constitutively expressed family member Max to function. Myc and Max form heterodimers via basic helix–loop–helix leucine zipper domains and bind to a regulatory sequence, called the E-box, in their target genes. Heterodimer formation and DNA binding are crucial for the oncogenic, mitogenic, and apoptotic effects of Myc. Other heterodimers, such as Max and Mad/Mxi, are inhibitory for Myc function. They can also bind to the E-box in gene promoters, but they repress transcription. Thus, the Myc family of transcription factors forms a network of interacting basic helix–loop–helix leucine zipper proteins, and the identity of the members within a heterodimer determines the biological effect elicited.

Self-test Close this book, and try to redraw Figure 4.2. Check your answer. Correct your work. Close the book once more, and try again.

Good! Now let us backtrack to illustrate how you can build levels of complexity on the foundation you have learned. RAS has been noted to be a "major player" for the integration of a growth factor signal with a number of crucial signaling pathways. In fact, all receptor tyrosine kinases activate RAS. Raf was described earlier as one effector protein that leads to the activation of the MAPK cascade, but there are several other effector proteins of RAS activation. Phosphatidylinositol 3-kinase (PI3K), a lipid kinase, is another effector protein downstream of RAS that can be introduced at this point (see "Pause and think").

Crystal studies have shown that RAS interacts directly with the catalytic structure of PI3K. Production of the second messenger, PIP3, recruits the serine/threonine kinase PDK-1 to the membrane. Akt, another serine/threonine kinase, is also recruited to the membrane where it is phosphorylated and activated by PDK-1. Activated Akt is involved in anti-apoptotic and survival roles by phosphorylating distinct target proteins. Mammalian target of rapamycin (m-TOR), a serine threonine kinase, is a downstream target of Akt that is involved in promoting anabolic programmes such as lipid and nucleotide synthesis. Also, activated Akt can be translocated into the nucleus where it phosphorylates nuclear substrates, including transcription factors such as forkhead box O (FOXO).

Self-test Draw a diagram of a growth factor signal transduction pathway, including two effector proteins of RAS activation. Check your answer with Figure 4.7. Correct your work.

The effect of cell signaling on cell behavior

We have seen earlier in this chapter that growth factor signaling can lead to cell proliferation via signaling to the nucleus. In addition, cell signaling can have effects on cell behavior. This is clearly demonstrated by the intracellular tyrosine kinase SRC (the gene *src* is discussed in the next section). In addition to having a role in cell proliferation, SRC plays an important role in the regulation of cell adhesion, invasion, and motility. SRC is a phosphoprotein which contains SRC homology domains, including a SH2 and SH3 domain. It also contains a negative regulatory

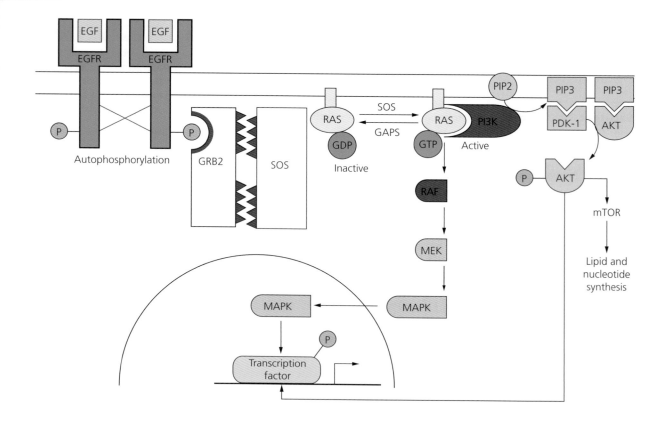

Figure 4.7 The EGF signal transduction pathway showing two effectors of RAS (shown in red). See text and Fruman and Rommel (2014) for details of PI3K signaling.

domain near the carboxy-terminus. When Tyr530 in this negative regulatory domain is phosphorylated, it binds to the internal SRC SH2 domain and results in an intramolecular association which represses the kinase domain and keeps SRC in an inactive state (Figure 4.8). One way in which SRC can be activated is via receptor tyrosine kinases such as the EGF receptor. Upon stimulation of the EGF receptor by growth factor, the autophosphorylated receptor can interact with the SH2 domain of SRC, disrupting its negative regulatory intramolecular conformation. Similarly, focal adhesion kinase (FAK) can also activate SRC by direct binding of FAK to the SH2 domain of SRC. The resulting SRC–FAK complex interacts with a wide range of target proteins, including focal adhesion proteins, adaptor proteins, and transcription factors. The regulation of cell matrix junctions, called focal adhesions, that associate with cytoskeletal fibers is particularly important for cell shape and motility. Assembly of focal adhesions facilitates cell adherence, while disassembly facilitates motility. Activation of SRC leads to disassembly of focal adhesions and thereby permits increased motility. SRC also regulates cell invasion by inhibiting E-cadherin (see Chapter 9), and influences proliferation and survival.

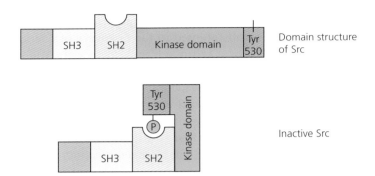

Domain structure of Src

Inactive Src

Figure 4.8 The protein domains of SRC and a negative regulatory intramolecular interaction.

Thus, the effects of growth factor cell signaling through tyrosine kinase receptors are indeed many fold, as is illustrated by the many downstream effects that occur upon SRC activation.

It is important to realize that, although the EGFR pathway is shown as a linear signaling pathway, it is more complicated; it is interconnected into a complex and dynamic signaling network. Remember the broad input from a range of ligands and the broad output from different effector proteins. Systems biology analysis has shown that EGFR influences 211 reactions! On top of the multiple components, there are regulatory signals from positive and negative feedback mechanisms that affect the amplitude and duration of responses. We are far from understanding quantitative aspects of the pathway, though mathematical modeling is becoming a useful tool (Kolch *et al.*, 2015).

4.2 Oncogenes

Cancer arises from mutations in genes that are involved in growth, differentiation, or death. There are two major classifications of mutated genes that contribute to carcinogenesis: oncogenes and tumor suppressor genes. A general description of an oncogene is a mutated gene whose protein product is produced in higher quantities or whose altered product has increased activity and therefore acts in a dominant manner. A mutation in only one allele is sufficient for an effect. Tumor suppressor genes (discussed in Chapter 6), however, are genes in which the mutation has caused a loss of function, and therefore most are recessive in nature because both alleles must be mutated. More than 100 oncogenes and at least 15 tumor suppressor genes have been identified.

Studies of retroviruses

Studies of retroviruses have led to great insights into cancer biology and have become the foundation of our knowledge of oncogenes (for a historical review, see Vogt, 2012). Several landmark experiments were performed, based on the early observation that viruses could cause cancer in animals, and the results pointed to the discovery of oncogenes. In 1911, Peyton Rous prepared a cell-free filtrate from a chicken sarcoma and demonstrated that he could induce sarcomas in healthy chickens with this filtrate. The causative agent was identified as the Rous Sarcoma Virus. Many decades later, oncogenic transformation by this virus was found to be caused by an "extra" gene contained in its genome that was not required for viral replication. The first so-called "oncogene" was identified as *v-src* (pronounced "v sark"). The oncogene product was characterized as a 60 kDa intracellular tyrosine kinase.

In 1976, a startling discovery was made. Michael Bishop and Harold Varmus found that there was a gene with a homologous sequence to *v-src* in uninfected chickens. Moreover, upon further investigation, this gene could be found in organisms from fruitflies to humans. Following further examination, a fundamental principle of cancer biology was revealed: almost all known oncogenes are altered forms of normal genes or proto-oncogenes.

> ● Exception: some protein products of DNA tumor viruses behave like oncogenes but are not viral versions of cellular oncogenes. They act by blocking the activity of tumor suppressor proteins. See Chapters 6 and 13.

The name proto-oncogene is sometimes used in cancer biology to distinguish the normal cellular (c) gene (e.g. *c-src*) from the altered form transduced by retroviruses (v) (e.g. *v-src*). The *v-src* sequence lacks the carboxy-terminal negative regulatory domain present in *c-src* and has point mutations throughout the gene.

A PIONEER IN THE FIELD . . . of oncogenes: Harold Varmus

Harold Varmus, along with J. Michael Bishop, received the Nobel Prize in Physiology or Medicine in 1989 for studies carried out at the University of California, San Francisco, that laid down the foundation for the role of mutations in carcinogenesis. They discovered that some genes of cancer-causing viruses were mutated forms of normal cellular genes, not viral genes. It meant that an oncogene was any cellular gene that, upon activation (i.e. by gain-of-function mutations), could transform cells. As we see in this chapter, this was the birth of the concept of proto-oncogenes and of the molecular biology of cancer.

Varmus was named by President Clinton to serve as the Director of the National Institutes of Health, a position he marked with many advancements during his 6 years' service. He was the President and Chief Executive Officer of Memorial Sloan-Kettering Cancer Center in New York City. After accepting President Obama's nomination in 2010, he was the Director of the National Cancer Institute for nearly 5 years.

At this point, a short review of the retroviral life cycle is necessary. The life cycle of retroviruses brands them as intracellular parasites in that they rely on their host cell for energy and to synthesize viral proteins. After injecting their infectious nucleic acid (RNA) into a host cell, the viral RNA is first reverse-transcribed into DNA and is called a provirus. Note that this is an exception to the central dogma of unidirectional flow of genetic information: DNA→RNA→protein. This provirus DNA is integrated randomly into the host chromosome where it will be replicated, transcribed, and translated as host DNA. The translation of viral RNA then produces viral proteins for the synthesis of new viral particles. During evolution, the virus can acquire fragments of genes from the host at integration sites, and this process may result in the creation of oncogenes. The Rous Sarcoma Virus acquired a truncated form of *c-src*. Alternatively, and depending on the integration site, viral DNA may be translated as a fusion protein, in conjunction with cellular DNA, resulting in a novel fusion protein, or host genes may fall under the regulation of viral regulatory sequences. The resulting disruptions to host gene expression are other mechanisms of virus-induced oncogenesis. This knowledge aids our general understanding of the mechanisms of carcinogenesis, because, although viruses are not the major cause of human cancers, the mechanism of oncogenic activation of proto-oncogenes is similar. For example, chromosomal translocations may have the same consequence as the integration of a virus into a host chromosome; a crucial gene may come under the influence of novel regulatory sequences and result in abnormal quantities of the gene product. The new gene configuration may serve as an oncogene.

It is important to become familiar with examples of oncogenes to bolster the lesson learned from viral studies: almost all known oncogenes are altered forms of normal genes.

There are examples of oncogenes for every type of protein involved in a growth factor signal transduction pathway. Several examples are described in the following sections.

Growth factors

The first evidence for the role of proto-oncogenes came from analysis of the viral oncogene *v-sis*. Its protein product was cytoplasmic and was found to be a truncated version of a growth factor normally secreted by platelets, called platelet-derived growth factor (PDGF). Thus, the identity of the product of the proto-oncogene, or cellular gene *c-sis*, is PDGF. It is a component of a wound response, and its normal role is to stimulate epithelial cells around the wound edge to proliferate and repair the damage. The significance of the oncogenic form may be its aberrant location (cytoplasmic rather than secreted) and the subsequent activation of the PDGF signal pathway at inappropriate times (e.g. other than in response to a wound), leading to unregulated growth.

Growth factor receptors

The oncogene *v-erbB* was originally identified from (and named after) the avian **erythroblastosis** leukemia virus. It is a truncated form of the epidermal growth factor receptor whereby the extracellular domain is deleted. Thus, the identity of the product of the proto-oncogene, or cellular gene *c-erbB*, is EGFR. The mutated receptor triggers cell division in the absence of EGF. Point mutations that accomplish the same effect of interfering with growth factor binding and inducing constitutive ("always on") activation have been identified in human cancers. Increasing the amount of normal *c-erbB* product by gene amplification is another mechanism that contributes to carcinogenesis, particularly breast cancer. Gene amplification involves multiple duplications of a DNA sequence, owing to errors at DNA replication forks.

RET signaling plays an important role in kidney development and neuronal differentiation during embryogenesis. The proto-oncogene *ret* codes for another growth factor tyrosine kinase receptor that heterodimerizes with cell surface co-receptors GFR-α1–4 in order to transduce the signal for glial-derived neurotrophic factor (GDNF) family ligands (for review, see Mulligan, 2014). The ligands do not bind to RET directly but rather forms a complex with the co-receptors first, which then can recruit RET. Within this complex, RET undergoes a conformational change, dimerizes, and undergoes autophosphorylation.

Papillary thyroid carcinoma cells often carry somatic chromosomal rearrangements involving the amino-terminal parts of numerous genes and the sequences of *ret* that code for the tyrosine kinase domain. The fusion protein products display cytoplasmic kinase activity that is independent of GDNF signaling. These RET chromosomal rearrangements are more frequent in locations of high environmental radiation exposure such as Chernobyl, Ukraine, or Hiroshima in Japan.

Germline mutations are associated with three familial autosomal dominant tumor syndromes: multiple endocrine neoplasia 2A (MEN2A), MEN2B, and familial medullary thyroid carcinoma. The mutations that have been identified illustrate different mechanisms for oncogenic activation. Almost all MEN2A patients have mutations in conserved extracellular cysteines. Resulting intermolecular disulfide bonds cause constitutive RET dimerization and aberrant activation. In MEN2B patients, oncogenic activation is achieved by altering the substrate-binding pocket of the tyrosine kinase domain. A conserved Met is characteristic of the substrate-binding domain of receptor tyrosine kinases whereby Thr is conserved for cytoplasmic tyrosine kinases. The characteristic mutation of MEN2B is a substitution mutation whereby Thr replaces this conserved Met residue (Met918Thr). This results in altered substrate access, leading to increased kinase activity and altered substrate specificity that is characteristic of cytoplasmic tyrosine kinases, instead of receptor tyrosine kinases. Thus, the signal transduction pathways are highly disrupted.

PAUSE AND THINK

What types of genes are usually involved in inherited predispositions to cancer? Although tumor suppressor genes are most often involved, here we see an unusual example of a heritable oncogene (*ret*) playing a role in cancer predisposition.

Oncogenic activation of receptor tyrosine kinases occurs through specific mutations that lead to constitutive tyrosine activation, dimerization, or altered substrate specificity.

Intracellular signal transducers

The oncogenic activation of *ras* is observed in about 30% of human tumors. It is the most commonly mutated oncogene in human cancers. The majority of mutations are located in codons 12, 13, and 61. The consequence of each of these mutations is a loss of GTPase activity of the RAS protein, normally required to return active RAS–GTP to inactive RAS–GDP. The effect is constitutive activation of RAS protein, even in the absence of mitogens. Some specific mutations in the *ras* gene are characteristic for specific cancers. A point mutation within codon 12 that results in the substitution of valine (GTC) for glycine (GGC), G12V, is characteristic of bladder carcinoma, while substitution of cysteine is a smoke-induced mutation and is common in lung cancer.

 HOW DO WE KNOW THAT?

Transformation assays and mutational analysis

The early classical paper by Reddy *et al.* (1982) illustrates fundamental experimental methods that were used to identify oncogenes and their genetic alterations. The ability of the *ras* oncogene (identified in a 6.6 *Bam*H1 fragment of T24 DNA) to transform cells was demonstrated in a prototypical (original) *in vitro* cell transformation assay (Figure 4.9a; see "Cancer cells can be distinguished from normal cells in cell culture conditions" on p. 5). This assay is based on the characteristic that cancer cells grow as foci against a monolayer of normal cells. To identify a smaller fragment of the gene that contained the genetic alteration, a series of deletion mutants were transfected into NIH3T3 cells and assayed for formation of foci

(transforming activity is indicated by a +/− sign next to each deletion construct shown in Figure 1 of Reddy *et al.*, 1982). The research team then used comparative sequence analysis of the small fragment identified by deletion analysis to identify a single point mutation within the small fragment in human bladder carcinoma cells. The method of sequence analysis was according to the procedures of Maxam and Gilbert (Figure 4.9b).

In vivo experiments supported the role of the *ras* oncogenes in carcinogenesis. Cells transfected with a plasmid capable of expressing high levels of *c-ras* showed the ability to form tumors in mice (Chang *et al.*, 1982), and a *ras* knock-out mouse exhibited decreased tumor formation during skin carcinogenesis (Ise *et al.*, 2000).

B-Raf is another transducer that may become oncogenic. Oncogenic activation of B-Raf is common in melanomas. The common mutated form of B-Raf, B-Raf (V600E), found in melanoma patients causes constitutive kinase activity and insensitivity to feedback mechanisms. Thus, growth signaling initiated from the transducer, and not from an external growth factor signal, occurs at inappropriate times and leads to abnormal cell growth.

Genes that code for cytoplasmic tyrosine kinases, such as SRC, serine/threonine kinases, such as RAF and MAPK, and nuclear kinases, such as

PAUSE AND THINK

Classify each type of evidence above according to the "show it," "block it," and "move it" criteria discussed in Section 1.2, "How do we know that? Types of evidence" on p. 9.

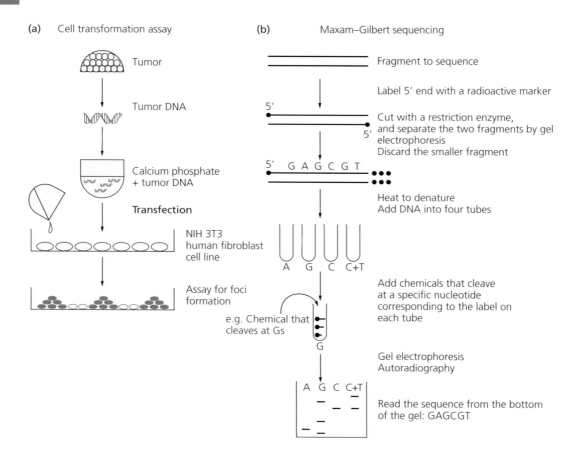

(a) Cell transformation assay

Tumor

Tumor DNA

Calcium phosphate + tumor DNA

Transfection

NIH 3T3 human fibroblast cell line

Assay for foci formation

(b) Maxam–Gilbert sequencing

Fragment to sequence

Label 5′ end with a radioactive marker

5′

Cut with a restriction enzyme, and separate the two fragments by gel electrophoresis
Discard the smaller fragment

5′

5′   G A G C G T

Heat to denature
Add DNA into four tubes

A   G   C   C+T

Add chemicals that cleave at a specific nucleotide corresponding to the label on each tube

e.g. Chemical that cleaves at Gs

G

Gel electrophoresis
Autoradiography

A   G   C   C+T

Read the sequence from the bottom of the gel: GAGCGT

Figure 4.9 (a) Cell transformation assay. (b) DNA sequencing by the Maxam–Gilbert method. Chemicals are used to cleave an end-labeled DNA segment at specific nucleotides, and fragments are analyzed by gel electrophoresis and autoradiography. Note that there is not a specific enzyme that cuts only at Ts but instead cuts at Cs + Ts. Thus, if there is a band in both the C lane and the C + T lane, the nucleotide is a C; if there is a band only in the C + T lane, the nucleotide is a T.

ABL, can also undergo oncogenic activation. As discussed earlier, intramolecular associations normally regulate c-SRC kinase activity; the SRC SH2 domain binds a carboxy-terminal phosphorylated tyrosine residue (Tyr530) and results in a conformation that blocks the SRC kinase active site. Repression of SRC kinase activity can be relieved by dephosphorylation of Tyr530 or by the binding of the SH2 domain to specific activated tyrosine kinase receptors (see "Pause and think" on p.95.). In colon cancer, the protein product of oncogenic *src* is characterized by a truncation at Tyr530. This aberrant protein is unable to adopt the inactive conformation described previously, and therefore kinase activity is constitutive ("always on").

Let us look at *abl* as another example. *c-abl* is a gene whose product is a nuclear tyrosine kinase that plays a role in DNA damage-induced apoptosis. It is normally activated by ionizing radiation and particular drugs via the serine/threonine kinase ATM. Oncogenic activation occurs through the chromosomal translocation t(9;22) whereby *abl* becomes juxtaposed to a breakpoint cluster region *bcr*. Thus, DNA sequences that are normally not next to each other are now fused and, upon transcription, give

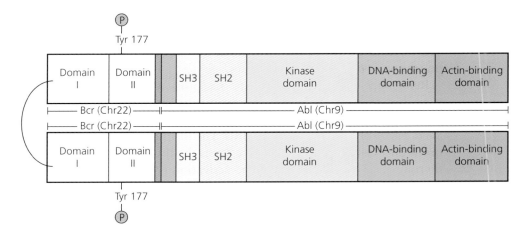

Figure 4.10 The BCR–ABL fusion protein.

rise to a fusion protein with novel features (Figure 4.10). Translocated BCR retains domains I and II (shown in white), while ABL retains the SH3 and SH2 domains, the kinase domain, the DNA-binding domain, and the actin-binding domain (shown in color). BCR–ABL molecules associate with each other, forming homo-oligomeric complexes, mediated by the coiled-coil motif in domain I of BCR. Oligomerization permits autophosphorylation at Tyr177 within domain II, and this triggers activation of the ABL tyrosine kinase. The fusion protein BCR–ABL is maintained in the cytoplasm. Consequently, a nuclear kinase is constitutively activated in the cytoplasm and has access to a range of novel substrates, interfering in the normal signal transduction pathways of the cell.

Transcription factors

It is not surprising that the transcription factor AP-1, which is an important regulator of cell growth, differentiation, and death, is also involved in transformation. Components of AP-1, Jun and Fos, are encoded by proto-oncogenes *c-jun* and *c-fos*, and several mechanisms of oncogenic activation of these genes exist. Normally, *c-fos* mRNA is short-lived so that the response to a mitogen is transient. Truncation of the 3′ end of *v-fos* eliminates a motif involved in mRNA instability (ATTTATTT) and produces an mRNA with a longer half-life. The aberrant expression of *v-fos* mRNA results in an increase in *v-fos* gene product and an inappropriate increase in the transcription of AP-1-regulated genes. Oncogenic activation may also involve the deletion of a regulatory promoter sequence, the serum response element, such that transcription of the *fos* gene occurs even in the absence of serum mitogens.

Oncogenic activation of *c-myc* occurs from constitutive expression and overexpression of the c-Myc protein. Chromosomal translocation of *myc* (chromosome 8) to a location that falls within the regulation of the strong promoter of immunoglobulin genes (chromosome 14) increases

PAUSE AND THINK

How does the role of the SH2 domain of SRC compare with that of the SH2 domain of Grb2? In both cases, the SH2 domain recognizes a phosphotyrosine residue. The SH2 domain of SRC regulates intramolecular interactions, but the SH2 domain of Grb2 regulates intermolecular interactions between itself and receptor tyrosine kinases.

the amount of expression from the *myc* gene. This mechanism of onco-genic activation of *c-myc* is commonly observed in Burkitt's lymphoma. The increase of Myc protein results in an inappropriate increase in the transcription of Myc-regulated genes.

Remember that steroid hormone receptors act as ligand-dependent transcription factors. In addition to *v-erbB* discussed earlier, another oncogene, *v-erbA*, was originally identified from (and named after) the avian erythroblastosis leukemia virus. The identity of the product of the proto-oncogene or cellular gene *c-erbA* is the thyroid hormone (triiodo-thyronine, T3) receptor. Oncogenic activation is achieved by mutations that prevent thyroid hormone binding and inhibit transcriptional activa-tion. This type of mutation is referred to as a **dominant negative** mutation because the product of this mutation codes for receptors that can bind to DNA and block access of wild-type receptors, including other steroid hor-mone receptor family members that may form heterodimers with thyroid receptors at their response elements. As the product of *v-erbA* can form **homodimers** (note that the product of the proto-oncogene *erbA* homodi-merizes poorly), it is thought that the homodimers mediate a dominant negative effect on the response elements. Somatic mutations of thyroid hormone receptors identified in human cancers give rise to products that have lost their ability to bind thyroid hormone and regulate transcription, and some result in dominant negative products, suggesting that they may be involved in human cancers. Silencing of the thyroid hormone receptor genes by hypermethylation is also associated with some cancers. Trans-genic mice models that lack thyroid hormone receptors develop thyroid cancer and support the critical role of thyroid receptors in cancer (Kim and Cheng, 2013) (see "Pause and think").

Mechanisms of oncogenic activation

As can be seen from the previous sections, several mechanisms can be used to activate proto-oncogenes to become oncogenes (Figure 4.11). Point mutations and deletions in the coding region are a common mechanism and often change the structure and/or function of the proto-oncogene prod-uct. Both mechanisms of oncogene activation were described for the *EGFR* gene. Mutations in the gene promoter region can lead to overexpression of a proto-oncogene. Chromosomal translocations, as well as insertional mutagenesis, cause the juxtaposition of sequences not normally next to each other, and often this configuration can cause altered expression. The translocation involving *c-myc* and immunoglobulin regulatory sequences mentioned under "Transcription factors" on p. 95 is one example. Alter-natively, fusion proteins can have novel characteristics. The Philadelphia chromosome t(9:22) activates and relocates the nuclear kinase, c-Abl, to the cytoplasm where it encounters novel substrates. **Gene amplification** is another mechanism for activation of *erbB2* and is observed in breast cancer.

PAUSE AND THINK

In the previous discussion regarding three specific tran-scription factors, you will notice that AP-1 and Myc transcrip-tional activity increases in some cancers, while thyroid hormone receptor transcriptional activ-ity decreases in some cancers. What can you postulate about the target genes of these tran-scription factors? The target genes for AP-1 and Myc are likely to code for proteins that promote growth, while the tar-get genes of thyroid hormone receptor are likely to inhibit growth.

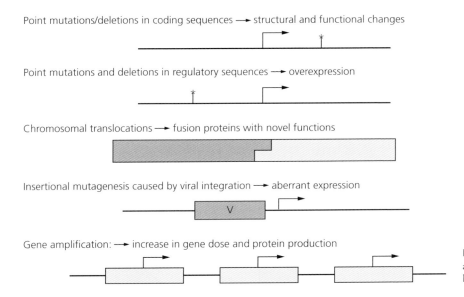

Point mutations/deletions in coding sequences ⟶ structural and functional changes

Point mutations and deletions in regulatory sequences ⟶ overexpression

Chromosomal translocations ⟶ fusion proteins with novel functions

Insertional mutagenesis caused by viral integration ⟶ aberrant expression

Gene amplification: ⟶ increase in gene dose and protein production

Figure 4.11 Mechanisms that activate proto-oncogenes to become oncogenes.

◎ Therapeutic strategies

Knowledge of the molecular details of the EGFR signal transduction pathway and related pathways has led to the launch of many new cancer therapeutics targeting individual components. There has been great success for some, and valuable lessons from others. The future is hopeful, as we continue to unravel the molecular biology of signal transduction pathways and move forward in the design of additional new therapeutics. The strategies of some therapeutics aimed at molecular targets within these pathways are described in the next sections.

4.3 Kinases as drug targets

Many types of kinases, including transmembrane tyrosine kinases, cytoplasmic kinases, and nuclear kinases, are implicated in cancer, as we have seen earlier. Therefore, they have become important targets for the design of new cancer therapeutics. In theory, one may predict that designing drugs with great specificity to a subset of kinases would be difficult because the structure of the catalytic domains of different kinases is very similar when the kinases are in the active state. However, the synthesis of specific kinase inhibitors has clearly been demonstrated, and examples are described in the following sections and illustrated in Figure 4.12.

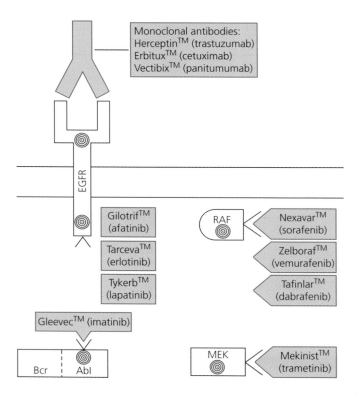

Figure 4.12 Kinase inhibitors as therapeutic strategies. Molecular targets are indicated with a target symbol (◎), and therapeutic agents are shaded in red. All drugs shown are approved for clinical use.

Anti-EGFR drugs

The *ErbB2* gene, a member of the EGFR family, is amplified in 30% of breast cancer patients. Overexpression of this gene in cell culture experiments and in **transgenic mice** (see Box "Analysis of gene function using transgenic mice") results in cell transformation and the induction of breast cancer, respectively. This suggests that ErbB2 has a causal role in breast cancer. Later, EGFR mutations were identified in a subset of lung cancers.

Several small molecule kinase inhibitor drugs directed against the tyrosine kinase activity of EGFR family members have been developed, including Iressa™ (gefitinib), Tarceva™ (erlotinib), and Tykerb™ (lapatinib) (Pao *et al.*, 2004; Sequist, 2015). Iressa™ (gefitinib) and Tarceva™ (erlotinib) are first-generation inhibitors, as their binding is competitive and reversible—that is ATP can compete for their binding. Gilotrif™ (afatinib) is a second-generation inhibitor that can form covalent bonds with EGFR, and this binding is irreversible. Tarceva™ (erlotinib) and Gilotrif™ (afatinib) have been approved for the treatment of advanced non-small-cell lung cancer; Iressa™ (gefitinib) is no longer approved in the USA (see Box "Lessons from pharmacogenomics" for further discussion). Although the response rate to both erlotinib and afatinib is quite high (70%), drug resistance commonly develops. A single mutation in EGFR, T790M, accounts for 50–60% of resistant cases. Even though afatinib showed promise against T790M in early pre-clinical trials, both drugs are not effective for this mutation. Second-generation mutant-selective covalent (irreversible) inhibitors, such as rociletinib (CO-1686) and AZD9291, that target T790M are being developed and have entered clinical trials (Jänne *et al.*, 2015; Sequist *et al.*, 2015). Their inability to bind to wild-type EGFR receptors promises to produce fewer side effects. Lapatinib has been approved for the treatment of breast cancer.

Analysis of gene function using transgenic mice

One strategy for investigating the function of a gene is to put it in a place where it is not normally located, or remove it from where it normally resides and observe any changes. Transgenic mice contain an additional or altered gene in all of their cells. Transgenic mice can be created by the direct injection of DNA into a fertilized egg. More commonly, foreign DNA is introduced into embryonic stem cells in culture prior to their transplantation into an early embryo (blastocyst). A series of breeding stages must follow to generate fully **homozygous** transgenic mice, as the founder animal is chimeric (not all cells of the organism contain the altered DNA) rather than transgenic. The cell culture step allows for selection and analysis of DNA integration to take place prior to the creation of a mouse. To create a knock-out mouse, a vector is designed, so that it will insert into a specific gene location by homologous recombination (see "Recombinational repair" on p. 41) and disrupt the endogenous gene such that gene function is repressed. Complex transgenic experiments may use tissue-specific or inducible promoters to induce the expression of the foreign gene in a particular location or specific time.

A different approach has also proved successful. Instead of targeting the tyrosine kinase domain of the ErbB2 receptor, the unique extracellular domain was targeted using monoclonal antibodies (see also Section 12.6, "Therapeutic antibodies" on p.286). Herceptin™ (trastuzumab) is a humanized (produced with human recombinant immunoglobulin genes) monoclonal antibody that binds the extracellular domain of ErbB2 with high affinity. Herceptin™ functions through a combination of mechanisms of action, including: enhanced receptor degradation, inhibition of angiogenesis, and recruitment of immune cells, resulting in antibody-dependent cellular cytotoxicity. Herceptin™ was approved by the US Food and Drug Administration (FDA) in 1998 for treatment of metastatic breast cancer whose tumors overexpress ErbB2, and thus it is the first genomics-based therapy administered selectively, based on the molecular profile of the tumor.

Erbitux™ (cetuximab) and Vectibix™ (panitumumab; ABX-EGF) are monoclonal antibodies that target EGFR and have been approved to treat colorectal cancers. Patients with *KRAS* mutations in their tumors do not respond to cetuximab (Erbitux™) and panitumumab (Vectibix™). Recalling the details of the EGF signaling pathway (see Section 4.1, "Epidermal growth factor signaling: an important paradigm" on p. 80 and Figure 4.2) helps to understand why patients with *KRAS* mutations will not respond. If the EGFR is the oncogenic signal, then blocking it will be a successful strategy. But if a molecule further downstream in the pathway is the oncogenic signal, then blocking EGFR will be in vain. The *KRAS* gene is mutated in about 40% of colorectal cancer patients. In 2009, the US FDA updated the cetuximab (Erbitux™) label to include a recommendation on screening for the *KRAS* mutation. This is one of the first cases where pharmacogenomic information is included on a drug label. Recommendation for *KRAS* screening is also included for the drug panitumumab (Vectibix™). The influence of genetic information on an individual's response to a drug, pharmacogenomics, helps doctors to choose the best treatment for an individual (see Box "Lessons from pharmacogenomics"). Genomic characterization of tumor DNA is elucidating subsets of molecularly distinct tumor types within a class of cancer (e.g. breast cancer) that will respond differently to targeted treatments. This is a significant step towards individually tailored treatment, whereby treatment is matched to the molecular make-up of the patient's tumor and promises to increase the success rate of a given drug.

Lessons from pharmacogenomics

Pharmacogenomics is the study of the influence of the genome on an individual's response to a drug. Gene variability in both the individual and in the tumor may lead to differences in drug response among individuals. Below are a few examples of mutations in the EGFR signaling pathway that are known to affect drug response. Prescribing drugs with respect to a tumor's genotype has begun the practice of "personalized" or "precision" medicine.

Gefitinib (Iressa™; AstraZeneca), a tyrosine kinase inhibitor targeted against the EGFR (see "Anti-EGFR drugs" on p. 98), was reported to have mixed clinical responses. Initially approved in the USA and Japan for the treatment of non-small-cell lung cancer, gefitinib (Iressa™) causes impressive and successful tumor regression but, curiously, in only a select 10% of patients. As non-small-cell lung cancer accounts for 80% of all lung cancers, the number of patients that respond to gefitinib (Iressa™) (10%) is significant. It has now been demonstrated that the patients who respond to gefitinib (Iressa™) carry specific mutations in the *EGFR* gene (Lynch *et al.*, 2004; Paez *et al.*, 2004).

Clinical trials showed that Japanese people are three times more likely to respond to gefitinib (Iressa™) than Americans. Scientists screened for mutations in the *EGFR* gene in two groups of non-small-cell lung cancer tumor samples: one set from a Japanese hospital and another set from an American hospital. Somatic mutations in the *EGFR* gene were found more often in Japanese patients. The mutations were either missense mutations or small deletions that were located in the coding region for the receptor tyrosine kinase active site, a region near the binding site for gefitinib (Iressa™). Scientists then asked ➔

PAUSE AND THINK

Why are these particular mutations in the *EGFR* gene more common in Japanese than Americans? It may be due to a lifestyle factor, such as exposure to a specific carcinogen, or a genetic predisposition, or a mixture of both.

→ whether the mutations correlated with treatment response. The analysis of a small set of patients showed that those who responded to gefitinib (Iressa™) also carried the identified mutations.

The mutations produce an altered protein that increases the amount and duration of EGF-induced signal transduction, compared with wild-type receptors. In other words, the receptors are hyperactive, but not constitutive. These mutations that underlie aberrant tyrosine kinase signaling and drive carcinogenesis also make the tumor more susceptible to gefitinib (Iressa™), because they cluster in the sequences that code for the region where gefitinib (Iressa™) binds. This suggests that knowing the particular molecular characteristics of a tumor (e.g. mutations of the *EGFR* gene) is important for selecting the best treatment for an individual.

In contrast to the *KRAS* mutation described in the text where the mutation blocks the ability of a tumor to respond to a drug, these *EGFR* mutations enable a drug response. The age of precision medicine has really begun.

Strategies against Ras and Raf

As the most commonly mutated oncogene in cancer, Ras is ideally a most important drug target. But the decades of failure to find small molecule inhibitors has caused the scientific community to label Ras as undruggable. Many strategies, including the creation of farnesyltransferase inhibitors that compete with the carboxy-terminal CAAX motif of RAS to prevent membrane insertion and inhibitors that disrupt RAS–RAF interactions, were unsuccessful. These failures were mainly due to gaps in our knowledge. For example, alternative enzymes that modify RAS compensated for farnesyltransferase inhibition. This trend may change, as the National Cancer Institute has launched a national Ras Program (see web sites listed at the end of this chapter) to focus research efforts on this problem. Also an inhibitor that is selective for the G12C K-Ras mutant has been identified (Ostrem *et al.*, 2013). As an allosteric inhibitor, it irreversibly binds to the mutant protein, selectively binds the GDP form of the oncoprotein, and interferes with Raf binding. Although these inhibitors need further development, they have opened another avenue for Ras targeting strategies.

Targeting downstream Ras effectors have proved valuable as a cancer treatment.

The strategy commonly used to target other kinases has been applied to the serine/threonine kinase Raf. A multi-kinase inhibitor that targets the ATP-binding site of Raf, called sorafenib (Nexavar™; previously called BAY43-9006), was approved for treatment of advanced renal cell carcinoma and hepatocarcinoma (USA 2005, Switzerland and Mexico 2006; Wilhelm *et al.*, 2006). In order to investigate whether there was modulation of the defined molecular target Raf, the phosphorylation of Raf targets was monitored. Data demonstrated a reduction in downstream

MAPK phosphorylation in the blood of patients receiving the well-tolerated oral treatment. Note that it is important to monitor molecular endpoints (MAPK activity), in addition to clinical endpoints (anti-tumor activity). As a multi-kinase inhibitor, sorafenib (Nexavar™) also blocks several receptor tyrosine kinases (VEGFR-1, -2, and -3, PDGFRb, c-Kit, and RET). The real contribution of Raf inhibition to clinical effectiveness, compared with the other sorafenib (Nexavar™) targets, will only be known by examining specific Raf inhibitors.

Two other Raf inhibitors approved for the treatment of melanoma are vemurafenib (Zelboraf™; PLX4032) and dabrafenib (Tafinlar™; GlaxoSmithKline). Half of all melanomas contain the V600E mutation that leads to constitutive kinase activity. Crystallography-guided approaches using the kinase domain of B-Raf (V600E) aided drug optimization. Vemurafenib (Zelboraf™) induced partial or complete tumor regression in 81% of patients who had melanoma with the B-Raf (V600E) mutation in an early clinical trial (Flaherty *et al.*, 2010). Although very promising, drug resistance develops for many of the patients. Resistance involves switching to other MEK kinases, rather than new mutations in the kinase domain, and this suggests that combination treatments may be more effective for durable responses. The combination of dabrafenib and an approved MEK inhibitor (trametinib; see "MEK inhibitors") has received FDA approval.

MEK inhibitors

Inhibiting MEK is a strategy that would be applicable to patients with mutations in either *ras* or *raf* genes. Mutations in *MEK* genes are rare; approximately 1% of tumors contain *MEK* mutations, and no mutations have been identified to date in *MAPK* (Fremin and Meloche, 2010). Several allosteric MEK inhibitors (inhibitors that do not compete with ATP) are in clinical trials (e.g. AZD6244 and RDEA119), and one, trametinib (Mekinist™; GlaxoSmithKline), has been approved.

Imatinib (Gleevec™; STI571)

Chronic myelogenous leukemia (CML) accounts for 15–20% of all leukemias. Bone marrow transplantation is the only hope for a cure but is not feasible for a variety of reasons (including donor matching) for the majority of patients. Interferon-α, accompanied by severe side effects, was, until recently, the standard treatment. The knowledge of the molecular biology of the disease has led to successful specific molecular targeting and the development of a most successful drug. Most CML patients (95%) carry the Philadelphia chromosome, the product of the chromosomal translocation t(9;22)(q34;q11) generating the BCR–ABL fusion protein. As a result of this translocation, the tyrosine kinase activity of ABL is constitutive and is retained in the cytoplasm, rather than the nucleus. As a result of aberrant kinase signaling, there is abnormal proliferation of white blood cells, the hallmark of leukemia. Transformation is dependent on the BCR–ABL kinase activity, and therefore it provides the perfect therapeutic target.

Imatinib (Gleevec™), a small-molecule tyrosine kinase inhibitor, has been successful in the treatment of CML, resulting in remission in 96% of early-stage patients. It is a paradigm for targeted cancer therapy, having flown through clinical trials and approval (2001) within 3 years (discussed further in Chapter 14). The compound was modeled and synthesized after related lead compounds (compounds that show a desired activity, e.g. kinase inhibition), called phenylaminopyrimidines, were identified from high-throughput screens of chemical libraries. The compound was originally optimized for inhibiting PDGF receptor (PDGFR) tyrosine kinase activity but was later found to inhibit ABL and c-Kit, as well. Imatinib (Gleevec™) binds to the ATP-binding pocket within the catalytic domain, but the fairly narrow specificity of the compound seems to be a result of preferential binding of the drug to the inactive state of the kinase, as evidenced by analysis of crystal structures (Schindler et al., 2000). Gleevec™ (imatinib) recognizes the auto-inhibitory conformation of the activation loop of the protein that regulates the kinase activity. The structure of the inactive state is distinctive between different kinases. The drug has a half-life of approximately 15 h and conveniently allows daily oral administration.

Pre-clinical data demonstrated inhibition of proliferation in cultured cells and in cells from CML patients with the Philadelphia chromosome, as well as tumor regression in mice. This evidence allowed progression to clinical trials. The threshold dose for significant therapeutic efficacy was found to be 300 mg in Phase I trials. Parameters of how well the drug works, efficacy endpoints, were measured by the degree of cytogenetic (chromosomal) and hematologic (blood count) response. A complete cytogenetic response was defined as 0% Philadelphia chromosome-positive cells in metaphase (partial, 1–35%; minor, 36–65%; minimal, 66–95%; or no response > 95% were additional classifications used). Hematologic response is simply graded by white blood cell counts. Importantly, molecular target inhibition was also analyzed. Quantification of the levels of phospho-CRKL, a BCR–ABL substrate found in neutrophils, allowed for the assessment of the inhibition of kinase activity and aided in the determination of effective dosage. As mentioned earlier, it is important to monitor the modulation of the defined molecular target (BCR–ABL).

Note that CML has three disease phases: chronic (lasting 3–5 years), accelerated (lasting 3–9 months), and blast crisis (lasting 3–6 months). Owing to an increase of cell proliferation, the number of white blood cells increases as the disease progresses. The effectiveness of Gleevec™ (imatinib) decreases with advanced disease phase (53% response in accelerated phase, and 30% response in blast crisis). Although only 9%

of early-stage patients relapsed, 78% of late-stage patients relapsed. The mechanism for the majority of these cases is owing to reactivation of the kinase activity due to mutation or *Bcr–Abl* amplification. Analysis of clinical samples showed that six out of nine patients had a single amino acid substitution at a contact residue identified in the crystal structure (Gorre *et al.*, 2001). These mechanisms suggest that the initial chromosomal translocation is important not only for initiation, but also for maintenance of the cancer phenotype, and supports the concept of oncogene addiction— the dependence of a cancer cell on a specific oncogene for its maintenance. The development of the imatinib and second- and third- generation drugs to treat resistance are discussed in Chapter 14 (see Sections 14.7 on p. 339 and 14.8 on p. 340). Imatinib has also been approved to target c-Kit in gastrointestinal stromal tumors, and additional studies are investigating its use against PDGFR in glioblastomas (see "Pause and think").

> **PAUSE AND THINK**
>
> What is the difference between the types of molecular targets described in this section? They include three different types of kinases: a transmembrane receptor tyrosine kinase, a cytoplasmic serine/threonine tyrosine kinase, and a nuclear tyrosine kinase, respectively.

Conclusion

Cancer is a disease characterized by uncontrolled growth. Therefore, a clear understanding of growth regulation has helped to reveal the mechanisms of carcinogenesis. This was illustrated by the elucidation of the existence of oncogenes, which include altered versions of normal genes involved in growth. Oncogenes often play a role in growth factor signal transduction. The knowledge of the intricacies of growth factor signal transduction pathways have been, and will be, essential to the design of successful, low-toxicity cancer therapeutics designed against molecular targets.

CHAPTER HIGHLIGHTS—REFRESH YOUR MEMORY

- Growth factors, growth factor receptors, intracellular signal transducers, and nuclear transcription factors play a role in growth factor signal transduction.

- Many growth factor receptors are tyrosine kinases. Kinases phosphorylate specific amino acid residues in target proteins.

- Phosphorylated proteins can be recognized by specific protein domains (e.g. SH2) and can thus serve as a recruitment platform.

- RAS plays a pivotal role in the EGFR pathway; it links activation of tyrosine kinase receptors with downstream signaling pathways.

- Raf, a serine/threonine kinase activated by RAS, initiates a cascade of phosphorylations by MEK and MAP kinases.

- One ultimate result of signaling initiated by growth factors is the regulation of transcription factors in the nucleus. Another is affecting cell behavior.

- Retroviruses have been instrumental in the elucidation of oncogenes.

- Most oncogenes are altered versions of normal genes.

- Constitutive kinase activation is a common consequence of oncogenic mutations of tyrosine kinase receptors.

- Aberrant subcellular localization is another consequence of oncogenic activation.

- Many molecular components of growth factor signal transduction pathways have been targets for new cancer therapeutics.

- Different domains of tyrosine kinase receptors have been targeted for the development of new cancer therapies.

- Iressa™ showed favorable responses in patients who carry specific mutations in the *EGFR* gene but is no longer approved in the USA.

- The study of the influence of the genome on a patient's response to a drug is called pharmacogenomics.

- The testing of new therapeutics should include an assay for the modulation of the defined molecular target.

ACTIVITY

1. Fibroblast growth factor receptors can function as an oncogene. Somatic mutations in fibroblast growth factor receptor 3 (FGFR3) have been identified in approximately 50% of bladder cancers. Identify the most common mechanism of the oncogenic activation and its consequences (what happens to the function of FGFR3?). Describe the components of the signal transduction pathway it activates, and compare this to the pathway of EGFR. Suggest a therapeutic strategy for designing a new anticancer drug for this molecular target. Refer to Touat *et al.* (2015) and Turner and Grose (2010).

2. Discuss the importance of protein–protein interactions in growth factor signal transduction pathways.

3. Suggest possible drug strategies and targets within the PI3K signaling pathway (see Figure 4.6). Search the literature for drugs that are being tested in the clinic or have been approved. Include everolimus, an approved drug that targets m-TOR. You may choose to begin with Fruman and Rommel, 2014.

FURTHER READING

Arteaga, C.L. and Engelman, J.A. (2014) ERBB receptors: from oncogene discovery to basic science to mechanism-based cancer therapeutics. *Cancer Cell* **25**: 282–303.

Brognard, J. and Hunter, T. (2011) Protein kinase signalling networks in cancer. *Curr. Opin. Genet. Dev.* **21**: 4–11.

Caunt, C.J., Sale, M.J., Smith, P.D., and Cook, S.J. (2015) MEK1 and MEK2 inhibitors and cancer therapy: the long and winding road. *Nat. Rev. Cancer* **15**: 577–592.

Druker, B.J. (2002) STI571 (Gleevec) as a paradigm for cancer therapy. *Trends Mol. Med.* **8**: S14–S18.

Grant, S. (2009) Therapeutic protein kinase inhibitors. *Cell Mol. Life Sci.* **66**: 1163–1177.

Greuber, E.K., Smith-Pearson, P., Wang, J., and Pendergast, A.M. (2013) Role of ABL family kinases in cancer: from leukaemia to solid tumours. *Nat. Rev. Cancer* **13**: 559–571.

Holdergield, M., Deuker, M.M., McCormick, F., and McMahon, M. (2014) Targeting RAF kinases for cancer therapy: BRAF-mutated melanoma and beyond. *Nat. Rev. Cancer* **14**: 455–467.

Krause, D.S. and Van Etten, R.A. (2005) Tyrosine kinases as targets for cancer therapy. *N. Engl. J. Med.* **353**: 172–187.

Lemmon, M. A., and Schlessinger, J. (2010) Cell signaling by receptor tyrosine kinases. *Cell* **141**: 1117–1134.

Maurer, G., Tarkowski, B., and Baccarini, M. (2011) Raf kinases in cancer—roles and therapeutic opportunities. *Oncogene* **30**: 3477–3488.

Ogunleye, F., Ibrahim, M., Stender, M., Kalemkerian, G., and Jaiyesimi, I. (2015) Epidermal growth factor receptor tyrosine kinase inhibitors in advanced non-small cell lung cancer. *Am. J. Hematol. Oncol.* **11**: 16–25.

Pylayeva-Gupta, Y., Grabocka, E., Bar-Sagi, D. (2011) RAS oncogenes: weaving a tumorigenic web. *Nat. Rev. Cancer* **11**: 761–774.

Sawyers, C.L. (2002) Rational therapeutic intervention in cancer: kinases as drug targets. *Curr. Opin. Genet. Dev.* **12**: 111–115.

Schlessinger, J. (2014) Receptor tyrosine kinases: legacy of the first two decades. *Cold Spring Harb. Perspect. Biol.* **6**: a 008912.

Seshacharyulu, P., Ponnusamy, M.P., Haridas, D., Jain, M., Ganti, A., and Batra, S.K. (2012) Targeting the EGFR signalling pathway in cancer therapy. *Expert Opin. Ther. Targets* **16**: 15–31.

Stephen, A.G., Esposito, D., Bagni, R.K., and McCormick, F. (2014) Dragging Ras back in the ring. *Cancer Cell* **25**: 272–281.

Walter, A.O., Sjin, R.T., Haringsma, H.J., Ohashi, K., Sun, J., Lee, K., *et al.* (2013) Discovery of a mutant-selective covalent inhibitor of EGFR that overcomes T790M-mediated resistance in NSCLC. *Cancer Discov.* **3**: 1404–1415.

Yeatman, T.J. (2004) A renaissance for Src. *Nat. Rev. Cancer* **4**: 470–480.

Zandi, R., Larsen, A.B., Andersen, P., Stockhausen, M.-T., and Poulsen, H.S. (2007) Mechanisms for oncogenic activation of the epidermal growth factor receptor. *Cell Signal.* **19**: 2013–2023.

Zhang, J., Yang, P.L., and Gray, N.S. (2009) Targeting cancer with small molecule kinase inhibitors. *Nat. Rev. Cancer* **9**: 28–39.

■ WEB SITES

Cancer Research UK. *Cancer growth blockers.* http://www.cancerresearchuk.org/about-cancer/cancers-in-general/treatment/biological/types/cancer-growth-blockers

The National Cancer Institute. Introducing the NCI Ras Program www.youtube.com/watch?v=wDhdJXJUDYo

American Association for Cancer Research. *Webcasts.* http://webcast.aacr.org/portal

■ SELECTED SPECIAL TOPICS

Bollag, G., Hirth, P., Tsai, J., Zhang, J., Ibrahim, P.N., Cho, H., *et al.* (2010) Clinical efficacy of a RAF inhibitor needs broad target blockade in *BRAF*-mutant melanoma. *Nature* **467**: 596–599.

Chang, E.H., Furth, M.E., Scolnick, E.M., and Lowy, D.R. (1982) Tumorigenic transformation of mammalian cells induced by a normal human gene homologous to the oncogene of Harvey murine sarcoma virus. *Nature* **297**: 479–483.

Fehrenbacher, N., Bar-Sagi, D., and Philips, M. (2009) Ras/MAPK signaling from endomembranes. *Mol. Oncol.* **3**: 297–307.

Flaherty, K.T., Puzanov, I., Kim, K.B., Ribas, A., McArthur, G.A., Sosman, J.A., *et al.* (2010) Inhibition of mutated, activated BRAF in metastatic melanoma. *N. Engl. J. Med.* **363**: 809–819.

Fremin, C. and Meloche, S. (2010) From basic research to clinical development of MEK1/2 inhibitors for cancer therapy. *J. Hematol. Oncol.* **3**: 8–19.

Fruman, D.A. and Rommel, C. (2014) PI3K and cancer: lessons, challenges and opportunities. *Nature Rev. Drug Discov.* **13**: 140–156.

Gorre, M.E., Mohammed, M., Ellwood, K., Hsu, N., Paquette, R., Nagesh Rao, P., *et al.* (2001) Clinical resistance to STI-571 cancer therapy caused by BCR-ABL gene mutation or amplification. *Science* **293**: 876–880.

Ise, K., Nakamura, K., Nakao, K., Shimizu, S., Harada, H., Ichise, T., *et al.* (2000) Targeted deletion of the *H-ras* gene decreases tumor formation in mouse skin carcinogenesis. *Oncogene* **19**: 2951–2956.

Jänne, P.A., Yang, J.C., Kim, D.W., Planchard, D., Ohe, Y., Ramalingam, S.S., *et al.* (2015) AZD9291 in EGFR inhibitor-resistant non-small-cell lung cancer. *N. Engl. J. Med.* **372**: 1689–1699.

Kim, W.G. and Cheng, S.Y. (2013) Thyroid hormone receptors and cancer. *Biochim. Biophys. Acta* **1830**: 3928–3936.

Kolch, W., Halasz, M., Granovskaya, M., and Kholodenko, B.N. (2015) The dynamic control of signal transduction networks in cancer cells. *Nat. Rev. Cancer* **15**: 515–527.

Lynch, T.J., Bell, D.W., Sordella, R., Gurubhagavatula, S., Okimoto, R.A., Brannigan, B.W., *et al.* (2004) Activating mutations in the epidermal growth factor receptor underlying responsiveness of non-small cell lung cancer to gefitinib. *N. Engl. J. Med.* **350**: 2129–2139.

Mulligan, L.M. (2014) RET revisited: expanding the oncogenic portfolio. *Nat. Rev. Cancer* **14**: 173–186.

Ostrem, J.M., Peters, U., Sos, M.L., Wells, J.A., and Shokat, K.M. (2013) K-Ras (G12C) inhibitors allosterically control GTP affinity and effector interactions. *Nature* **503**: 548–551.

Paez, J.G., Janne, P.A., Lee, J.C., Tracy, S., Greulich, H., Gabriel, S., *et al.* (2004) EGFR mutations in lung cancer: correlation with clinical response to Gefitinib therapy. *Science* **304**: 1497–1500.

Pao, W., Miller, V.A., and Kris, M.G. (2004) 'Targeting' the epidermal growth factor receptor tyrosine kinase with gefitinib (Iressa) in non-small cell lung cancer (NSCLC). *Semin. Cancer Biol.* **14**: 33–40.

Reddy, E.P., Reynolds, R.K., Santos, E., and Barbacid, M. (1982) A point mutation is responsible for the acquisition of transforming properties by the T24 human bladder carcinoma oncogene. *Nature* **300**: 149–152.

Schindler, T., Bornmann, W., Pellicena, Miller, W.T., Clarkson, B., and Kuriyan, J. (2000) Structural mechanism for STI-571 inhibition of Abelson tyrosine kinase. *Science* **289**: 1938–1942.

Sequist, L.V. (2015) The role of third-generation epithelial growth factor receptor inhibitors in non-small cell lung cancer. *Clin. Adv. Hematol. Oncol.* **13**: 147–149.

Sequist, L.V., Soria, J.-C., Goldman, J.W., Wakelee, H.A., Gadgeel, H.M., Varga, A., *et al.* (2015) Rociletinib in EGFR-mutated non-small-cell lung cancer. *N. Engl. J. Med.* **372**: 1700–1709.

Touat, M., Ileana, E., Postel-Vinay, S., Andre, F., and Soria, J.C. (2015) Targeting FGFR signaling in cancer. *Clin. Cancer Res.* **21**: 2684–2694.

Turner, N. and Grose, R. (2010) Fibroblast growth factor signaling: from development to cancer. *Nat. Rev. Cancer* **10**: 116–129.

Vogt, P.K. (2012) Retroviral oncogenes: a historical primer. *Nat. Rev. Cancer* **12**: 639–648.

Wilhelm, S., Carter, C., Lynch, M., Lowinger, T., Dumas, J., Smith, R.A., *et al.* (2006) Discovery and development of sorafenib: a multikinase inhibitor for treating cancer. *Nat. Rev. Drug Discov.* **5**: 835–844.

Chapter 5

The cell cycle

Introduction

Cancer is characterized by abnormal cell proliferation. Cell proliferation involves the reproduction of a cell to form two daughter cells. Each daughter cell can reproduce to form two daughter cells, and so on; thus cell reproduction is cyclic. The sequence of stages through which a cell passes between one cell division and the next is called the **cell cycle** (Figure 5.1) and is made up of four stages: G_1, **S phase**, G_2, and **M phase**. G_1, S, and G_2 make up the part of the cycle called interphase. The genetic material of a cell is replicated in S phase (DNA synthesis). M phase involves the partitioning of the cell to produce two daughter cells and includes **mitosis** and cytokinesis. G_1 and G_2 are "gaps" preceding the S and M phases, respectively, during which time the cell prepares for the next phase.

Figure 5.1 The cell cycle. The four phases, G_1, S, G_2, and M phase (including cytokinesis), are shown in the center of the diagram. Cells are illustrated around the phases of the cell cycle, and their chromosome content in relation to the four phases of the cell cycle is depicted. Cells in different phases of the cell cycle have different DNA content: G_1: $2n$ (two copies of every chromosome); S: begins with $2n$ and becomes $4n$; G_2: $4n$; M: begins with $4n$ and produces $2n$ daughter nuclei. Figure from *The Cell Cycle: Principles of Control* by David O. Morgan with the permission of Oxford University Press.

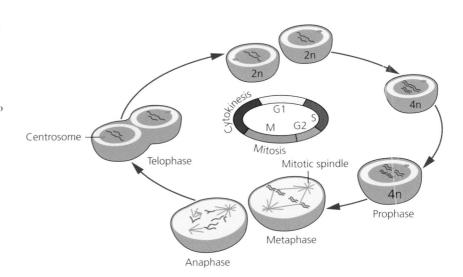

In Chapter 4, we examined the molecular mechanisms involved in growth factor signal transduction pathways. We mentioned that the pathway culminates by regulating the expression of target genes, including the induction of genes whose products are essential in cell proliferation. In this chapter we will introduce the cyclins and cyclin-dependent kinases, main players in the cell cycle, some of which are encoded by genes regulated by growth factor signaling pathways. We will also examine the molecular mechanisms involved in the regulation of the cell cycle. Specific mutations that affect the cell cycle and play a role in carcinogenesis will be described. Lastly, we will discuss therapeutic strategies that target molecules of the cell cycle.

5.1 Cyclins and cyclin-dependent kinases (cdks)

The average length of the cell cycle is 16 h (15 h for interphase and 1 h for mitosis) as shown in Figure 5.2, but note that this can vary depending on the cell type. Cells in interphase can be distinguished microscopically from cells in mitosis because chromosomes are not visible in interphase and can only be observed during mitosis because of chromosome condensation.

Starting from the beginning of G_1 (Figure 5.2, top), progression of the cell cycle is illustrated in a clockwise manner. Most cells in an adult are not in the process of cell division. They are quiescent and enter an inactive period called G_0, a phase outside of the cell cycle. **Mitogens** or growth factors can, however, induce cells in G_0 to re-enter the cell cycle and pass a control point called the G_1 restriction point (find this point in Figure 5.2). Before the passage of the restriction point, cell division is dependent on mitogens;

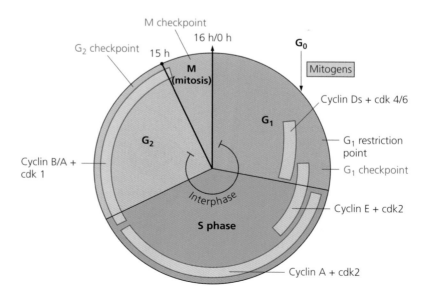

Figure 5.2 The pattern of cyclin–cdk activity during the cell cycle (shown by red bars) and the cell cycle checkpoints (G_1, G_2, and M).

afterwards, cells are irreversibly committed to progress through the cycle without the need for growth factors. The passage of the cell through the different phases of the cell cycle is coordinated and regulated by a set of proteins called cyclins and their associated cyclin-dependent kinases (cdks). Cyclins were so named because of the cyclical changes in their concentrations that occur over a series of cell divisions. The concentration of cyclin protein is dependent on the transcription of its gene and by subsequent regulated protein degradation. The pairing of cyclins to the cdks is highly specific. Cyclins are regulatory subunits of their cdks. Upon binding of a cyclin to its cdk partner, cyclin induces a conformational change in the catalytic subunit of the cdk, revealing its active site. Note that the concentration of cdks does not fluctuate during the cell cycle.

Different cyclin–cdk complexes are present at specific points in the cell cycle and are important regulators of irreversible phase transitions (depicted by the red bars in Figure 5.2). In Chapter 4, we saw that the cyclin D gene is one of the final targets of the EGF signaling pathway. This is an important molecular link between growth factors and how they actually stimulate cell proliferation. Cyclin D is the first cyclin to be synthesized and, together with cdks 4/6, drives progression through G_1. As we will see later in this chapter, cyclin D plays a role in the regulation of expression of the cyclin E gene, whose product is important for the G_1 to S phase transition. Cyclin A–cdk2 is important for S phase progression. Cyclins A, B–cdk1 directs G_2 and the G_2 to M phase transition.

Cell cycle checkpoints (Figure 5.2), a series of biochemical signaling pathways that sense and induce a cellular response to DNA damage, are important for maintaining the integrity of the genome. The G_1 checkpoint leads to the arrest of the cell cycle in response to DNA damage, ensuring that DNA damage is not replicated during S phase. The G_2 checkpoint leads to the arrest of the cell cycle in response to damaged and/or unreplicated DNA to ensure proper completion of S phase. The M checkpoint leads to the arrest of chromosomal segregation in response to misalignment on the mitotic spindle. The components of the checkpoints are proteins that act as DNA damage sensors, signal transducers, or effectors. Disruption of checkpoint function leads to genomic and chromosomal instability, leading to mutations that can induce carcinogenesis.

The cyclin–cdk complexes exert their effect by phosphorylating target proteins. As we saw in Chapter 4, phosphorylation is an important mechanism for regulating the activity of proteins. The targets of cyclin–cdk complexes include a diverse set of proteins, including transcriptional regulators, cytoskeletal proteins, nuclear pore and envelope proteins, and histones. Specific examples include condensins, nuclear lamins, GM130 of the Golgi apparatus, the famous transcriptional regulator retinoblastoma protein (RB or pRB), and transcription factors E2F and Smad 3. Consequently, essential events of the cell cycle, including chromosomal condensation, nuclear breakdown, fragmentation of the Golgi apparatus, regulated gene expression, and mitotic spindle assembly, are facilitated. Note that dephosphorylation is an important mechanism for resetting the cell for another round of the cell cycle.

? HOW DO WE KNOW THAT?

Protein cross-linking and immunopurification

(See Sanchez and Dynlacht, 2005.)

There are several approaches that can help identify the substrates of cdks. One common method is to cross-link proteins, carry out immunopurification of protein kinases, and use mass spectroscopic analysis to identify associated proteins. The identification of RB as a cdk substrate was demonstrated by the immunoprecipitation of RB using a cdk2 antibody.

Another method for identifying cdk substrates is to screen cDNA expression libraries using a solid-phase phosphorylation assay. Bacterial expressed cDNA libraries are transferred to

filters and probed with a solution of a functionally active cdk of interest and labeled ATP. Proteins are produced from the cDNA libraries, some of which interact with cdk. When cdk phosphorylates a target protein using the labeled ATP, a radioactive signal is generated. The signal is visualized by autoradiography at the position of proteins phosphorylated during the assay.

Candidate molecules must be validated by several criteria, including: (i) the phosphorylation state must be the same *in vitro* and *in vivo*; (ii) phosphorylation must be shown to be cell cycle-dependent; (iii) phosphorylation should have a functional consequence in the cell cycle.

PIONEERS IN THE FIELD ... of the cell cycle: Tim Hunt, Lee Hartwell, and Paul Nurse

The award of the Nobel Prize in Physiology or Medicine in 2001 to Tim Hunt, along with Lee Hartwell and Paul Nurse, indicates the great contributions that these men have made to science. The prize was received for studies on the regulation of the cell cycle during which the cyclins were discovered. Each of these three scientists worked independently on three different model systems: Hunt worked on sea urchins, Hartwell on budding yeast, and Nurse on fission yeast. One early experiment seemed to serve as the foundation for future work on the cell cycle: the discovery that the cytoplasm of a hormone-treated frog oocyte was able to induce maturation (including the first meiotic division) in a recipient untreated oocyte. The substance in the cytoplasm was termed maturation-promoting factor (MPF). It was later identified in mitotic cells and was then also called mitosis-promoting factor. Similarities between MPF and cyclins–cdks uncovered by these scientists led to the discovery that MPF was, in fact, a cyclin–cdk complex. The mechanism by which the cell cycle is regulated, as elucidated by the work of Hunt, Hartwell, and Nurse, offered new insights into the molecular biology of cancer.

5.2 Mechanisms of cdk regulation

In an adult, more than 25 million cells undergo cell division per second. The magnitude of this number suggests the need for precise regulation. Control of cell cycle progression is essentially controlled by two processes: protein phosphorylation and dephosphorylation by kinases and phosphatases, and specific proteolytic degradation targeted by the addition of ubiquitin to the proteasome. Cdks are serine/threonine kinases that, sequentially, regulate progression through the phases of the cell cycle via phosphorylation. Therefore, the regulation of cdk activity is crucial for precise cell reproduction.

There are four mechanisms of cdk regulation: association with cyclins, association with cdk inhibitors, addition of phosphate groups that activate cdk activity, and addition of phosphate groups that inhibit cdk activity (Figure 5.3). Because of the precise window of time for which regulators of the cell cycle are required, the "disappearance" of a factor is as important as its appearance. That is, precise protein degradation also plays an important role in the control of the cell cycle (Teixeira and Reed, 2013). The mechanisms of cdk regulation are discussed in the following sections.

Association with cyclins

The binding of cyclins to their partner cdk causes a crucial conformational change in the cdk that allows binding of a protein substrate and correct positioning of ATP. The inactive cdk molecule has a conformation that blocks the binding of the protein substrate and correct alignment of

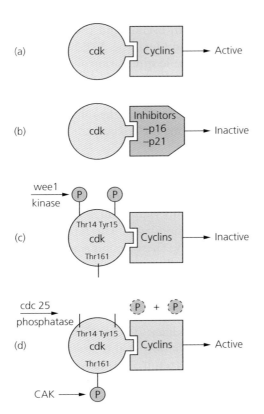

Figure 5.3 Mechanisms of cdk regulation: (a) association with cyclins activates cdks; (b) association with inhibitors inactivates cdks; (c) inhibitory phosphorylation at Thr14 and Tyr15 by wee1 kinase inactivates cdks; (d) the removal of inhibitory phosphates by cdc phosphatase and activating phosphorylation by cdk-activating kinases (CAK) activate cdk.

ATP. Some cyclins are also involved in increasing the affinity of cdks to specific substrates. As mentioned earlier, the amounts of cyclin protein vary through the cell cycle. Protein levels are modified by transcriptional regulation of the cyclin genes and by protein degradation. As we saw in Chapter 4, the signaling pathway for the growth factor EGF results in the transcriptional activation of the cyclin D gene and allows progression through the restriction point. Degradation of cyclin proteins is carried out by the proteasome, a complex of proteases. The covalent addition of ubiquitin, a small polypeptide, to the lysine amino acids of the cyclin flags the protein for degradation by the proteasome. The enzyme that catalyzes the transfer of ubiquitin to the target protein is called a ubiquitin-protein ligase. Ubiquitin-mediated proteolysis of cyclins prevents the constitutive activity of cdks. Note that ubiquitination is also important in stem cell maintenance and angiogenesis, and will be discussed in Chapters 8 and 10.

Association with inhibitors

Two families of inhibitors are involved in regulating cyclin–cdk activity: the $p16^{ink4a}$ (INK) family and the p21 (Cip/Kip) family. Members of the $p16^{ink4a}$ family include $p16^{ink4a}$, $p15^{ink4b}$, $p18^{ink4c}$, and $p19^{ink4d}$. The INK proteins bind cdks 4/6 and interfere with the binding of cdks 4/6 to cyclin D. The p21 family members include $p21^{cip1}$, $p27^{kip1}$, and $p57^{kip2}$. These inhibitors interact with both cyclins and their associated cdks (mainly with cdk2 and cyclin E) and block the ATP-binding site, thus disabling kinase activity. Upon mitogenic stimulation and subsequent cyclin D synthesis, cyclin D-dependent kinases sequester inhibitors of the Cip/Kip family, facilitating cyclin E–cdk2 activation. Again, ubiquitin-mediated degradation of inhibitors ensures that the inhibitors are present during a defined window of time during the cell cycle.

Regulation by phosphorylation

Regulation of cdk activity by phosphorylation involves both activation and inhibition. Two phosphorylation sites on the amino-terminal end are inhibitory when phosphorylated. The tyrosine kinase, wee1, phosphorylates Thr14 and Tyr15. These amino acids are located deep within the ATP-binding site of the cdk, and phosphorylation of these sites physically interferes with ATP binding. Two steps are required for cdks to become active: dephosphorylation of the inhibitory phosphate groups by cdc25 phosphatases, and phosphorylation of a central threonine residue, Thr161, by cdk-activating kinase (CAK). Note that complete activation of cdks requires phosphorylation at this site and that association with cyclins alone does not lead to full activation. In addition, this phosphorylation event is not temporally regulated with respect to the phases of the cell cycle, as CAK activity is constant throughout the cycle.

5.3 Progression through the G₁ checkpoint

A key substrate of the cyclin D–cdk 4/6 complex is the RB protein (discussed further in Chapter 6). RB, an important tumor suppressor protein, serves as a molecular link for the G₁–S phase transition. RB does not bind to specific DNA sequences but instead regulates the activity of the E2F transcription factor family, which is crucial for the expression of genes needed for S phase. It does this by physically interfering with the transactivation domain of E2Fs. Note that, to date, there are eight E2Fs and two associated subunits (called DP). The activity of RB is regulated by sequential phosphorylation events by cyclin–cdks. Let us now examine the details of the RB protein.

Structure of the RB protein

The nuclear RB protein, along with two other related proteins, p107 and p130, is a member of the "pocket proteins" and contains conserved structural and functional domains that bind to various cell proteins. The pocket comprises the A domain and the B domain joined by a linker region. The binding of the two main cellular effector proteins, histone deacetylase (HDAC) and the E2F transcription factor, to the pocket region (Figure 5.4) is important for its function. HDACs contain the LXCXE motif, an amino acid sequence (X represents any amino acid) that is required to bind to the

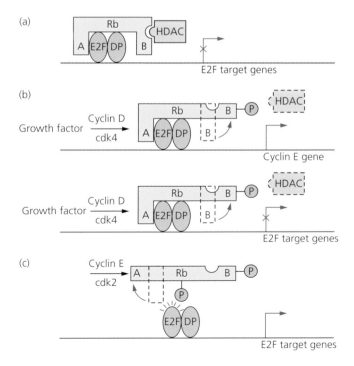

Figure 5.4 Structure and function of the RB protein. (a) Hypophosphorylated RB sequesters E2F/DP and recruits HDAC. Transcription is repressed. (b) Partial phosphorylation of RB by cyclin D–cdk4 causes a conformational change, and release of HDAC but not E2F/DP. Repression is relieved for some genes, such as cyclin E (top), but not for E2F target genes (bottom). (c) Additional phosphorylation by cyclin E–cdk2 causes an additional conformational change, release of E2F/DP, and transcription of E2F target genes.

B domain of the pocket of the RB protein. E2F and its associated subunit DP can bind to the pocket of RB simultaneously with HDAC, because E2F and DP recognize a different conserved sequence at the interface of the A and B domains of the pocket. The next question to examine is how binding of HDAC and E2F contributes to the function of RB.

Molecular mechanisms of the effects of RB

The major point of control for RB protein is the transition from G_1 phase of the cell cycle to S phase (Figure 5.2). It executes this control by its interactions with the transcription factor E2F and HDACs. (Recall that HDACs regulate gene expression via an epigenetic mechanism; see Chapter 3.) The interactions between RB and E2F and HDACs are regulated by serine/threonine phosphorylation. In the absence of a growth signal, RB is in a hypophosphorylated state (i.e. it does not have many phosphates attached) and binds to both E2F and HDAC (Figure 5.4a). By binding to E2F, RB sequesters it and blocks its transactivation domain, preventing E2F from interacting with the general transcription factors (e.g. TATA-binding protein). RB also inhibits the expression of E2F target genes by recruiting HDACs, enzymes that deacetylate histones and increase chromatin compaction. Thus, the trimeric complex of RB with HDAC and E2F regulates transcription, and consequently cell cycle progression; genes, such as *cyclin E*, *cyclin A*, and *cdk 2*, whose products are required for progression through the cell cycle are not expressed.

It is the cyclin D and E families and their cdks that phosphorylate RB in a progressive manner, in response to a growth signal. Figure 5.4 shows how phosphorylation leads to conformational changes in the RB protein and causes the sequential release of HDAC and E2F. HDAC is no longer localized to repress transcription, and the transcription factor E2F is free to activate genes necessary for proliferation. Phosphorylation of RB is carried out in two steps. First, cyclin D–cdk4 phosphorylates carboxy-terminal residues of RB upon growth factor stimulation. The increase in negative charge causes intramolecular interactions with lysine residues (positively charged amino acids) near the LXCXE domain. The resulting conformational change releases HDAC, an LXCXE-bound protein, but not E2F (Figure 5.4b). RB-mediated transcriptional repression of some genes, and not others, is relieved in the absence of HDAC. The cyclin E gene (Figure 5.4b, top), but not other E2F target genes (Figure 5.4b, bottom), is expressed upon the release of HDAC from RB. The cyclin E–cdk2 complex then phosphorylates additional amino acid residues of RB, including Ser567 close to the linker region. This results in a conformational change of the RB pocket domain, causing the release of E2F and subsequent expression of its target genes, such as cyclin A, thymidylate synthase, and dihydrofolate reductase, that are important for S phase (Figure 5.4c).

Activation of E2F also influences other later cell cycle events, in addition to regulation of the G1 transition. For example, another E2F target gene produces the spindle assembly checkpoint protein, MAD2 which has a role in mitosis.

In conclusion, the sequential action of these two cyclin–cdk complexes is important. Phosphorylation of RB by cyclin D–cdk4 is a prerequisite for cyclin E–cdk2 phosphorylation in that it induces the expression of the *cyclin E* gene and uncovers cyclin E–cdk2 phosphorylation sites. Subsequent phosphorylation of RB by cyclin E–cdk causes the release of E2F. It is speculated that additional cyclin–cdk complexes may also be involved. Note that there are many phosphorylation sites within RB which are involved in its regulation, but Figure 5.4 is a simplified version used to illustrate the concept of sequential phosphorylation. For example, recently phosphorylation of the carboxy-terminal domain has been demonstrated to be involved in the release of E2F (Rubin *et al.*, 2005).

Self-test Close this book, and try to redraw Figure 5.4. Check your answer. Correct your work. Close the book once more, and try again.

5.4 The G$_2$ checkpoint

The G$_2$ checkpoint blocks entry into M phase in cells that have incurred DNA damage in previous phases or have not correctly completed S phase (reviewed in Medema and Macurek, 2013). This allows DNA repair. Specific Cdc25 tyrosine phosphatases are important in the G$_2$–M phase transition. DNA damage activates either of two kinases, ATR or ATM. These kinases then phosphorylate and activate checkpoint kinases Chk1 and Chk2, respectively. Note full activation of Chk1 requires interaction with a mediator protein, claspin. One target of Chk1 is Cdc25 tyrosine phosphatase that regulates cdk activity by removing inhibitory phosphates. Activation of the G$_2$ checkpoint activates Chk1 and inhibits Cdc25. This blocks the removal of an inhibitory phosphate on cdk, and cdk remains inactive, preventing mitosis. After successful DNA repair, cells re-enter the cell cycle in a process called checkpoint recovery. Checkpoint recovery is dependent on the protein Plk1. Plk1 targets claspin and Wee1 for degradation and directly inhibits Chk2.

There is also a decatenation G$_2$ checkpoint that is involved in detangling intertwined daughter chromatids after DNA synthesis. This process enables chromatid separation during anaphase of mitosis. Topoisomerase II, an enzyme that can release torsional stress by making double-strand DNA breaks to allow unwinding, is key in the decatenation G$_2$ checkpoint.

PAUSE AND THINK

Try to draw a diagram that illustrates this pathway. Compare your diagram with Figure 5.5.

PAUSE AND THINK

One of the types of classical chemotherapies discussed in Chapter 2 targets topoisomerase II. Can you remember the name of this drug? The answer is doxorubicin, a fungal anthracycline antibiotic.

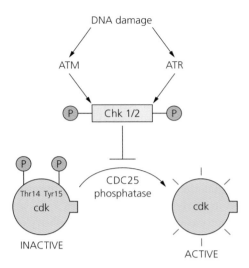

Figure 5.5 Mechanism of the G$_2$ checkpoint. DNA damage activates ATM or ATR. The ATM and ATR kinases phosphorylate and activate Chk1/2 kinases. The Chk1/2 kinases inhibit Cdc25. Cdc25 is then unable to remove inhibitory phosphate groups and activate cdks. Cdks remain inactive in response to DNA damage.

5.5 The mitotic checkpoint

The mitotic checkpoint (also known as the spindle assembly checkpoint) is a signaling cascade that ensures correct chromosomal segregation during mitosis and the production of two genetically identical nuclei.

A little lesson about the stages of mitosis . . .

There are four stages of mitosis: prophase, metaphase, anaphase, and telophase (Figure 5.6).

Prophase is marked by the appearance of the chromosomes as a result of condensation, nuclear membrane breakdown, separation of duplicated centrosomes, and assembly of mitotic checkpoint proteins at the centromeres.

Metaphase is characterized by the aligning of the chromosomes on the metaphase plate and the assembly of microtubules to form the mitotic spindle. Microtubule capture of both centromere regions of a chromatid pair results in checkpoint silencing. When the last pair is attached to the spindle, metaphase is completed and anaphase begins.

Anaphase is marked by the spindle pulling apart and separating chromatid pairs.

Telophase includes the accumulation of chromosomes at their respective poles, re-forming of the nuclear membrane, chromosome decondensation, and cytokinesis (separation into two separate cells).

 Spindle microtubules attach to the centromere regions of chromosomes during metaphase, such that sister chromatids can be pulled to opposite poles during anaphase. Unattached chromatid pairs recruit several checkpoint proteins that produce inhibitors of the anaphase-promoting complex. This complex functions as a ubiquitin-protein ligase and targets specific proteins for degradation in order for anaphase to begin. After each

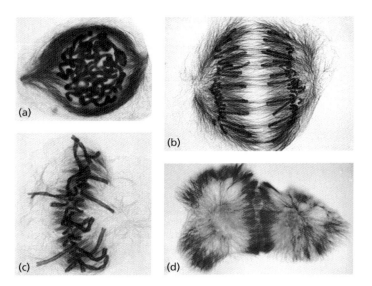

Figure 5.6 Cells in the four stages of mitosis: (a) prophase, (b) metaphase, (c) anaphase, (d) telophase (all magnified about 2700 times). Courtesy of www. micro.utexas.edu.

chromatid pair is attached to the spindle, the inhibition of the anaphase-promoting complex stops. One crucial target protein of the anaphase-promoting complex is securin; upon its degradation, the protease separase is activated. Separase cleaves a protein link between sister chromatids, and this allows them to separate during anaphase. Cyclins are also targets of the anaphase-promoting complex. Thus, the mitotic checkpoint plays an important role, preventing mis-segregation of single chromosomes.

Aurora kinases

The Aurora kinases (A, B, and C) regulate important aspects of mitosis, including chromosome segregation and the spindle checkpoint. (Note that Aurora kinases A, B, and C are also known as STK15, STK12, and STK13, respectively.) They are serine/threonine kinases that phosphorylate target proteins, many of which play a role in chromosome structure and spindle assembly. Histone H3 is one such target. The activities of the Aurora kinases are coordinated to specific cellular locations and events of the cell cycle (Figure 5.7). Aurora kinase A localizes to centrosomes during interphase. It is upregulated at the beginning of mitosis and relocates to the spindle poles and spindle microtubules, suggesting a role in centrosome maturation and assembly of the spindle apparatus. Aurora kinase B activity is highest later in mitosis and localizes first with centromeres, then with the middle of the spindle, and later between the dividing cells, suggesting a role in bipolar spindle attachment to chromosomal centromeres, and the spindle checkpoint and monitoring of chromosomal segregation and cytokinesis. Aurora kinase C is active during late mitosis and localizes to spindle poles. The precise temporal regulation of the Aurora kinases is regulated by phosphorylation, protein inhibitors, and targeted degradation.

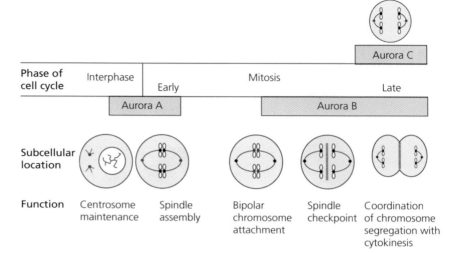

Figure 5.7 Subcellular location of Aurora kinase A, B, and C (illustrated in red) in relation to the cell cycle. Implicated functions are noted.

5.6 The cell cycle and cancer

Genes encoding cell cycle regulators are frequently mutated in human tumors, leading to aberrant regulation of the cell cycle, unscheduled proliferation, and carcinogenesis. Recall from Chapter 1 that growth signal autonomy is one of the eight hallmarks of cancer. Mutations in genes of the molecular components of the growth factor signaling pathway (see Chapter 4) and in genes that code for regulation of the cell cycle can lead to growth signal autonomy. Mutations in RB will be discussed in Chapter 6 which focuses on tumor suppressor proteins.

Mutations in cdk genes found in cancer cells have been well characterized. For example, a miscoding mutation (Arg24Cys) in cdk4 blocks binding to INK4 inhibitors in a subset of melanoma patients. Expression of this mutation in mice leads to the induction of several types of tumors. Chromosomal translocations cause overexpression of cdk6 in some leukemias. In addition, recent data suggest that the cdk profiles described in the classical model of the cell cycle (Figure 5.2) may be modified in specific cell types and tumor cells. For example, cdk4 is dispensable for mammary gland development but is required for the development of tumors of the mammary gland. Thus, cdk4 inhibition may be a drug strategy for breast cancer.

Alterations in cell cycle regulation in cancer include overexpression of cyclins (e.g. cyclin D and E) by gene amplification. DNA amplification of the cyclin D gene occurs in one in six breast cancers (about 15%). One study reported 20% of squamous cell carcinomas (SCCs) of the skin contain additional cyclin D gene copies.

? HOW DO WE KNOW THAT?

Fluorescence *in situ* hybridization (FISH)

(See Utikal *et al.*, 2005.)

FISH is used to detect gene amplification. Serial sections of tissue or cells in culture are placed on a microscope slide and incubated at a high temperature in a solution that causes the DNA strands to separate. A single-stranded DNA probe that consists of a section of the sequence that is being searched for is either directly fluorescently labeled or labeled with a small molecule (usually biotin), and allowed to hybridize to the chromosomal DNA within the tissue. Molecules that have not hybridized are washed away. Directly labeled fluorescent probes may be viewed immediately under a fluorescence microscope. Alternatively, a fluorescently labeled molecule that binds biotin (e.g. avidin and/or antibodies) may be used to "visualize" the DNA of interest. Fluorescence can be used to quantify gene copy number. Controls may include unaffected healthy tissue and probes to additional unrelated genes. An example of cells analyzed by FISH is shown in Figure 5.8. When looking at this image, one can see that there is a higher number of cyclin D gene copies (red) in relation to the number of centromeric regions of chromosome 11 (blue) in many of the nuclei of the carcinoma.

Cyclin D mRNA and protein levels are overexpressed in 50% of breast cancers. Experimental evidence, including the induction of hyperplasia and adenocarcinoma of the breast in transgenic mice, suggests that the gene for cyclin D is a proto-oncogene. Both EGFR and estrogen exert their mitogenic effect by transcriptional activation of cyclin D. The cyclin D gene does not contain an estrogen response element in its promoter region, and thus the estrogen receptor (ER) may act as a transcriptional co-activator (see Chapter 3). However, the mechanism by which cyclin D exerts its oncogenic effects is not clear and may even involve a cdk-independent mechanism (Roy and Thompson, 2006). Cyclin D enhances ER-mediated transcription by binding to the hormone-binding domain of ER and increasing protein interactions with ER co-activators.

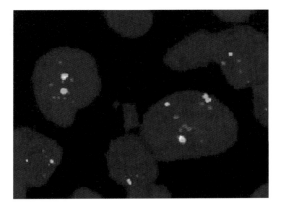

Figure 5.8 Fluorescence *in situ* hybridization (FISH) analysis used to detect amplification of the cyclin D1 gene. A tissue section of non-melanoma skin cancer analyzed by FISH. Directly labeled DNA probes were used to detect copies of the cyclin D gene (red) in relation to the number of centromeric regions of the chromosome 11 (blue). Reprinted from Figure 1A in Utikal, J., *et al.* (2005) Numerical abnormalities of the Cyclin D1 gene locus on chromosome 11q13 in non-melanoma skin cancer. Cancer Lett. **219**: 197–204. Copyright (2005), with permission from Elsevier.

Deletion of the gene coding for the inhibitor p16^{ink4a} (frequent in mesothelioma, a cancer linked to asbestos exposure and pancreatic carcinomas) is also common. Experimental findings from p16^{ink4a} knock-out mice demonstrated that loss of p16^{ink4a} results in increased incidence of spontaneous and carcinogen-induced cancers (Sharpless *et al.*, 2001). Mutations in RB will be discussed in Chapter 6.

Defects in the decatenation G$_2$ checkpoint are associated with chromosome breakage and may lead to genetic instability. This has been supported by evidence from cancer cell lines (see references within Kaufmann (2006) such as the work of Doherty *et al.*, 2003 and Nakagawa *et al.*, 2004).

It is a current debate about whether aneuploidy, the condition of having an abnormal chromosome number and content, facilitates or drives tumorigenesis. Whatever the answer, aneuploidy is the most common characteristic of human solid tumors. Aneuploidy may be caused by defects in centrosomes, the organizers of the mitotic spindle, or cytokinesis. Aberrations of the mitotic checkpoint may also lead to aneuploidy. The mitotic checkpoint is not all or none but rather can be weakened by depleted individual components. This is because many proteins are involved, so a deletion of one component may allow the mitotic checkpoint to function, albeit at a reduced efficiency. Large numbers of missegregated chromosomes lead to cell death, but a weakened checkpoint may lead to some abnormal number of chromosomes that is not sufficient to induce cell death.

Evidence suggests that mutations in genes that code for components of the mitotic spindle are not common in human tumor cells. However, a decreased quantity of mitotic checkpoint proteins has been observed in aneuploidy tumor cells with an aberrant mitotic checkpoint. Tumor suppressors or oncoproteins may transcriptionally regulate these proteins, causing a decrease in protein levels. A rare recessive disorder called mosaic variegated aneuploidy, caused by mutations in a gene encoding one of the checkpoint proteins, is characterized by aneuploidy and an increased risk of childhood cancers. The evidence described supports a link between a weakened mitotic checkpoint and the process of carcinogenesis.

The Aurora kinases are frequently overexpressed in several types of human tumors, including lymphomas. The Aurora-A gene lies within a region of chromosome 20q13, that is amplified in many epithelial malignant tumors, including breast, colon, and pancreatic cancers. Overexpression of Aurora-A in mice yields mitotic defects and mammary hyperplasia, suggesting it is important early in carcinogenesis. A single nucleotide polymorphism T91A (amino acid Phe31→Ile) present in the coding region of human Aurora-A is associated with susceptibility to multiple cancer types. These findings suggest that Aurora kinases play a role in cancer and may be good drug targets.

◎ Therapeutic strategies

Therapeutic strategies that target components of the cell cycle are in development, and many are in clinical trials. Because kinases play a central role in the cell cycle and have been implicated in carcinogenesis, identification of kinase inhibitors is an important drug strategy. Proteins involved in the mitotic spindle are other important drugs targets. Several strategies are discussed in the following sections.

5.7 Cyclin-dependent kinase inhibitors

As we have seen earlier (see Section 5.1, "Cyclins and cyclin-dependent kinases (cdks)" on p. 110), phosphorylation by cdks is a key step in the regulation of the cell cycle. These serine/threonine kinases are overexpressed and/or amplified in some cancers, making them possible molecular targets for cancer therapies. First-generation cdk inhibitors were non-selective. One of these, a semi-synthetic flavonoid called flavopiridol, acts as a competitive inhibitor of many cdks (1, 2, 4, 6, 7, and 9) by targeting their ATP-binding site. Interestingly, it is related to a compound isolated from a plant found in India that is known to have medicinal properties. Flavopiridol induces cell cycle arrest at G_1/S and G_2/M phases. It also affects cdk family members that have a role in transcriptional control and inhibits gene expression of cyclin D1 and D3. Flavopiridol was the first cdk inhibitor to be tested in clinical trials. Owing to its poor oral bioavailability, it is administered intravenously. Anti-tumor activity was demonstrated in some patients with lymphoma, but, in general, flavopiridol failed to demonstrate significant clinical activity as a single agent for many solid tumors in Phase II studies. However, investigations into its potential, in combination with existing chemotherapeutic agents, are ongoing. Pre-clinical evidence suggests that cdk inhibitors are synergistic with cytotoxic drugs (e.g. cisplatin, 5-FU), possibly because cells that are arrested or synchronized with respect to the cell cycle are more sensitive to cdk inhibitors (Musgrove *et al.*, 2011). Next-generation cdk inhibitors were designed to be more selective for the ATP-binding site of specific cdks. Both seliciclib (R-roscovitine) that targets cdk 2, 7, and 9 and dinaciclib that targets cdk 1, 2, 5, and 9 are currently in clinical trials (Figure 5.9). Palbociclib (Ibrance™; Pfizer) (Cadoo *et al.*, 2014), a selective cdk 4 and cdk 6 inhibitor, has been approved for clinical use. Clinical trial results include a multiyear Phase II study that found a significantly higher progression-free survival rate for patients with advanced ER⁺/HER2⁻ breast cancer who were given palbociclib, in addition to letrozole, a standard anti-estrogen treatment,

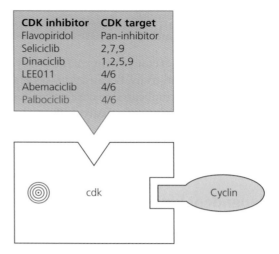

CDK inhibitor	CDK target
Flavopiridol	Pan-inhibitor
Seliciclib	2,7,9
Dinaciclib	1,2,5,9
LEE011	4/6
Abemaciclib	4/6
Palbociclib	4/6

Figure 5.9 Cyclin-dependent kinase inhibitors in clinical trials; only palbociclib has been approved. Inhibitors are shown in the red box.

compared with women who received letrozole alone. LEE011 and abemaciclib are two other cdk inhibitors that target cdks 4/6 and are being tested in clinical trials (Figure 5.9).

5.8 Other cell cycle kinase targets

Cell cycle checkpoint kinase inhibitors against Chk1 and Chk2 are also being identified and used as an anticancer strategy (Ma *et al.*, 2011). The ATP-competitive inhibitors AZD7762 (equipotent Chk1 and Chk2 inhibitor) and SCH900776 (Chk1 inhibitor) have reached clinical trials. These agents prevent cell cycle arrest and may potentiate the effects of classical chemotherapeutics that cause DNA damage and subsequent apoptosis. This approach may be especially effective in p53-deficient cells that are unable to cause G_1 arrest and are thus dependent on activation of the G_2 checkpoint by ATR/Chk1 after drugs that damage DNA. In this case, Chk1 inhibitors make the p53-deficient cancer cells sensitive to chemotherapies.

Inhibitors of the Aurora kinases (many of which are ATP-competitive inhibitors) are in early-phase clinical trials (Choudary *et al.*, 2015). Their selectivity ranges from pan-Aurora inhibitors (danusertib, Nerviano Pharm.; tozasertib, Merck and Vertex Pharm.) to dual Aurora inhibitors (Aurora A and Aurora B: AT-9283) to selective Aurora inhibitors (Aurora A-specific: alisertib, MLN8237; Millennium Pharm.). We await the results of this first wave of clinical trials for these Aurora inhibitors.

5.9 Inhibitors of the mitotic spindle

As discussed under therapeutic strategies in Chapter 2 (see "Organic drugs" on p. 46), some conventional chemotherapies interfere with microtubule formation and spindle formation. Paclitaxel/taxol stabilize microtubules, while the vinca alkaloids (vinblastine, vincristine) inhibit microtubule assembly. These drugs result in chromatid pairs that are not attached to spindle fibers and thus activate the mitotic checkpoint. It is thought that the mechanism of action of these drugs is cytostatic, but induction of apoptosis may also be a consequence. A molecule called KSP, an ATP-dependent microtubule-motor protein, is required for spindle pole separation and is a new molecular target for the development of cancer therapies. A small molecule inhibitor of KSP called ispinesib (Cytokinetics) prevents mitotic spindle pole separation and also leads to chronic mitotic checkpoint activation, and is being tested in multiple Phase I and II trials. Thus far, it shows a favorable safety profile and a 9% response rate in patients with advanced or metastatic breast cancer.

▓ CHAPTER HIGHLIGHTS—REFRESH YOUR MEMORY

- The cell cycle is made up of four phases: G_1, S, G_2, and M.
- There are three cell cycle checkpoints: the G_1, G_2, and M checkpoints.
- The progression of a cell through the different phases of the cell cycle is highly regulated by cyclins and cyclin-dependent kinases (cdks).
- Cdks are regulated by association with cyclins and inhibitors, and by activating and inhibitory phosphorylation.
- Proteolysis is important for regulating the activity of key regulators of the cell cycle.
- The retinoblastoma (RB) protein is an important target of cyclin D–cdks 4/6 and a key regulator of the G_1 to S phase transition.
- RB exerts its effects by protein–protein interactions with the E2F transcription factor and HDACs.
- The activity of RB is regulated by phosphorylation via different cyclins–cdks.
- Hypophosphorylated RB inactivates E2F and recruits HDACs.
- Phosphorylated RB releases E2F and HDACs, which facilitates transcription and cell cycle progression into S phase.
- The G_2 checkpoint is induced by DNA damage and aberrant DNA synthesis, and blocks entry into M phase.
- The mitotic checkpoint prevents mis-segregation of chromosomes during anaphase.
- Aurora kinases are important for centrosome and mitotic spindle function.
- Aberrant regulation of the cell cycle can lead to cancer.
- Cyclin D amplification often occurs in breast cancer and SCC.
- Several cdk inhibitors have entered clinical trials, and one has been approved.
- Several conventional chemotherapies exert their effects by activating the mitotic checkpoint.

▓ ACTIVITY

1. Critically discuss your views on whether you would carry out research on cdk inhibitors and, if so, what strategy you would use. Support your view with pre-clinical and clinical evidence.

2. It has been stated in this chapter that ubiquitin-mediated proteolysis is crucial for regulation of the cell cycle. Find evidence that supports the statement that unregulated proteolysis in the cell cycle can lead to cancer. Begin with the paper by Teixeira and Reed (2013).

■ **FURTHER READING**

Bryere, C. and Meijer, L. (2013) Targeting cyclin-dependent kinases in anti-neoplastic therapy. *Curr. Opin. Cell Biol.* **25**: 772–779.

Burkhart, D.L. and Sage, J. (2008) Cellular mechanisms of tumor suppression by the retinoblastoma gene. *Nat. Rev. Cancer* **8**: 671–682.

Chinnam, M. and Goodrich, D.W. (2011) RB1, development, and cancer. *Curr. Topics Devel. Biol.* **94**: 129–169.

Collins, I. and Garrett, M.D. (2005) Targeting the cell division cycle in cancer: CDK and cell cycle checkpoint kinase inhibitors. *Curr. Opin. Pharmacol.* **5**: 366–373.

Dick, F.A. and Rubin, S.M. (2013) Molecular mechanisms underlying RB protein function. *Nat. Rev. Mol. Cell Biol.* **14**: 297–306.

Dickson, M.A. (2014) Molecular pathways: CDK4 inhibitors for cancer therapy. *Clin. Cancer Res.* **20**: 3379–3383.

Kitzen, J., de Jonge, M., and Verweij, J. (2010) Aurora kinase inhibitors. *Crit. Rev. Oncol. Hematol.* **73**: 99–110.

Kops, G.J.P.L., Weaver, B.A.A., and Cleveland, D.W. (2005) On the road to cancer: aneuploidy and the mitotic checkpoint. *Nat. Rev. Cancer* **5**: 773–785.

Lapenna, S. and Giordano, A. (2009) Cell cycle kinases as therapeutic targets for cancer. *Nat. Rev. Drug Discov.* **8**: 547–566.

Malumbres, M. and Barbacid, M. (2009) Cell cycle, CDKs and cancer: a changing paradigm. *Nat. Rev. Cancer* **9**: 153–166.

Manning, A.L. and Dyson, N.J. (2012) RB: mitotic implications of a tumour suppressor. *Nat. Rev Cancer* **12**: 220–226.

Massagué, J. (2004) G1 cell-cycle control and cancer. *Nature* **432**: 298–306.

Meraldi, P., Honda, R., and Nigg, E.A. (2004) Aurora kinases link chromosome segregation and cell division to cancer susceptibility. *Curr. Opin. Genet. Dev.* **14**: 29–36.

Shah, M.A. and Schwartz, G.K. (2006) Cyclin dependent kinases as targets for cancer therapy. *Update Cancer Ther.* **1**: 311–332.

Sherr, C.J. and Roberts, J.M. (2004) Living with or without cyclins and cyclin-dependent kinases. *Genes Dev.* **18**: 2699–2711.

Swanton, C. (2004) Cell-cycle targeted therapies. *Lancet Oncol.* **5**: 27–36.

Weaver, B.A.A. and Cleveland, D.W. (2005) Decoding the links between mitosis, cancer and chemotherapy: the mitotic checkpoint, adaptation, and cell death. *Cancer Cell* **8**: 7–12.

Zhu, L. (2005) Tumour suppressor retinoblastoma protein Rb: a transcriptional regulator. *Eur. J. Cancer* **41**: 2415–2427.

■ **SELECTED SPECIAL TOPICS**

Cadoo, K.A., Gucalp, A., and Traina, T.A. (2014) Palbociclib: an evidence-based review of its potential in the treatment of breast cancer. *Breast Cancer (Dove Med Press)* **6**: 123–133.

Choudary, I., Barr, P.M., and Friedberg, J. (2015) Recent advances in the development of Aurora kinases inhibitors in haematological malignancies. *Ther. Adv. Hematol.* **6**: 282–294.

Kaufmann, W.K. (2006) Dangerous entanglements. *Trends Mol. Med.* **12**: 235–237.

Ma, C.X., Janetka, J.W., and Piwnica-Worms, H. (2011) Death by releasing the breaks: CHK1 inhibitors as cancer therapeutics. *Trends Mol. Med.* **17**: 88–96.

Medema, R.H. and Macurek, L. (2013) Checkpoint control and cancer. *Oncogene* **31**: 2601–2613.

Morgan, D. (2006) *The Cell Cycle: Principles of Control*. Oxford University Press, Oxford.

Musgrove, E.A., Caldon, C.E., Barraclough, J., Stone, A., and Sutherland, R.L. (2011) Cyclin D as a therapeutic target in cancer. *Nat. Rev. Cancer* **11**: 558–572.

Roy, P.G. and Thompson, A.M. (2006) Cyclin D1 and breast cancer. *Breast* **15**: 718–727.

Rubin, S.M., Gall, A.-L., Zheng, N., and Pavletich, N.P. (2005) Structure of the Rb C-terminal domain bound to E2F-D P1: a mechanism for phosphorylation-induced E2F release. *Cell* **123**: 1093–1106.

Sanchez, I. and Dynlacht, B.D. (2005) New insights into cyclins, CDKs, and cell cycle control. *Semin. Cell Dev. Biol.* **16**: 311–321.

Sharpless, N.E., Bardeesy, N., Lee, K.H., Carrasco, D., Castrillon, D.H., Aguirre, A.J.,*et al.* (2001) Loss of p16Ink4a with retention of p19Arf predisposes mice to tumorigenesis. *Nature* **413**: 86–91.

Teixeira, L.K. and Reed, S.I. (2013) Ubiquitin ligases and cell cycle control. *Annu. Rev. Biochem.* **82**: 387–414.

Utikal, J., Udart, M., Leiter, U., Kaskel, P., Peter, R.U., and Krahn, G. (2005) Numerical abnormalities of the Cyclin D1 gene locus on chromosome 11q13 in non-melanoma skin cancer. *Cancer Lett.* **219**: 197–204.

Chapter 6

Growth inhibition and tumor suppressor genes

Introduction

The human body has mechanisms exerted by tumor suppressor genes that normally "police" the processes that regulate cell numbers and ensure that new cells receive DNA that has been precisely replicated. Recall from Chapter 1 that the balance between cell proliferation, differentiation, and apoptosis maintains appropriate cell numbers. Many tumor suppressor gene products act as stop signs to uncontrolled growth and therefore may inhibit the cell cycle, promote differentiation, or trigger apoptosis. If both copies of a tumor suppressor gene become inactivated by mutation or epigenetic changes, the inhibitory signal is lost, and the result may be unregulated cell growth, a hallmark of cancer. Other tumor suppressor gene products are involved in DNA repair. If inactivated, DNA repair may be defective, and failure to repair DNA may give rise to mutations that lead to cancer. Two alleles of every gene are present in the human genome (except those on sex chromosomes), and, in most cases, loss of tumor suppressor gene function requires inactivation of both copies. This often happens by mutation in one copy and loss of the remaining wild-type allele (**loss of heterozygosity**, LOH). Note there are exceptions to this rule.

6.1 Definitions of tumor suppressor genes

Hereditary syndromes that predispose individuals to cancer can be explained by the inheritance of a germline mutation (passed on from egg/sperm DNA and thus present in all cells of an individual) in one tumor suppressor allele and the acquisition of a somatic mutation or other inactivating alteration in the second allele later in life. This was first proposed by Knudson and is known as Knudson's two-hit hypothesis. It states a strict definition of a tumor suppressor gene: a gene in which a germline mutation predisposes an individual to cancer. Examples of tumor suppressor genes that fit this definition are shown in Table 6.1.

Although this hypothesis describes the mechanism by which mutation of most tumor suppressor genes has an effect, exceptions and additional complexities exist and will be mentioned later.

Let us look at the breast and ovarian cancer susceptibility genes *BRCA1* and *BRCA2* as examples. Some families are prone to increased risk of developing breast and ovarian cancers. The hereditary breast and ovarian cancer susceptibility genes *BRCA1* and *BRCA2* are well-known tumor suppressor genes that play a role in this hereditary syndrome, which make up about 5–7% of all breast cancer cases (Roy *et al.*, 2012). The mechanism of these tumor suppressors follows Knudson's hypothesis in that one germline mutation results in hereditary breast and ovarian cancer syndrome and predisposes individuals to breast and ovarian cancer. Individuals with this syndrome have a 50–80% lifetime risk of developing breast cancer and a 30–50% lifetime risk of developing ovarian cancer. Breast and ovarian tumors that develop in these individuals exhibit a loss of heterozygosity. In some cases of sporadic (non-hereditary) breast cancer, BRCA1 protein levels are reduced, not because of mutation, but rather as a result of epigenetic mechanisms. Note that, although there is no clear homology of the coding region between BRCA1 and BRCA2, both BRCA proteins are involved in homologous recombination and double-stranded break repair (Figure 2.10), and therefore help maintain the integrity of the genome. The primary function of BRCA2 is recruitment of the recombinase RAD51 to double-strand breaks, whereas BRCA1 has diverse roles in homologous recombination, including recruitment of RAD51. BRCA1 also has a role in the regulation of transcription. Mutations in the *BRCA* genes may lead to defective homologous recombination that destabilizes the genome, leading to chromosomal rearrangements and mutation. Another proposal regarding a role for BRCA proteins in estrogen signaling is discussed in Chapter 11.

Historically, tumor suppressor genes were called "anti-oncogenes", as some of them seemed to "undo" pathways of oncogene activation. Although the term is no longer used, it can be a helpful tool for illustrating the function of some tumor suppressor genes.

The role of aberrant phosphorylation by kinases during carcinogenesis was emphasized in Chapter 4. It is therefore predictable that some genes that encode phosphatases which antagonize kinase activity could act as "anti-oncogenes." Inactivation of these phosphatase genes by mutation removes the inhibitory signal, and the kinase activity becomes unregulated. Many, but not all, phosphatases are tumor suppressors.

> **PAUSE AND THINK**
>
> Why does the loss of BRCA1 or BRCA2 cause breast and ovarian cancer? One hypothesis (discussed in Roy *et al.*, 2012) is that, during each menstrual cycle, growth in these tissues induced by hormones produces reactive oxygen species and oxidation of DNA. This damage may cause replicative stress and double-stranded breaks that require homologous recombination, and thus BRCA1 and BRCA2.

> **PAUSE AND THINK**
>
> As kinases are enzymes that phosphorylate, what types of enzymes "undo" kinases? Phosphatases are enzymes that remove phosphate groups.

Table 6.1 Tumor suppressor genes

Tumor suppressor gene	Human chromosomal location	Gene function	Human tumors associated with sporadic mutation	Associated cancer syndrome	Tumor phenotype of knock-out mouse mutants (hetero-/homozygote)
RB1	13q14	Transcriptional regulator of cell cycle	Retinoblastoma, osteosarcoma	Familial retinoblastoma	MTC, pituitary adenocarcinoma, pheochromocytomas
Wt1	11p13	Transcriptional regulator	Nephroblastoma	Wilms tumor	None
p53	17q11	Transcriptional regulator/ growth arrest/apoptosis	Sarcomas, breast/brain tumors	Li–Fraumeni	Lymphomas, sarcomas
NF1	17q11	Ras-GAP activity	Neurofibromas, sarcomas, gliomas	von Recklinghausen neurofibromatosis	Pheochromocytomas, myeloid leukemia, neurofibromas in DKO chimeras
NF2	22q12	ERM protein/cytoskeletal regulator	Schwannomas, meningiomas	Neurofibromatosis type 2	Sarcomas: metastases on p53 background
VHL	3p25	Regulates proteolysis	Hemangiomas, renal, pheochromocytoma	von Hippel–Lindau	None
APC	5q21	Binds/regulates β-catenin activity	Colon cancer	Familial adenomatous polyposis	Intestinal polyps in Apc^{Min}
INK4a	9p21	$p16^{ink4a}$ cdki for cyclin D–cdk (4/6); $p19^{ARF}$ binds mdm2, stabilizes p53	Melanoma, pancreatic	Familial melanoma	Lymphomas, sarcomas
PTC	9q22.3	Receptor for sonic hedgehog	Basal cell carcinoma, medulloblastoma	Gorlin syndrome	Medulloblastomas
BRCA1	17q21	Transcriptional regulator/ DNA repair	Breast/ovarian tumors	Familial breast cancer	None
BRCA2	13q12	Transcriptional regulator/ DNA repair	Breast/ovarian tumors	Familial breast cancer	None
DPC4	18q21.1	Transduces TGF-β signals	Pancreatic, colon, hamartomas	Juvenile polyposis	Cooperates with $Apc^{\Delta716}$ in colorectal carcinoma

PAUSE AND THINK

Recall that stimulation of cell membrane receptors recruits PI3 kinase to the membrane where it phosphorylates PIP2 to generate phosphatidyl-inositol-3 phosphate (PIP3), a potent second messenger that activates a cascade of proteins important for cell division and inhibition of apoptosis. This signal must be tightly regulated to prevent uncontrolled growth.

One gene encoding a phosphatase that is frequently mutated in many cancers is *PTEN* (phosphatase and tensin homolog on chromosome 10). *PTEN* codes for a phosphatase with dual specificity—it can act as both a protein and lipid phosphatase. Its role as a lipid phosphatase in tumor suppression is best known. PTEN dephosphorylates the membrane lipid phosphatidyl-inositol-3 phosphate (PIP3) to form PIP2. This antagonizes (shown by the reversed red arrow) the PI3 kinase pathway (Figure 6.1).

Loss of the inhibitory dephosphorylation activity of PTEN in the *PTEN* mutant phenotype may result in a constitutively active PI3 kinase pathway, involving activation of protein kinases Akt and m-TOR (mammalian target of rapamycin). The net result is the inhibition of apoptosis and induction of cell proliferation. This favors oncogenesis. Note that this gene also fits the tumor suppressor definition given earlier, as a germline mutation of *PTEN* causes Cowden syndrome which predisposes patients

Table 6.1 Tumor suppressor genes (*Continued*)

Tumor suppressor gene	Human chromosomal location	Gene function	Human tumors associated with sporadic mutation	Associated cancer syndrome	Tumor phenotype of knock-out mouse mutants (hetero-/homozygote)
FHIT	3p14.2	Nucleoside hydrolase	Lung, stomach, kidney, cervical carcinoma	Familial clear cell renal carcinoma	Not reported
PTEN	10q23	Dual-specificity phosphatase	Glioblastoma, prostate, breast	Cowden syndrome, BZS, Ldd	Lymphoma, thyroid, endometrium, prostate
TSC2	16	Cell cycle regulator	Renal, brain tumors	Tuberous sclerosis	Not reported
NKX3.1	8p21	Homeobox protein	Prostate	Familial prostate carcinoma	Not reported
LKB1	19p13	Serine/threonine kinase	Hamartomas, colorectal, breast	Peutz–Jeghers	Not reported
E-cadherin	16q22.1	Cell adhesion regulator	Breast, colon, skin, lung carcinoma	Familial gastric cancer	Dominant negative, promotes invasion/metastasis
MSH2	2p22	*mut S* homolog, mismatch repair	Colorectal cancer	HNPCC	Lymphoma, colon/skin carcinoma
MLH1	3p21	*mut L* homolog, mismatch repair	Colorectal cancer	HNPCC	Lymphoma, intestinal adenoma/carcinoma
PMS1	2q31	Mismatch repair	Colorectal cancer	HNPCC	None
PMS2	7p22	Mismatch repair	Colorectal cancer	HNPCC	Lymphoma, sarcoma
MSH6	2p16	Mismatch repair	Colorectal cancer	HNPCC	Lymphoma, intestinal adenomas/carcinomas

This table does not include the susceptibility genes associated with ataxia telangiectasia (*ATM/ATR*), xeroderma pigmentosum (nucleotide excision repair genes), Bloom's syndrome (*BLM*), Werner's syndrome (*WRN*), or Fanconi's anemia (*FAA, FAC, FAD*), although mutation of these genes is associated with cancer predisposition. Nor does it include putative tumor suppressor genes which are subverted by chromosomal translocation, for example, *PML*. Genes, such as *MADR2*, *TGF-β receptor 2*, *IRF-1*, *p73*, *p33^{ING1}*, *PPARγ*, *BUB1*, and *BUBR1*, have been shown to be mutated in certain human tumors but are not included here, because germline mutation of these genes is not yet associated with any hereditary human cancer syndrome.

BZS, Bannayan–Zonana syndrome; HNPCC, hereditary non-polyposis colorectal cancer; Ldd, Lhermitte–Duclos syndrome; MTC, medullary thyroid cancer.

From Macleod, K. (2000) Tumor suppressor genes. *Curr. Opin. Genet. Dev.* **10**: 81–93, Copyright (2000). Reprinted with permission from Elsevier.

to cancer. Many of the mutations found in patients with Cowden disease have mutations in the phosphatase core motif, but some mutations can be found in the promoter region or splice junctions, leading to reduced expression or truncated products, respectively.

Another protein-tyrosine phosphatase, receptor PTPRT, is frequently mutated in cancers (i.e. in 26% of colorectal cancers). Knock-out mice are highly susceptible to colon cancer induced by a carcinogen, and this provides *in vivo* evidence for its role as a tumor suppressor. Several other protein-tyrosine phosphatases also act as tumor suppressor genes.

Note that not all kinases are oncogenic, and not all phosphatases are tumor suppressors. Ataxia telangiectasia mutated (ATM) kinase functions in DNA repair, as mentioned in Chapter 2, and plays a role in tumor suppression. Many other examples exist. *In vivo* evidence suggests that one protein-tyrosine phosphatase, PTPN1, may act as an oncoprotein or

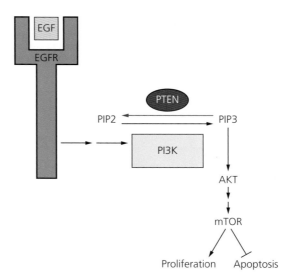

Figure 6.1 PTEN antagonizes the PI3 kinase pathway.

tumor suppressor, depending on tissue type and cellular context. Accelerated development of lymphomas has been reported upon analysis of *PTPN1/Trp53* knock-out mice (tumor suppressor activity), while mammary tumors are induced by PTPN1 overexpression in mammary glands (oncogenic activity). The role of different protein-tyrosine phosphatases as either oncogenes or tumor suppressor genes is reviewed in Julien *et al.* (2011). Also see "How do we know that?" below.

An examination of two "star players" in the world of tumor suppressor genes, the retinoblastoma (*Rb*) gene (also discussed in Chapter 5) and the *p53* gene, is central to this chapter. The roles of both gene products during carcinogenesis are described in the following sections.

 HOW DO WE KNOW THAT?

Enzyme kinetics and cell growth assays

(See data reported in Wang *et al.*, 2004.)

Mutations in several protein-tyrosine phosphatase genes were identified in human colorectal tumors. This was done by amplifying the exons of 87 different protein-tyrosine phosphatase genes from tumor genomic DNA and sequencing these exons. DNA from normal tissue of matching patients was used as a control to identify somatic (tumor-specific) mutations. Biochemical analysis demonstrated that these mutations gave rise to proteins that had reduced phosphatase activity. This was done by expressing mutant protein-tyrosine phosphatase catalytic domains in bacteria.

After purification, enzyme kinetics was studied. The rate of substrate hydrolysis was plotted against substrate concentration, and the Michaelis–Menten equation was used to determine K_m and K_{cat}. Growth of cells in culture was suppressed by transfection of the wild-type protein, but not upon transfection of the mutant proteins. Growth was examined by staining cells with crystal violet 2 weeks post-transfection, and colonies were counted. This genetic, biochemical, and cellular evidence demonstrates that these protein-tyrosine phosphatase genes are mutated in tumors and produce loss-of-function proteins, suggesting that they act as tumor suppressors.

6.2 The retinoblastoma gene

Retinoblastoma is a rare childhood cancer with a worldwide incidence of 1 in 20,000. There are two forms of the disease, a familial (inherited) form and a sporadic form (Figure 6.2). About 40% of all retinoblastoma cases are familial, and about 60% are sporadic. In the familial form of the disease, one germline mutation in the *Rb* gene is passed to the child and is present in all cells. A second mutation is acquired in a particular retinoblast that consequently gives rise to a tumor in the retina. One inherited mutated gene results in a sufficiently high probability that a second mutation may occur. The second mutation most often results from somatic mitotic recombination, during which the normal gene copy is replaced with a mutant copy. In sporadic retinoblastoma, both mutations occur somatically in the same retinoblast. As there are approximately 10^8 retinoblasts in the retina, there is a low chance that sporadic retinoblastoma will occur more than once in an individual; hence, sporadic cases usually only affect one eye, while familial cases are often bilateral. The disease demonstrates Knudson's two-hit hypothesis: two separate mutations—one in each of the two retinoblastoma alleles—are needed to inactivate the two copies of the *Rb* allele and prevent expression of the RB protein. A mutation in one *Rb* allele is insufficient to knock out functional RB, and so cancer-causing mutations are recessive. An understanding of the molecular mechanisms of the gene product underlying this disease elucidates important principles about tumor suppressor proteins.

The retinoblastoma protein (RB, sometimes called pRB) is the product of the retinoblastoma tumor suppressor gene (*Rb*). RB is a multi-functional

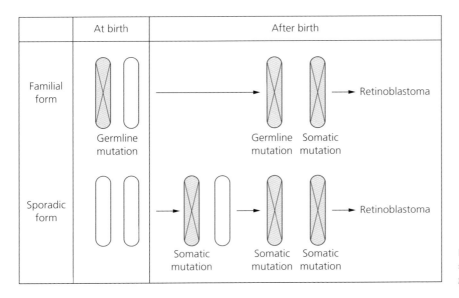

Figure 6.2 The familial and sporadic forms of retinoblastoma: germline versus somatic mutations.

PAUSE AND THINK

As a tumor suppressor protein, do you suppose RB inhibits or activates the transcription factors needed for cell proliferation? It inhibits the transcriptional activity of factors needed for cell cycle progression. Therefore, target genes important for cell growth are not expressed. However, loss of the tumor suppressor protein RB results in the loss of inhibition and, consequently, uncontrolled cell cycle progression and division. Think about the role of RB in differentiation. Do you suppose it inhibits or activates transcription factors that are responsible for turning on cell type-specific genes? As a tumor suppressor, it stimulates the activity of transcription factors, such as Myo D, that activate genes involved in differentiation. Loss of RB leads to an increase in cell number and to the failure of differentiation.

protein that has over a hundred known protein-binding partners. It is considered as a transcriptional co-factor that can bind to transcription factors and either inhibit or induce transcription factor activity. As we discussed in Chapter 5, its main role is to regulate the cell cycle by inhibiting the G_1 to S phase transition. Cell proliferation is dependent on the transcription of a set of target genes to produce proteins that are needed for cell division (e.g. thymidylate synthase, dihydrofolate reductase). RB is an indirect regulator of transcription for specific gene expression that affects cell proliferation and differentiation. Protein–protein interactions facilitate the function of RB as a transcriptional regulator; RB binds to, and modulates the activity of, a critical transcription factor (E2F) and chromatin-remodeling enzymes (Figure 5.4). Cell cycle arrest can also be induced by RB via stabilization of the Cdk inhibitor p27. RB acts as a scaffold to recruit an inhibitor of p27, S phase kinase-associated protein 2 (SKP2), and proteins that target SKP2 for degradation; thus p27 remains free.

6.3 Mutations in the RB pathway and cancer

Retinoblastoma is initiated by the loss of both *Rb* alleles. The types of mutations identified are mostly deletions, frameshift, or nonsense mutations that result in the abrogation of RB function, as would be predicted from Knudson's two-hit hypothesis. In addition, missense mutations that lie within the pocket domain have been reported. It is of interest that mutations of Ser567 (described earlier) have been found in human tumors, because normally phosphorylation of this amino acid causes the release of E2F. Mutation of Ser567 may disrupt the regulation usually observed at this site. As the RB pathway is central in cell cycle regulation, tumor initiation may be induced via any mutation that blocks RB function and causes E2F to be available to activate transcription, regardless of the presence or absence of a growth signal.

Although the *Rb* gene is expressed in all adult tissues, only retinoblastoma and a very few other types of cancer, such as small-cell lung carcinomas and osteosarcomas, are initiated by loss of RB. Yet this pathway is still inactivated in most human tumors and is targeted by human tumor viruses (see Section 6.6, "Interaction of DNA viral protein products with RB and p53" on p. 145). These observations suggest that loss of RB may have roles that are necessary for cancer progression, in addition to cancer initiation. For example, inactivation of RB promotes chromosomal instability and angiogenesis via increased expression of E2F target genes (e.g. MAD2, VEGF, respectively). Some studies suggest that RB may have a role in differentiation, as well as cell cycle progression, in the developing retina. We have more to learn about the different tumor suppressive roles of RB and the requirements of different cell types.

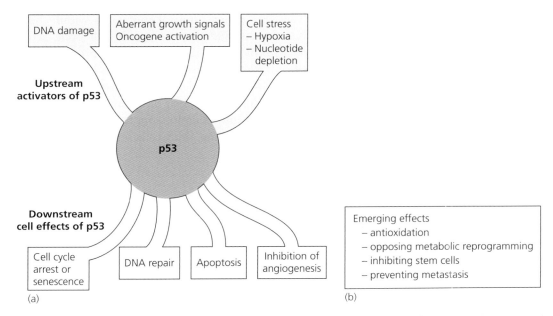

Figure 6.3 (a) Upstream activators and downstream effects of p53—a classical view. (b) Emerging effects.

6.4 The p53 pathway

The *p53* gene was the first tumor suppressor gene to be identified, and, since its discovery, scientists have found that the p53 pathway is altered in most human cancers. Two *p53* homologs, *p73* and *p63*, have also been identified, but mutations in cancer cells are rare. Its protein product, p53, is at the heart of the cell's tumor suppressive mechanism and thus has been nicknamed the "guardian of the genome." In the absence of cellular stress, low levels of p53 induce antioxidant activity which decreases the levels of reactive oxygen species (ROS) and subsequent DNA damage (Budanov, 2014). As mentioned in Chapter 2, normal cell metabolism produces ROS that can react with DNA. It has been estimated that endogenous ROS modify approximately 20,000 bases of DNA per day in a single cell. p53 protects against ROS by upregulating genes whose products have anti-oxidant functions such as glutathione peroxidase 1 and sestrins—proteins involved in hydrogen peroxide metabolism. This antioxidant activity guards against mutation and may help prevent cancer.

Many types of "danger signals," such as cell stress and DNA damage, can activate p53 and trigger several crucial cellular responses that suppress tumor formation (Figure 6.3a). Upstream stress activators include radiation-, drug-, or carcinogen-induced DNA damage, oncogenic activation, hypoxia, and low ribonucleotide pools. These conditions may nurture tumor initiation. In response to these stress signals, p53 can elicit downstream cellular effects, including transient or permanent cell cycle arrest, DNA repair, apoptosis, and inhibition of angiogenesis (see Chapter 10). The ability to cause the cell cycle to pause allows for the repair of mild DNA damage and prevents the propagation of mutations within the genome. More severe DNA damage induces irreversible cell cycle arrest called senescence. Apoptosis is another means of preventing propagation of mutations; cell suicide benefits the organism as a whole if DNA damage cannot be repaired. Mutated cells are better dead. Apoptosis is the critical biological function mediating the tumor suppressor function of p53. Other emerging downstream effects include antioxidation, opposing metabolic reprogramming (see Chapter 11), inhibiting stem cells, and preventing metastasis (Figure 6.3b).

Self-test Close this book, and try to redraw Figure 6.3. Check your answer. Correct your work. Close the book once more, and try again.

The overall regulation of the p53 pathway possesses an extraordinary complexity that compels us to try to unravel each layer. Let us begin by examining the structure of the p53 protein and its interactions with its inhibitors, and then move on to dissecting how its activity is switched on and how it exerts its effects.

Structure of the p53 protein

The *p53* gene, located on chromosome 17p13, contains 11 exons that encode a 53 kDa phosphoprotein. The p53 protein is a transcription factor containing four distinct domains: the amino-terminal transactivation domain, the DNA-binding domain containing a Zn^{2+} ion, a tetramerization domain, and a carboxy-terminal regulatory domain (Figure 6.4). The p53 protein binds as a tetramer to a p53 response element containing two direct repeats, as shown: 5'-RRRCWWGYYY(spacer of n = 0 – 14) RRRCWWGYYY-3' where R is a purine (A or G), Y is a pyrimidine (T or C), and W is adenine or thymine, in order to regulate transcription

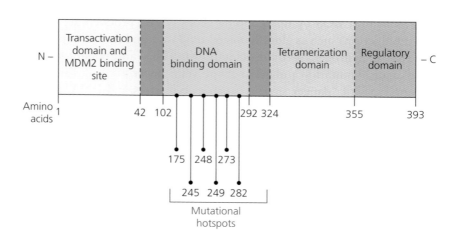

Figure 6.4 Domains of the p53 protein and location of mutational hotspots (marked in red).

of its target genes. Oligonucleotide array experiments have demonstrated that p53 binds to approximately 300 different gene promoter regions, thus suggesting that p53 has a powerful regulatory role. The p53 target genes code for proteins and miRNAs. Remember miRNAs are potent post-transcriptional regulators. Phosphorylation patterns of p53, as well as interactions with binding partners, are correlated with distinct transcriptional programs. Several specific p53 target genes and the mechanism of how they exert their effect will be discussed later in the chapter.

Regulation of p53 protein by MDM2

Normally, the level of p53 protein in a cell is low. The activity of p53 in a cell is regulated at the level of protein degradation, not at the level of expression of the *p53* gene. The MDM2 protein, a ubiquitin ligase, is its main regulator. Ubiquitin ligases are enzymes that attach a small peptide called ubiquitin to proteins, flagging it for proteolysis (enzymatic protein degradation involving cleavage of peptide bonds) in proteosomes. MDM2 modifies the carboxy-terminal domain of p53 and thus targets it for degradation by proteosomes in the cytoplasm. In addition, MDM2 modifies the activity of p53 as it binds to, and inhibits, the p53 transactivation domain at the amino-terminal and transports the protein into the cytoplasm, away from nuclear DNA. Thus, the activity of p53 as a transcription factor is out of reach. The binding of MDM2 to p53 is part of an autoregulatory feedback loop (Figure 6.5, shown by red arrows). The

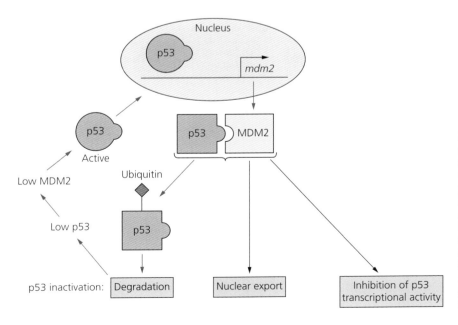

Figure 6.5 The p53–MDM2 feedback loop. p53 induces the transcription of the *Mdm2* gene, whose product negatively regulates p53. MDM2 is a ubiquitin ligase that adds ubiquitin to p53 and targets it for degradation. Low amounts of p53 leads to a reduction of *Mdm2* gene expression, and thus a reduction in the negative regulation of p53. Active p53 can bind the *Mdm2* gene promoter, completing the cycle.

Mdm2 gene is a transcriptional target of p53. Therefore, p53 stimulates the production of its negative regulator MDM2 that causes the degradation of p53. Low amounts of p53 will reduce transcription of the *Mdm2* gene and the amount of MDM2 protein, and this will result in an increase of p53 activity, thus completing the loop. Note that there are additional regulators of p53, including the negative regulator MDMX, a protein that does not contain intrinsic ligase activity, and a stabilizing factor, HAUSP, that removes ubiquitin from p53.

Upstream: molecular pathways of p53 activation

The mechanism by which p53 becomes activated depends on the nature of the stress signal. Stress is "sensed" by cellular proteins, many of which are kinases that convey the danger signals to p53 via phosphorylation. Disruption of the p53–MDM2 interaction is fundamental to the activation of p53 by its upstream factors.

The upstream activators of p53 utilize three main independent molecular pathways to signal cellular distress (Figure 6.6). DNA damage caused by ionizing radiation is signaled by two protein kinases. The first kinase, ATM, stimulated by DNA double-stranded breaks, phosphorylates and activates a second kinase, Chk2. Both ATM and Chk2 kinases phosphorylate amino-terminal sites of p53, and this phosphorylation interferes with binding of MDM2. A second molecular pathway that signals cellular stress to p53 is executed by two different kinases, ATR and casein kinase II. These also phosphorylate p53 and disrupt its interaction with

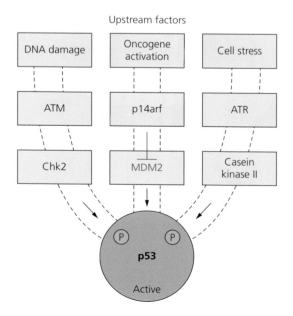

Figure 6.6 Upstream activators of p53. The first and the last pathways involve kinases and result in the phosphorylation of p53 (shown as P). All pathways disrupt the interaction of p53 with MDM2.

MDM2. Lastly, activated oncogenes, such as Ras, induce the activity of the protein p14arf, another modulator of the p53–MDM2 complex. p14arf is one of two translational products of the *INK4a/CDKN2A* gene (p16, a cyclin kinase inhibitor, is the other product). p14arf does not bind to the interface of p53–MDM2 but functions by sequestering MDM2 to the nucleolus of the cell. All three pathways prevent degradation of p53 by MDM2.

Downstream: molecular mechanisms of p53 cellular effects

The main mechanism by which p53 exerts its tumor-suppressing effects is by inducing the expression of specific target genes. Let us examine how the resulting network of proteins triggers these responses (Figure 6.7).

Inhibition of the cell cycle

One of the central functions of p53 is to cause either transient cell cycle arrest or senescence in response to DNA damage, so that there is either an opportunity to repair the damage prior to the next round of replication

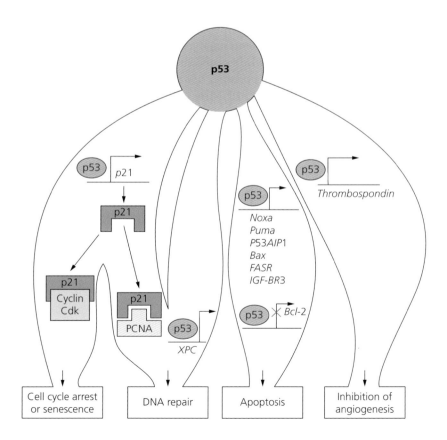

Figure 6.7 Downstream effects of p53. p53 exerts many of its effects by regulating target genes, as shown.

PAUSE AND THINK

Why would an inhibitor of cyclin–cdk complexes cause a pause in the G₁ to S transition? Recall the role of cyclin–cdk complexes in the cell cycle; importantly, they act as kinases. As kinases, they phosphorylate. What do they phosphorylate? RB. Failure of the cdk complex to phosphorylate RB prevents the release of the transcription factor E2F and blocks the transition into S phase.

or a complete restraint of cell division, respectively; thus, damaged DNA will be prevented from being replicated and passed on to daughter cells, and maintenance of the genome will be facilitated. The molecular mechanism responsible for this cellular response involves the transcriptional induction of the *p21* gene (also known as the *Cdkn1a* gene). Its product, the p21 protein, inhibits several cyclin–cdk complexes and causes a pause in the G_1 to S (and G_2 to M) transition of the cell cycle (see "Pause and think").

In addition, p21 also binds PCNA (proliferating cell nuclear antigen), a protein that has a role in DNA synthesis and DNA repair. The interaction with p21 is such that it inhibits the role of PCNA in DNA replication, but not in DNA repair. Therefore, p21 is an important part of the molecular mechanism that facilitates the ability of p53 to bring about a pause in the cell cycle and at the same time allow DNA repair. An miRNA regulated by p53, miR-34a, can also induce cell cycle arrest and senescence (not shown in Figure 6.7).

Apoptosis

The expression of several mediators of apoptosis is transcriptionally regulated directly by p53 (Table 6.2). The targets include genes that code for proteins involved in two apoptotic pathways that respond to external and internal signals, respectively. (Apoptosis will be described in Chapter 7.) In general, genes encoding proteins that promote apoptosis, pro-apoptotic proteins, are induced, while genes encoding proteins that antagonize apoptosis, anti-apoptotic proteins, are repressed. The mitochondrial pro-apoptotic proteins NOXA, PUMA, and p53AIP1, that cause the release of cytochrome *c* and activate the apoptosome, are induced. Also, p53 tips the balance regulated by the Bcl-2 protein family towards apoptosis by inducing gene expression of the pro-apoptotic protein Bax

Table 6.2 p53-inducible apoptotic target genes

Gene	Location of gene product
Bax	Intrinsic pathway
NOXA	Intrinsic pathway
PUMA	Intrinsic pathway
P53AIP1	Intrinsic pathway
FAS	Extrinsic pathway
IGF-BP3	Extrinsic pathway
DR5	Extrinsic pathway
PIDD	Extrinsic pathway
PERP	Endoplasmic reticulum

and repressing the expression of the anti-apoptotic protein Bcl-2. Fas receptor (FASR) is a transmembrane receptor that receives extracellular stimuli to stimulate apoptosis. Expression of the *FASR* gene is induced by p53. Apoptosis is also triggered when survival signaling is blocked by the induction by p53 of IGF-BP3 (insulin-like growth factor-binding protein 3). IGF-BP3 blocks the signaling of insulin-like growth factor 1 (IGF-1) to its receptor. Activation of these different pathways in concert is required for a full apoptotic response. Transcription-independent mechanisms for the induction of apoptosis by p53 also exist and will be discussed in Chapter 7.

DNA repair and angiogenesis and other emerging roles

Both DNA repair and angiogenesis are covered in depth elsewhere in this volume (see Chapters 2 and 9, respectively). In general, a role for the transcriptional regulation of important genes in these processes by p53 has been established. For example, the gene *XPC* that is involved in nucleotide excision repair is regulated by p53 through a p53 response element in its promoter. Thrombospondin, an inhibitor of angiogenesis, is also transcriptionally regulated by p53. This further supports the role of p53 as a transcriptional regulator in different biological responses.

Transcriptional activation of genes involved in the emerging roles of p53 have been identified, and some examples follow. A role in anti-oxidation is supported by the induction of the *glutathione peroxidase* gene (Gpx1) and *sestrin* gene (Sesn). Blocking metabolic reprogramming and inhibiting glycolysis are supported by induction of *Tigar*. The induction of miRNAs is important for inhibiting qualities of stem cells (mir-145) and preventing metastasis (mir 34). mir 34 reduces the expression of transcription factors (Snail) important for cancer cell migration.

Factors that decide downstream outcome: cell cycle arrest or apoptosis

As the guardian of the genome, p53 prevents damaged DNA from being passed on to daughter cells either by inhibiting the cell cycle or by inducing apoptosis. Cell cycle arrest and apoptosis are two independent effects of p53. There are several molecular factors that determine the biological outcome of whether cell cycle arrest or apoptosis takes place (Murray-Zmijewski *et al.*, 2008). Promoter selectivity and p53 binding affinity are mechanisms for influencing cell fate. This may be determined by the nucleotide sequence of the p53 response element (note the variable length of the spacer within this element), its context within the promoter, or post-translational modifications of p53. For example, phosphorylation of Ser46 results in a preference of p53 to induce pro-apoptotic genes. Absolute levels of p53 is another factor that influences outcome. Evidence suggests that a threshold is needed to enable apoptosis to occur. Subcellular localization of p53 also plays a role in its downstream effects. As we discussed earlier, although p53 is a transcription factor and must reside in the nucleus for this function, it has other transcription-independent roles which are carried out elsewhere in the cell. And lastly, the presence of factors that interact with p53, such as different combinations of transcription factors, influence the biological response. In summary, target gene selectivity, level of p53 protein, subcellular localization, and co-factors are important factors that affect the downstream effect of p53.

Let us examine one example of how p53 may be directed towards either cell cycle arrest or apoptosis. Oncogene activation (e.g. Myc) is an upstream inducer of p53 that triggers apoptosis. The mechanism of this stress signal acts via transcriptional regulators of the gene coding for the cyclin–cdk inhibitor p21, an important effector of cell cycle arrest but also an inhibitor of cell death (Figure 6.8). The regulation of the *p21* gene (also known as *CDKN1A*) is a pivotal point in the p53 decision-making process. Both p53 and a transcription factor called Miz-1 are required for *p21* gene expression. Now enter the oncogene Myc, which competes with p53 for binding with Miz-1. Myc interacts with Miz-1 and inhibits the transcription of *p21*. Through this mechanism of preventing expression of *p21*, Myc not only overrides the p53-regulated block to cell cycle progression, but also blocks p21-mediated inhibition of apoptosis. p53 is

Figure 6.8 An example of molecular factors that mediate the downstream effects of p53: cell cycle arrest or apoptosis? The regulation of a gene that codes for a cyclin–cdk inhibitor, the *p21* gene, plays a pivotal role. Top: without competition from Myc, p53 and Miz-1 bind to the promoter of *p21* and induce transcription, resulting in cell cycle inhibition. Bottom: upon oncogenic activation, Myc competes with the binding of p53. Myc and Miz-1 bind to the *p21* promoter, inhibit transcription, and block cell cycle inhibition. ASPP binds to p53 and facilitates activation of apoptotic genes to induce apoptosis.

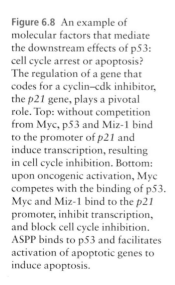
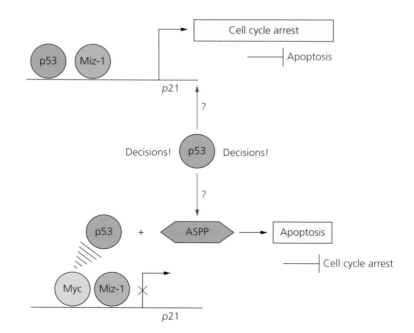

not altered and is free to induce the expression of pro-apoptotic targets. Additional events are also required for full activation of apoptosis, as p53 phosphorylation and apoptotic co-factors are required for the induction of some apoptotic genes. Revealing the mechanisms behind other modes of upstream stress inducers of p53, such as oxidative stress, requires further studies.

The apoptosis-stimulating proteins of p53 (ASPP) family are apoptotic co-factors that play a role in p53 decision-making. These proteins bind to the p53 DNA-binding domain and have been shown specifically to enhance the ability of p53 to activate genes involved in apoptosis and not cell cycle arrest. The selection of apoptotic genes (such as *Bax*), and not growth arrest genes (such as *CDKN1A*), could potentially be accomplished by specific promoter sequences that serve to distinguish the functionally distinct classes of genes. The opposite outcome is seen from the binding of p53 to an inhibitor of ASPP protein (iASPP). Mutations in the ASPP binding site of the *p53* gene and epigenetic silencing of the *ASPP* gene have been identified in tumor cells. These tumor cells may have been initiated, because they escaped from the apoptotic program normally augmented by ASPP.

6.5 Mutations in the p53 pathway and cancer

Owing to the central role of p53 as a tumor suppressor and guardian of the genome, cell transformation is less likely to occur in cells that maintain a functional p53 pathway. p53 mutant cells are characterized by genomic instability, as mutations are more likely to be maintained in dividing cells, providing an environment that is permissive for tumor initiation. The high frequency of p53 pathway mutations found in tumor cells is most likely to be the result of selective pressure favoring mutant cells that escape tumor suppression. Over 75% of all *p53* mutations are missense mutations and result in single amino acid substitutions. More than 90% of the missense mutations are located in the DNA-binding domain (amino acids 102–292), and more than 30% of these affect only six codons and are therefore referred to as "hotspots" (Figure 6.4).

Thus, p53 differs from many "classical" tumor suppressor genes in that "classical" tumor suppressors are usually characterized by nonsense or frameshift mutations that lead to inactivated truncated proteins. Many mutant p53 molecules are more stable than wild-type p53 protein and can accumulate in cells and show "gain of function." That is, some mutant p53 proteins show oncogenic capabilities, and this effect may be driven by altered target gene profiles or inappropriate protein–protein interactions. Some gain-of-function mutant p53 proteins induce histone-modifying enzymes that epigenetically increase transcription of some genes that lead to enhanced cell growth (Zhu *et al.*, 2015).

In addition to mutation of the *p53* gene, there are other ways to interfere with the p53 pathway. Defects in pathways that lead to the activation of p53 in response to stress, such as mutations in the *chk2* gene, have been identified in cancer cells that do not contain mutations in the *p53* gene. In these cases, the stress signal would not be transmitted to p53 via phosphorylation as in normal cells (Figure 6.6). Also, overexpression of the MDM2 protein has been demonstrated to alter the regulation of p53, leading to a "p53-inactivated" phenotype. Inactivation of downstream effectors, such as Bax and FASR, perturbs the apoptotic response. Some of these other p53 pathway disruptions may not lead to effects as severe as *p53* mutation, as the p53 protein is a central node for receiving and eliciting many stress signals and biological responses, respectively, but may mimic a partial aspect of p53 inactivation and be permissive for tumor formation.

Li–Fraumeni syndrome

Li–Fraumeni syndrome is predominantly characterized by a germline mutation of the *p53* gene and leads to a predisposition to a wide range of cancers. It is an autosomal dominant disease, so an affected individual has a 50% chance of passing the mutation to each offspring. Patients have a 25-fold increased risk of developing cancer before they are 50 years old, compared with the general population. The young age at which individuals develop cancer and the frequent occurrence of multiple primary tumors in individuals are characteristic features of the syndrome. The types of cancer seen within families that carry the mutation include sarcomas, breast cancer, leukemia, and brain tumors. Cancer develops at an earlier age over several generations. It is noteworthy that tumors in Li–Fraumeni syndrome patients do not always exhibit loss of the wild-type *p53* allele, suggesting that haploinsufficiency of p53 is sufficient for tumor formation.

More complex than Knudson's two-hit hypothesis

The mechanisms of some tumor suppressor genes are more complex than Knudson's two-hit hypothesis suggests. This is particularly clear for *p53*. As stated earlier, loss of heterozygosity is not commonly observed in tumors from Li–Fraumeni syndrome patients, suggesting that reduced amounts (haploinsufficiency) of p53 can cause transformation. In addition, specific mutations may result in varied amounts of tumor suppressor gene expression, and therefore tumor suppressor "dose" may play a role in the cancer outcome. Recent experiments which generated p53 hypomorphs (animals that exhibit reduced levels of p53 expression) using RNA interference (see "Analysis of gene function by small interfering RNAs (siRNAs)" on p. 18) support this mechanism (Hemann *et al.*, 2003).

Unlike most other tumor suppressor genes, some *p53* mutations do not lead to loss of function. Some missense mutations form an altered protein that interacts with the product of a normal p53 allele via dimerization to inactivate its function. This type of effect is referred to as dominant negative whereby the mutated gene product dominates to inactivate the wild-type gene product. In this situation, the autoregulatory loop is affected because p53 fails to induce its inhibitor MDM2, and, as a result, p53 mutant protein accumulates. Other mutations can lead to a newly acquired "gain-of-function" phenotype that can accumulate in, and transform, cells. In this instance, mutant *p53* (not wild-type *p53*) can be considered an oncogene having an active role in carcinogenesis.

The continuum model of tumor suppression

A new model, called the continuum model, for tumor suppression that integrates the two-hit hypothesis with a broader concept that also explains exceptions to this rule has been proposed (Berger *et al.*, 2011). It states that subtle dosage effects of tumor suppressors, either as a result of changes in levels of expression or protein activity, can play an important role in carcinogenesis. As described earlier for p53, a single dominant negative mutation over the wild-type protein is a possible mechanism of tumor suppression that does not fit Knudson's two-hit hypothesis. Tumor suppressor dosage plays a role for other tumor suppressors. It has been reported that PTEN haploinsufficiency is actually more tumorigenic in some contexts, compared with complete loss, because complete loss triggers p53-induced senescence. Mutations of the tumor suppressor *Pax5* gene most often affect one allele and seem to function as hypomorphs. This concept of dosage is not new for oncogenes: gene amplification of mutated *EGFR* is one example of altered oncogene dosage in cancer. Also, it is easy to envisage that single nucleotide polymorphisms may affect regulation by microRNAs and lead to altered protein levels. Further research about microRNAs and aberrant regulation in cancer may help to further elucidate our understanding of gene dosage of tumor suppressors in cancer.

6.6 Interaction of DNA viral protein products with RB and p53

Viruses are cellular parasites that hijack host cell proteins to maintain their life cycle. Coercing the host cell into S phase is essential for viral propagation. Several DNA viruses have been shown to be carcinogenic to humans. Most notable are papovaviruses, adenoviruses, herpes viruses, and hepatitis B. Interestingly, several of the DNA viruses share a common oncogenic mechanism that involves the interaction of viral proteins with the two important regulators of tumor suppression, RB and p53 (Figure 6.9). The viral proteins adenovirus E1A, papillomavirus E7, and

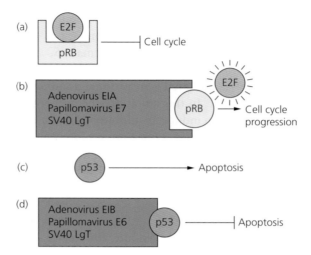

Figure 6.9 Viral protein products (red) interact with RB (pRB) and p53 (parts b and d, respectively), and inhibit their normal function (shown in a and c).

SV40 large T antigen inactivate RB (Figure 6.9b); adenovirus E1B, papillomavirus E6, and SV40 large T antigen inactivate p53 (Figure 6.9d).

The ability of both E6 and E7 to degrade p53 and RB, respectively, using the ubiquitin–proteasome system, correlates with their oncogenic potential. The biochemical events involved in p53 degradation by E6 are as follows: E6 binds to a ubiquitin–protein ligase (E6-AP) and forms a dimer that subsequently binds to p53. p53 is then ubiquitinated and tagged for recognition by the proteasome for degradation. A similar mechanism is likely for E7-targeted degradation of RB and E1B-targeted degradation of p53.

A PIONEER IN THE FIELD…Apoptosis: David Lane

David Lane's contributions to cancer research were acknowledged when he received a knighthood from Queen Elizabeth II in January 2000. He discovered the p53 protein–SV40 T antigen complex that helped characterize the mechanisms of viral transformation and tumor suppressor gene function. He was instrumental in publicizing the importance of p53 as the "guardian of the genome."

Human papillomavirus (HPV) causes cervical cancer

The *p53* gene is rarely mutated in cervical cancers, suggesting that the causative agent, HPV, may be functionally equivalent to *p53* mutation. This is indeed the case, as the binding of HPV protein E6 to p53 causes subsequent degradation of p53 which is functionally equivalent to the loss of the wild-type p53 allele. A polymorphism in the *p53* gene at amino acid 72 in humans leads to differences in the risk of cervical cancer, following exposure to HPV. Patients with two alleles coding for Arg at this position have a seven times higher risk of cervical cancer than those with alleles coding for Pro at this site. The Arg-containing p53 protein is more susceptible to degradation by HPV E6 (probably because of altered protein conformation). As a result of these cells having decreased p53 activity, they are likely to have an increased mutation rate and an increased potential to form tumors.

Therapeutic strategies

6.7 Targeting of the p53 pathway

The role of p53 as a "star player" in suppressing tumorigenesis and the high occurrence of mutations in the *p53* gene found in tumors draw attention to the p53 pathway as a promising cancer therapeutic target. As a result, many different strategies that target the p53 pathway have been developed (Bykov and Wiman, 2014; Chen *et al.*, 2010; Hoe, Verma and Lane, 2014; Muller and Vousden, 2014). Several are described in the following sections. The variety of p53 pathway aberrations, including loss-of-function and gain-of-function *p53* gene mutations and defective regulation of p53, suggests that the future success of these therapies will be dependent on knowing the genotype of tumors in patients prior to treatment—are there mutations in the *p53* gene itself or in its regulators? If there are mutations in *p53*, what type of mutations are present? Different mutations of *p53* have different functions and are currently grouped into two categories: those that contact DNA (contact mutants) or those that alter the protein structure (conformational mutants). Therapeutics may strive to correct a *p53* mutation or potentiate normal p53 protein function in cases where other alterations in the p53 pathway affect its function. Several different strategies are described below, and some are illustrated in Figure 6.10 (therapeutics are shaded red).

Strategies that aim to correct a *p53* mutation

Gene therapy is one of the most obvious approaches to correct for a *p53* mutation. In fact, many different vectors have been examined in pre-clinical and clinical settings (Bouchet *et al.*, 2006). Retroviruses have been used in several pre-clinical studies and have shown anti-tumor activity. However, only one Phase I clinical trial using retrovirus-mediated *p53* gene therapy has been published to date (discussed in Bouchet *et al.*, 2006), and expression of *p53* could not be detected, although some clinical responses in lung cancer patients were observed. Unlike retroviruses

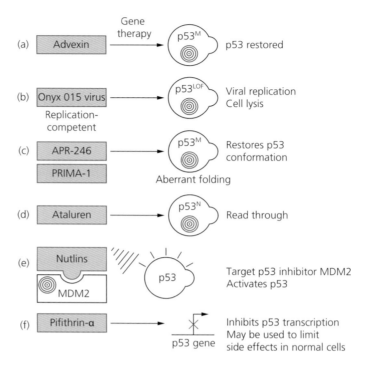

Figure 6.10 Therapeutic strategies that target the p53 pathway. Cell targets are indicated with a ◎ symbol; therapeutic agents (red); p53^{LOF}, loss-of-function mutant p53; p53^{M}, missense mutant p53; p53^{N}, nonsense mutant p53; nutlins compete (////) with p53.

that integrate into the genome, and therefore carry a risk of causing insertional mutagenesis, adenoviruses, which are DNA viruses, do not cause damage to the genome. Frequently used adenoviral vectors can carry large amounts of DNA and can infect a range of cell types via coxsackievirus and adenovirus receptors with high efficiencies. Adenoviruses that have been modified to be replication-defective have been popular vectors. Advexin™ (Introgen Therapeutics; Genedicine™, Shenzhen SiBonoGene Tech Co. Ltd) is a replication-defective adenoviral vector that contains the human *p53* gene, driven by a viral promoter (the cytomegalovirus (CMV) promoter), within the region of an E1 deletion (Figure 6.10a). Many clinical trials have demonstrated safety and low toxicity. It has been reported recently that Advexin™ was used as an adenoviral-mediated *p53* gene therapy to treat one individual with Li–Fraumeni syndrome. Genedicine™ obtained a drug license in China and is the world's first commercial gene therapy for cancer.

Replication-competent adenoviruses provide another strategy as the basis of new cancer therapeutics. As mentioned previously, wild-type adenovirus can replicate in cells by inactivating p53 and RB. The Onyx 015 replication-selective adenovirus was designed to selectively kill cancer cells that contain *p53* mutations (Larson *et al.*, 2015). It takes advantage of the fact that interference with the RB and p53 pathways is exploited both by viruses and cancer cells. The Onyx 015 virus contains a deletion of the *E1B* gene. Because it lacks E1B, the Onyx 015 virus can trigger a

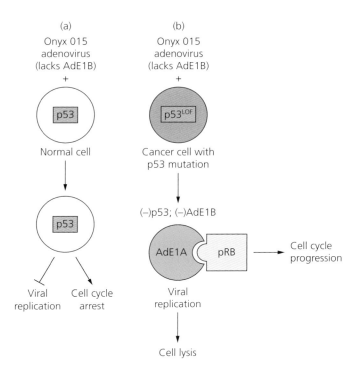

Figure 6.11 The Onyx 015 adenovirus can selectively kill cancer cells with *p53* mutations. (a) Normal cell. (b) Cancer cell with loss-of-function p53.

p53 response (apoptosis and growth arrest) in normal cells, resulting in the interference of viral replication (Figure 6.11a). Cancer cells that have an inactive p53 pathway cannot induce apoptosis or growth arrest, but the E1A product present in the Onyx 015 virus binds to, and inactivates, RB in p53 mutant cancer cells and, as a result, induces the G_1 to S phase transition of the cell cycle, allowing for viral replication and cell destruction. Thus, Onyx 015 can only replicate within, and subsequently lyse, cells that have an inactive p53 pathway (Figure 6.11b). Such selective replication produces a treatment with minimal side effects. Phase I and II clinical trials have given encouraging results. Intratumoral injection proved safe, and specific anti-tumor effects were demonstrated. Although further development of Onyx 015 has been halted, H101, an oncolytic adenovirus similar to Onyx 015, was approved by the Chinese State Food and Drug Administration in 2005 for use in combination with chemotherapy for the treatment of nasopharyngeal cancers.

Small molecules that can restore wild-type function to products of *p53* gene mutations have been investigated. Many missense mutations of *p53* result in aberrant protein conformation and subsequently interfere with the DNA-binding function of the p53 protein. As a result, the p53 inhibitor MDM2 and target genes essential for apoptosis are not induced. Note that tumors that carry such mutations are more likely to be resistant to conventional chemo- and radiotherapies, as these drugs work via the induction of apoptosis. Reactivation of *p53* mutants by restoring wild-type protein conformation aims to eliminate tumor cells via the induction of apoptosis. PRIMA-1 (and its derivative APR-246) (Figure 6.10c), identified from a chemical library, is thought to act as a chaperone that aids in the refolding of the protein. It demonstrates restoration of the wild-type p53 conformation and DNA binding in a number of mutant p53 proteins, including those with mutations in the hotspots (Figure 6.4). Both drugs can be converted to bind covalently to cysteines in p53 mutant proteins. PRIMA-1 showed mutant p53-dependent antitumor effects in human tumor xenografts in animal models, and APR-246 demonstrated safety in a Phase I clinical trial (Lehmann *et al.*, 2012). Also, APR-246 can restore unfolded wild-type protein and mutant p63 (due to high sequence homology between p53 and p63).

Nonsense mutations that give rise to truncated p53 protein account for 8% of the cancer-associated p53 mutations. The strategy to use "read-through" drugs, such as ataluren (PTC Therapeutics) that is in Phase III clinical trials for cystic fibrosis patients, may be investigated for cancer patients with p53 nonsense mutations.

Strategies that aim to activate endogenous p53

In many tumors, wild-type *p53* is expressed, but the regulation of the p53 protein is defective and results in an altered p53 pathway. As mentioned earlier, overexpression of the p53 inhibitor MDM2 leads to defective regulation and p53 inactivation.

An approach on the horizon to activate wild-type p53 protein is the development of inhibitors of the p53–MDM2 interaction (see "Pause and Think").

Detailed structural information about the p53–MDM2 complex has been obtained by nuclear magnetic resonance and X-ray crystallography, and has revealed that MDM2 has a well-defined binding site for p53. The opposite is not true. Thus, inhibitors are best designed to mimic amino acids of p53. In addition, the binding interface was found to be relatively small, which suggested that it was possible to design small inhibitors that could be taken orally. Following a strategy of high-throughput screening of synthetic chemicals and computer modeling, drugs called nutlins (Figure 6.10d) were identified and showed promising results in pre-clinical tests (Hoe, Verma, and Lane, 2014). These results include triggering p53 activation and its biological responses in cancer cells containing wild-type p53. In addition, inhibition of tumor growth by 90% was demonstrated in animal models. Nutlins support the idea that protein–protein interactions are good targets for cancer therapeutics, and second-generation nutlin RG7112 (Roche) is currently in early-phase clinical trials.

There has been a push to develop dual inhibitors of both MDM2 and MDMX. One such drug candidate is the stapled peptide ATSP-7041 (Aileron Therapeutics). Stapling is a technique that is used to stabilize the

PAUSE AND THINK

What type of information is required to begin to design inhibitors of the p53–MDM2 interaction? Structural studies of the interactions between p53 and MDM2 produced important information for drug design. The crystal structure of a p53 peptide complexed to MDM2 revealed a deep hydrophobic pocket in MDM2 and three amino acids of the p53 peptide (Phe19, Trp23, and Leu26) that are critical for the interaction. Combined with molecular modeling, structure-based screening allowed for new drugs to be developed. Many bind to MDM2 by mimicking the key amino acid residues of p53.

helical motif that is important for binding. It introduces a hydrocarbon linker between two non-adjacent amino acids. This class of molecules shows excellent pharmaceutical properties.

PAUSE AND THINK

When and where would suppression of p53 be clinically beneficial?

Strategies that aim to suppress endogenous p53

It is clear that p53 is important for the disposal of tumor cells. The success of chemo- and radiotherapy is often limited by side effects in normal tissues. Many of the side effects of chemo- and radiotherapy are, in part, mediated by p53. There is normally high expression of p53 in the tissues that are sensitive to these conventional therapies, such as the hematopoietic organs and intestinal epithelia, and the DNA damage caused by these agents induces p53 to elicit apoptosis, the mechanism behind the side effects. Therefore, temporary and reversible suppression of p53 in normal tissue may help alleviate the side effects of conventional therapies only in patients with tumors that have lost p53 function. This concept to use drugs to protect normal tissues from a cytotoxic drug is called cyclotherapy. A chemical screen has identified pifithrin-α (**p** fifty **thr**ee **in**hibitor) as a potential agent to test this approach (Figure 6.10e). Pifithrin inhibits *p53* gene transcription. Prevention of hair loss and an increase in tolerated dose in irradiated mice promise future clinical applications.

■ CHAPTER HIGHLIGHTS—REFRESH YOUR MEMORY

- Tumor suppressor genes act as stop signals for uncontrolled growth or may play a role in DNA repair.

- Knudson's two-hit hypothesis states that a germline mutation in one tumor suppressor allele predisposes an individual to cancer. Acquiring a second mutation in the second allele later in life triggers carcinogenesis.

- Germline mutations in *BRCA1* and *BRCA2* predispose individuals to breast and ovarian cancer.

- *PTEN* is a tumor suppressor gene that codes for a phosphatase regulating the activity of a potential oncogenic kinase.

- The *Rb* tumor suppressor gene in retinoblastoma follows Knudson's two-hit hypothesis: a germline mutation predisposes the individual to cancer, and mutations in both alleles are necessary for tumor initiation.

- The tumor suppressor *p53* has been nicknamed the "guardian of the genome" because of its central role in maintaining the integrity of the cell's DNA.

- The p53 protein is a transcription factor that regulates genes involved in inhibition of the cell cycle, DNA repair, apoptosis, and angiogenesis.

- Additional emerging roles of p53 include antioxidant responses, blocking metabolic reprogramming, inhibiting stem cell characteristics, and blocking metastasis.

- More than 90% of *p53* missense mutations are located in the DNA-binding domain.

- MDM2 is a main regulator of p53 protein activity. As a ligase, it targets p53 for degradation. MDMX is an additional regulator.

- The protein product of the *p21* gene, a cdk inhibitor, is key for eliciting the p53 response of cell cycle inhibition.

- Several gene products, including Bax and IGF-BP3, are important for eliciting the apoptotic response of p53.

- The biological response exerted by p53, either inhibition of the cell cycle or apoptosis, is mediated by several factors and may include the regulation of *p21* and binding to the ASPP family.

- Li–Fraumeni syndrome is a disease that is characterized by an inherited mutation of the *p53* gene. Patients have a predisposition to a variety of cancers.

- The continuum model for tumor suppression has recently been proposed, and it states that subtle dosage effects of tumor suppressors, either because of expression levels or protein activity, can play an important role in carcinogenesis.

- Viral proteins from adenovirus, papilloma virus, and SV40 virus inactivate p53 and RB as a common oncogenic mechanism; several utilize the ubiquitin–proteasome system.

- Both the RB pathway and the p53 pathway provide molecular targets for the design of new cancer therapeutics.

▨ ACTIVITY

1. Choose a genetic syndrome that leads to a predisposition to cancer (excluding familial breast cancer and Li–Fraumeni syndrome). Describe in detail the molecular mechanisms involved.
2. Find evidence of specific tumor suppressor genes that exemplify the spectrum of the continuum model for tumor suppression (see Berger *et al.*, 2011).

▨ FURTHER READING

Bieging, K.T., Mello, S.S., and Attardi, L.D. (2014) Unravelling mechanisms of p53-mediated tumor suppression. *Nat. Rev. Cancer* **14**: 359–370.

Brown, C.J., Lain, S., Verma, C.S., Fersht, A.R., and Lane, D.P. (2009) Awakening guardian angels: drugging the p53 pathway. *Nat. Rev. Cancer* **9**: 862–873.

Burkhart, D.L. and Sage, J. (2008) Cellular mechanisms of tumour suppression by the retinoblastoma gene. *Nat. Rev. Cancer* **8**: 671–682.

Goh, A.M., Coffill, C.R., and Lane, D.P. (2011) The role of mutant p53 in human cancer. *J. Pathol.* **223**: 116–126.

Goldstein, I., Marcel, V., Olivier, M., Oren, M., Rotter, V., and Hainaut, P. (2011) Understanding wild-type and mutant p53 activities in human cancer: new landmarks on the way to targeted therapies. *Cancer Gene Ther.* **18**: 2–11.

Levitt, N.C. and Hickson, I.D. (2002) Caretaker tumour suppressor genes that defend genome integrity. *Trends Mol. Med.* **8**: 179–186.

Manning, A.L. and Dyson, N.J. (2012) RB: mitotic implications of a tumour suppressor. *Nat. Rev. Cancer* **12**: 220–226.

Sherr, C.J. (2004) Principles of tumor suppression. *Cell* **116**: 235–246.

Song, M.S., Salmena, L., and Pandolfi, P.P. (2012) The functions and regulation of PTEN tumor suppressor. *Nat. Rev. Mol. Cell Biol.* **13**: 283–296.

Volgelstein, B., Lane, D., and Levine, A. (2000) Surfing the p53 network. *Nature* **408**: 307–310.

Vousden, K.H. and Prives, C. (2009) Blinded by the light: The growing complexity of p53. *Cell* **137**: 413–431.

≡ **WEB SITES**

International Agency for Research on Cancer. Database of p53 mutations
  http://p53.iarc.fr/

A web resource for tumor suppressor genes http://bioinfo.mc.vanderbilt.edu/TSGene/
  (see Zhao, M., Sun, J., and Zhao, Z. (2013) TSGene: a web resource for tumor
  suppressor genes. *Nucleic Acids Res*. **41**: D970–976. doi:10.1093/nar/gks937)

≡ **SELECTED SPECIAL TOPICS**

Berger, A.H., Knudson, A.G., and Pandolfi, P.P. (2011) A continuum model for tumor
  suppression. *Nature* **476**: 163–169.

Bouchet, B.P., de Fromentel, C.C., Puisieux, A., and Galmarini, C.M. (2006) P53 as a
  target for anti-cancer drug development. *Crit. Rev. Oncol. Hematol*. **58**: 190–207.

Budanov, A.V. (2014) The role of tumor suppressor p53 in the antioxidant defense and
  metabolism. *Subcell. Biochem*. **85**: 337–358.

Bykov, V.J.N. and Wiman, K.G. (2014) Mutant p53 reactivation by small molecules
  makes its way to the clinic. *FEBS Lett*. **588**: 2622–2627.

Chen, F., Wang, W., and El-Deiry, W.S. (2010) Current strategies to target p53 in
  cancer. *Biochem. Pharmacol*. **80**: 724–730.

Hemann, M.T., Fridman, J.S., Zilfou, J.T., Hernando, E., Paddison, P.J., Cordon-
  Cardo, *et al*. (2003) An epi-allelic series of p53 hypomorphs created by stable RNAi
  produces distinct tumor phenotypes. *Nat. Genet*. **33**: 396–400.

Hoe, K.K., Verma, C.S., and Lane, D.P. (2014) Drugging the p53 pathway:
  understanding the route to clinical efficacy. *Nat. Rev. Drug Discov*. **13**: 217–236.

Julien, S.G., Dubé, N., Hardy, S., and Tremblay, M.L. (2011) Inside the human cancer
  tyrosine phosphatome. *Nat. Rev. Cancer* **11**: 35–49.

Kang, M.Y., Kim, H.-B., Piao, C., Lee, K.H., Hyun, J.W., Chang, I.Y., and You, H.J.
  (2013) The critical role of catalase in prooxidant and antioxidant function of p53.
  Cell Death Differ. **20**: 117–129.

Larson, C., Oronsky, B., Scicinski, J., Fanger, G.R., Stirn, M., Oronsky, A., *et al*. (2015)
  Going viral: a review of replication-selective oncolytic adenoviruses. *Oncotarget* **6**:
  19976–19989.

Lehmann, B.D., Bykov, V.J., Ali, D., Andrén, O., Cherif, H., Tidefelt, U., *et al*. (2012)
  Targeting p53 in vivo: a first in human study with p53-targeting compound APR-
  246 in refractory hematologic malignancies and prostate cancer. *J. Clin. Oncol*. **30**:
  3633–3639.

Macleod, K. (2000) Tumor suppressor genes. *Curr. Opin. Genet. Dev*. **10**: 81–93.

Muller, P.A.J. and Vousden, K.H. (2014) Mutant p53 in cancer: new functions and
  therapeutic opportunities. *Cancer Cell* **25**: 304–317.

Murray-Zmijewski, F., Slee, E.A., and Lu, X. (2008) A complex barcode underlies the
  heterogeneous response of p53 to stress. *Nat. Rev. Mol. Cell Biol*. **9**: 702–712.

Roy, R., Chun, J., and Powell, S.N. (2012) BRCA1 and BRCA2: different roles in a
  common pathway of genome protection. *Nat. Rev. Cancer* **12**: 68–78.

Sablina, A.A., Budanov, A.V., Ilyinskaya, G.V., Agapova, L.S., Kravchenko, J.E., and Chumakov, P.M. (2005) The antioxidant function of the p53 tumor suppressor. *Nat. Med.* **11**: 1306–1313.

Vassilev, L.T. (2006) MDM2 inhibitors for cancer therapy. *Trends Mol. Med.* **13**: 23–31.

Vu, B.T. and Vassilev, L.T. (2011) Small molecule inhibitors of the p53-MDM2 interaction. *Curr. Topics Microbiol. Immunol.* **348**: 151–172.

Wang, Z., Shen, D., Parsons, D., Bardelli, A., Sager, J., Szabo, S., *et al.* (2004) Mutational analysis of the tyrosine phosphatome in colorectal cancers. *Science* **304**: 1164–1166.

Zhu, J., Sammons, M.A., Donahue, G., Dou, Z., Vedadi, M., Getlik, M., *et al.* (2015) Gain-of-function p53 mutants co-opt chromatin pathways to drive cancer growth. *Nature* **525**: 206–211.

Chapter 7

Apoptosis

Introduction

Apoptosis is a highly regulated process of cell death that not only plays a role in developmental morphogenesis, but also controls cell numbers and gets rid of damaged cells. It therefore plays an important role in tumor suppression. As described in Chapter 1, the balance between cell growth, differentiation, and apoptosis affects the net number of cells in the body, and aberrant regulation of these processes can give rise to tumors. Apoptosis is a crucial tumor suppression mechanism within the body, because it gets rid of cells that have extensive DNA damage and the potential to lead to cancer. The peeling of your skin after sunburn is, in fact, a result of apoptosis of cells that have had extensive DNA damage after UV exposure. This process is an important defense against skin cancer. Elimination of cells that have damaged DNA helps protect the entire organism from cancer. Defects in apoptosis also influence the effectiveness of those conventional therapies that mainly exert their effect by inducing apoptosis. In this chapter, we will describe the molecular mechanisms of apoptosis and examine specific mutations that affect the apoptotic pathway and play a role in carcinogenesis. We will also investigate how mutations in the apoptotic pathway can lead to resistance to chemotherapeutic drugs. Lastly, strategies for the design of new cancer therapeutics that target apoptosis will be presented. Let us begin with a description of apoptosis.

Apoptosis is a type of "cell suicide" that is intrinsic to the cell. It is an active process requiring the expression of a genetic program that every cell is capable of executing. The apoptotic process is organized, neat, and tidy, leaving behind little evidence of the pre-existing cell. The cell undergoing apoptosis is swept clean during **phagocytosis** by macrophages and neighboring cells that recognize molecular flags (e.g. phosphatidylserine) exhibited by the apoptosing cell. Apoptosis is characterized by cell shrinkage, membrane blebbing and budding, and chromatin condensation and precise fragmentation, all of which contribute to the neat disposal of the cell. This sharply contrasts with the "sloppy" process of **necrosis** whereby cells swell, cell membranes become leaky, and cells spill out their contents into the surrounding

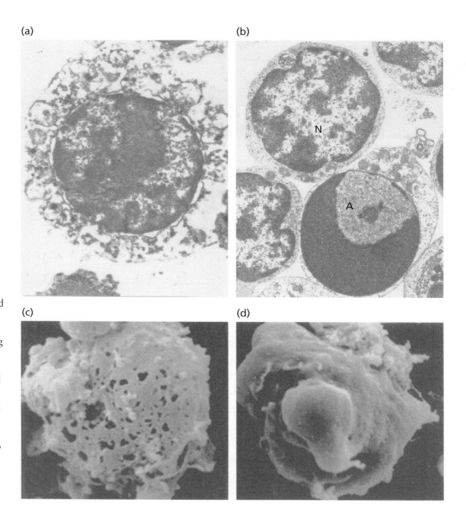

Figure 7.1 (a) Transmission electron microscopy (TEM) of a necrotic cell: the disruption of the plasma membrane and organelles is observable. A relative preservation of nuclear morphology appears (original magnification ×10,000). (b) TEM of an apoptotic (A) and a normal (N) cell. The characteristic chromatin rearrangement appears in A, strongly different from its normal organization (N). The good preservation of membrane and organelles is also evident (original magnification ×8000). (c) Scanning electron microscopy (SEM) of a necrotic cell. Numerous lesions appear on the cell surface (original magnification ×5000). (d) SEM of an apoptotic cell. Surface blebbing is evident (original magnification ×5000). From Watson, J. (1997) *The Purdue Cytometry CD-ROM*, Vol. 4, guest ed. J. Paul Robinson. Purdue University Cytometry Laboratories, West Lafayette, IN (ISBN 1-890473-03-0).

tissue and cause inflammation. Morphological differences between cells undergoing necrosis and apoptosis can be seen in Figure 7.1.

Similar to the central role that kinases have in growth factor signaling pathways, particular **proteases**, called **caspases**, play a central role in apoptosis. Proteolysis, catalyzed by caspases, helps to break down cellular components for the neat disposal that is characteristic of apoptosis. For example, the shrinkage of the nucleus is aided by caspases that degrade the protein network of lamins that underlie the structure of the nuclear envelope.

7.1 Molecular mechanisms of apoptosis

Cells may be induced to undergo apoptosis by extracellular signals, so-called "death factors," or by internal physical or chemical insults such as DNA damage or oxidative stress. Subsequently, two non-exclusive

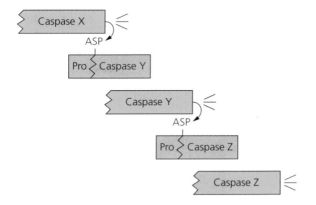

Figure 7.2 A simple caspase cascade.

molecular pathways, the extrinsic and the intrinsic, respectively, may be activated. Caspases are specific proteases that act like molecular scissors to cleave intracellular proteins at aspartate residues (one of the 20 amino acids). Caspases are central to both apoptotic pathways. The term "caspases" derives from three of the characteristics of the enzymes: they are **c**ysteine-rich **asp**artate prote**ases**. More than 13 mammalian caspases have been identified. They are synthesized as inactive enzymes called procaspases that need to be cleaved at aspartate residues in order to be activated. Although, for the most part, procaspases are considered inactive, procaspases possess some activity—about 2% of the proteolytic activity of fully activated caspases. This may seem insignificant at the moment, but, as we will see later, it is an important feature for some pathways of caspase activation. Moreover, as caspases cleave at aspartate residues, and procaspases are themselves activated by cleavage at aspartate residues, caspases participate in a cascade of activation whereby one caspase can activate another caspase in a chain reaction (Figure 7.2). This mechanism, whereby caspases activate procaspases, leads to amplification of an apoptotic signal—only a few initially activated caspase molecules can produce the rapid and complete conversion of a pool of procaspases. Let us now examine both the extrinsic and intrinsic apoptotic pathways.

The extrinsic pathway: mediated by membrane death receptors

The extrinsic pathway for triggering cell death (Figure 7.3) shares some common features with pathways involved in triggering cell growth (see Chapter 4). A death factor, such as Fas ligand (shown as a blue diamond) or tumor necrosis factor (TNF) (shown as a blue circle), is received by a transmembrane death receptor such as Fas receptor or TNF receptor, respectively. TNF is a soluble factor, while Fas ligand is bound to the plasma membrane of neighboring cells. When ligands bind to the death receptors, the receptors undergo a conformational change and form homotrimers in order to transduce the signal into the cell. The

Figure 7.3 The extrinsic pathway of apoptosis. Death signals, TNF and Fas, activate their death receptors TNF receptor and Fas receptor, respectively. Binding causes a change in shape and trimerization of the receptors. Adaptor proteins recognize the activated receptors and lead to the aggregation of procaspase-8. Procaspase aggregation leads to caspase-8 activation (shown in red). This initiator caspase initiates a caspase cascade, proteolysis, and apoptosis. The protein c-Flip (shaded yellow) can inhibit the interaction of procaspase-8 with the adaptors.

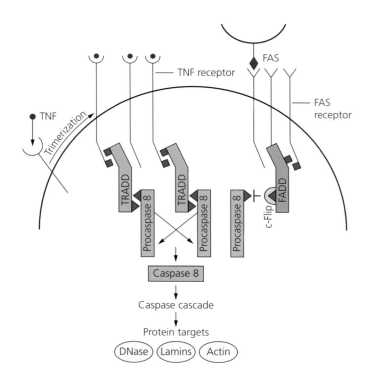

conformational change exposes so-called death domains (blue squares) that are located on the cytoplasmic tail of the receptors and enable intracellular adaptor proteins, such as FADD (Fas-associated death domain protein) and TRADD (TNF receptor-associated death domain protein), to bind via their death domains (see "Pause and Think").

The function of adaptor proteins is to transduce the death signal from the receptor to caspases. The adaptors recruit several molecules of procaspase-8 via death effector domains (DEDs; blue triangles). Molecules of procaspase-8, now in close proximity to each other, become activated by self-cleavage, as procaspases have low enzymatic activity. Caspase-8 is known as an initiator caspase, as it is the first link between the receptor and the apoptotic proteases, and it is key to the extrinsic pathway. Together, the death ligands, receptors, adaptors, and initiator caspase are called the **d**eath-**i**nducing **s**ignaling **c**omplex (DISC). Caspase-8 initiates a cascade of caspase activation: one activated caspase cleaves and activates other caspases, called executioner caspases (caspase-3, -6, and -7). The cascade ultimately causes the cleavage of specific protein targets and results in apoptosis.

This process can be inhibited by c-Flip (shown in yellow in Figure 7.3), a catalytically inactive caspase-8 homolog. c-Flip can bind to adaptor FADD or caspase-8 via a DED and inhibit caspase-8 recruitment and activation.

PAUSE AND THINK

What do you think is the function of a death domain? A death domain is part of a protein, approximately 70 amino acids, that allows for specific protein–protein interactions to occur and is analogous to the SH2 domain characteristic of growth factor signal transduction pathways.

(a) Mouse B lymphoma cells

Control | Heat-shocked

(b)

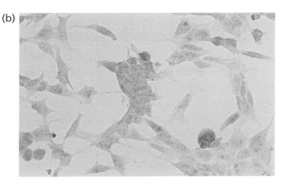

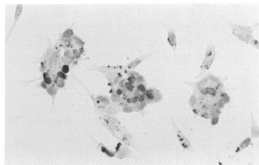

Figure 7.4 Methods for the detection of apoptotic cells. (a) A typical DNA ladder, characteristic of cells undergoing apoptosis. Detection by agarose gel electrophoresis and visualized by ethidium bromide staining under UV light. Reprinted from Otsuki, Y., Li, Z., and Shibata, M.A. (2003) Apoptotic detection methods—from morphology to gene. *Progress in Histochemistry and Cytochemistry* **38**: 275–339, Figure 13a. Copyright (2003), with permission from Elsevier. (b) TUNEL staining. The induction of apoptosis in a human neuroblastoma cell line was analyzed by TUNEL staining (described in Box "Analysis of apoptosis by the TUNEL technique"). Control (top) and induced (bottom) cells. Apoptotic cells (red) are detected using an alkaline phosphatase-conjugated anti-fluorescein antibody. Reprinted from Lui, X.-H., Yu, E.Z., Li, Y.-Y., Rollwagen, F.M., and Kagan, E. (2006) RNA interference targeting Akt promotes apoptosis in hypoxia-exposed human neuroblastoma cells. *Brain Research* **1070**: 24–30, Figure 1. Copyright (2006) with permission from Elsevier.

The breakdown of the cell results from the proteolysis of the target proteins. Target proteins include nuclear lamins allowing for nuclear shrinkage, cytoskeletal proteins such as actin and intermediate filaments for rearranging cell structure, specific kinases for cell signaling, and other enzymes such as caspase-activated DNase for the cleavage of chromatin. The caspase-activated DNase cuts DNA between nucleosomes and generates a DNA ladder (corresponding to multiples of 180 bp—the distance between nucleosomes; Figure 7.4a) that can be detected experimentally and used by scientists as a molecular marker of apoptosis. The TUNEL technique, described in Box "Analysis of apoptosis by the TUNEL technique," is another procedure used by scientists to detect apoptosis. Caspases also cleave the tumor suppressor protein RB (discussed in Chapters 5 and 6), and this cleavage results in the degradation of RB protein. This event is required for apoptosis induced by TNF and points to a role for RB in the inhibition of apoptosis.

Self-test Close this book, and try to redraw Figure 7.3. Check your answer. Correct your work. Close the book once more, and try again.

The intrinsic pathway: mediated by the mitochondria

The intrinsic pathway of apoptosis (Figure 7.5) does not depend on exter-nal stimuli (e.g. death factors). Stimuli from inside the cell, such as DNA damage and oxidative stress, induce the intrinsic pathway of apoptosis through the Bcl-2 family of proteins that act at the outer mitochondrial membrane. There are two groups within the Bcl-2 family that have oppos-ing functions: one group of Bcl-2 proteins inhibits apoptosis, and another group promotes apoptosis (Table 7.1). It is the permeabilization of the outer mitochondrial membrane and the release of stored pro-apoptotic factors that are critical in triggering the intrinsic pathway of apoptosis.

The Bcl-2 family consists of approximately 25 members, all of which contain at least one Bcl-2 homology (BH) domain that mediates pro-tein–protein interactions. Most family members share three or four BH domains. Within the group of pro-apoptotic molecules is a subset referred to as the BH3-only proteins (e.g. Bim) because they only share one BH domain, BH3. The BH3-only proteins function by either inducing the activity of the pro-apoptotic molecules (BH3-only activators) or bind-ing and inhibiting the anti-apoptotic Bcl-2 proteins (BH3-only sensitiz-ers). The pro-apoptotic members of the Bcl-2 family (e.g. Bax, Bak) act by forming pores in the outer mitochondrial membrane. The anti-apo-ptotic members of the Bcl-2 family (e.g. Bcl-2, Mcl-1) bind and seques-ter pro-apoptotic proteins to inhibit apoptosis. It is the balance of these two activities that regulates the release of important molecular apoptotic mediators from the mitochondria. The members of this family can associ-ate by protein–protein interactions, and it appears that it is the ratio of activity that determines function and hence the outcome. For example, if the activity of the pro-apoptotic factors is high owing to low inhibition

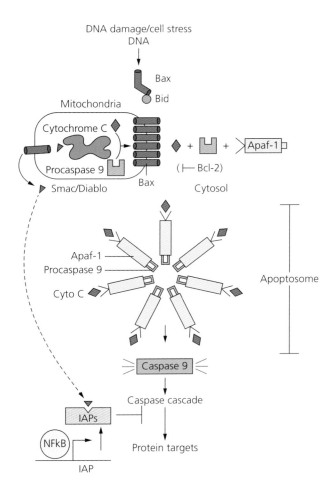

Figure 7.5 The intrinsic pathway of apoptosis, based on current knowledge. Cell stress triggers the BH3-only protein Bid to transiently bind to, and activate, Bax. Bax undergoes a conformational change, inserts into the outer mitochondrial membrane, and oligomerizes (6–8 molecules). Important regulators are released from the intermembrane space (shown in red and pink). Cytochrome *c* joins to the adaptor protein, apoptotic protease-activating factor-1 (Apaf-1), and recruits procaspase-9 to form the apoptosome. Caspase aggregation leads to the activation of procaspase-9 and finally a caspase cascade. Smac/DIABLO (red triangle), also released from the mitochondria, inhibits IAPs that normally act to block caspases.

Table 7.1 Members of the Bcl-2 family

Anti-apoptotic members	Pro-apoptotic members	Pro-apoptotic members— BH3-only members
Bcl-2	Bax	Bad
Bcl-x_L	Bok/Mtd	Bik/Nbk/Blk
Bcl-w	Bcl-x_s	Bid
A1	Bak	Hrk/DP5
Mcl-1	Bcl-G_L	Bim/Bod
Boo		Bmf
		Noxa
		Puma/Bbc3, BNIP3, BNIP3L

from anti-apoptotic factors, apoptosis is triggered. The activity of the proteins of the Bcl-2 family can also be regulated by phosphorylation.

The intermembrane space between the two mitochondrial membranes acts as a supply cabinet for apoptotic mediators. The pro-apoptotic Bcl-2 members regulate the release of the apoptotic mediators from this mitochondrial compartment in a process sometimes referred to as **m**itochondrial **o**uter **m**embrane **p**ermeabilization (MOMP). Upon activation by an apoptotic signal, BH3-only proteins, Bid and Bim, bind to, and activate, Bax. This interaction is transient ("kiss and run") and induces a conformational change in Bax, as it translocates from the cytoplasm to the mitochondria. Bax monomers insert into the outer mitochondrial membrane and first dimerize and later oligomerize further in the membrane (Subburaj *et al.*, 2015) (six are shown in Figure 7.5). This new conformation within the mitochondrial membrane increases the permeability of the outer mitochondrial membrane by forming membrane pores and allows the release of apoptotic mediators. The BH3 domain of Bax is required for its apoptotic activity and interactions with anti-apoptotic proteins. Anti-apoptotic molecules, such as Bcl-x$_L$, cause the dissociation of Bax oligomers.

As shown in Figure 7.5, cytochrome *c*, which also functions in the electron transport chain of aerobic respiration, and procaspase-9 (shaded in red and pink) are released into the cytoplasm and assemble into a complex called an apoptosome, along with dATP bound to Apaf-1. The binding of cytochrome *c* to cytosolic Apaf-1 triggers the formation of a wheel-like heptameric complex that facilitates the recruitment of procaspase-9 via protein domains, called CARD domains, present on both Apaf-1 and procaspase-9. Structural studies suggest that the CARD domains, and therefore procaspase-9 molecules, reside in a central ring. Further, cytochrome *c* binds to Apaf-1 within clefts formed by a pair of β propellers at the end of the spoke-like helical domain of Apaf-1 with a 1:1 stoichiometry (see "How do we know that?" below).

Apaf-1 is a protein co-factor that is required for activation of procaspase-9. Caspase-9 is an initiator caspase activated by procaspase aggregation that begins another caspase cascade activating downstream caspase-3, -6, and -7. Thus, caspase-9 is key to the intrinsic pathway.

 HOW DO WE KNOW THAT?

Electron microscopy

(See Yu *et al.*, 2005.)

Yu *et al.* (2005) used electron cryomicroscopy and single-particle methods to elucidate the structure of the human apoptosome at 12.8 Å resolution. Based on the scientific evidence available at the time, the first edition of this book included a diagram in which the structure of the apoptosome was shown inverted; procaspase-9 was illustrated on the outside of the wheel-like heptameric structure, and cytochrome *c* was drawn at the center. This was a model put forth based on stoichiometry. Structural data were not available at the time. The update presented in Figure 7.5 illustrates the progress made in understanding of the apoptosome.

Other factors, such as inhibitors of apoptosis proteins (IAPs; eight mammalian IAPs have been identified) and Smac (**s**econd **m**itochondria-derived **a**ctivator)/DIABLO (shown as a gray box and red triangles, respectively, Figure 7.5), play a role in modulating the process. The X chromosome-linked member XIAP is one member of the IAP family that directly binds to, and inhibits, the activity of caspase-3 and caspase-7, after they have been processed, by binding to their active site. XIAP also inhibits caspase-9 but does so by binding to monomeric caspase-9 and locking the active site in an aberrant conformation. A transcription factor called NF-κB, a major player in inflammation (see Chapter 13), is a potent inhibitor of apoptosis. It induces the transcription of IAPs.

Smac/DIABLO, another regulator released from the mitochondria, eliminates inhibition by IAPs. Smac/DIABLO competes with activated caspase-9 for binding to XIAP. Both caspase-9 and Smac contain a similar tetrapeptide domain that binds to XIAP. Thus, opposing effects on caspase activity are regulated by conserved IAP-binding motifs in caspase-9 and Smac.

Self-test Study Table 7.1. It is crucial to remember that Bcl-2 blocks apoptosis, and this fact will help you to remember the role of the other Bcl-2 family members. Close this book, and try to redraw Figure 7.5. Check your answer. Correct your work. Close the book once more, and try again.

Cross-talk between extrinsic and intrinsic pathways

Note that there is cross-talk between the extrinsic and intrinsic pathways, and the two converge at the activation of downstream caspases. For example, caspase-8, a key regulator of the extrinsic pathway, can proteolytically cleave and activate Bid, a pro-apoptotic Bcl-2 family member (Figure 7.6). Bid can then stimulate the intrinsic pathway of apoptosis by

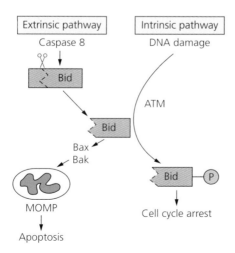

Figure 7.6 Bid links the intrinsic and extrinsic pathways of apoptosis.

directly activating Bax and Bak, facilitating the release of cytochrome *c* from the mitochondria and inducing the subsequent activation of downstream caspases. In addition, Bid links the intrinsic pathway with the regulation of cell cycle progression in response to DNA damage. Recall from Chapter 2 that ATM kinase is activated upon DNA damage. Some data demonstrate that phosphorylation of Bid by ATM kinase is required for cell cycle arrest in response to DNA damage (Figure 7.6; Kamer *et al.*, 2005; Zinkel *et al.*, 2005).

A PIONEER IN THE FIELD ... of apoptosis: in memory of Stanley Korsmeyer

Korsmeyer did pioneering work on the regulation of apoptosis, including the identification of Bax, one of the first pro-apoptotic members of the Bcl-2 family reported, and structural analysis of the Bcl-2 homology domains and how they allow for family member heterodimerization. He and his research group have helped to elucidate the molecular mechanisms of the Bcl-2 family and their role in cancer. The basic research carried out by Korsmeyer and his research group has created a foundation for the development of future cancer therapeutics.

---

p53 and apoptosis

As we saw in Chapter 6, the tumor suppressor protein p53 accomplishes its role as the guardian of the genome, in part, by inducing apoptosis in response to DNA damage and cellular stress. It does this by both transcription-dependent and transcription-independent means. As a transcription factor, p53 induces the expression of genes that code for death receptors and pro-apoptotic members of the Bcl-2 family (Table 6.2). Examples include *Fas* receptor, *Bax*, and *Bak*. These genes contain a consensus p53-binding site in their promoter regions. p53 can also repress the expression of anti-apoptotic factors such as Bcl-2 and Bcl-x and IAPs. Recent evidence demonstrates that a member of the Bcl-2 family called PUMA (**p**53 **u**pregulated **m**odulator of **a**poptosis), a target of p53, is essential for apoptosis induced by p53. Apoptosis induced by DNA-damaging drugs, irradiation, oncogenic activation, and cell stress was blocked in *PUMA* gene knock-out mice. p53 can exert transcription-independent regulation of apoptosis. This has been demonstrated by the induction of apoptosis with p53 mutants incapable of regulating transcription; these p53 mutants included one that lacked the DNA-binding domain disabling interaction with p53 target genes, one without a nuclear localization signal preventing p53 from reaching the target genes in the nucleus, and one mutated in the p53 transactivation domain preventing the transcriptional activation function. It was also demonstrated that UV-induced apoptosis could be triggered by wild-type p53 strictly

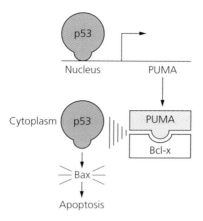

Figure 7.7 Transcription-dependent and transcription-independent functions of p53 linked to PUMA.

from the cytoplasm in cells that were treated with wheatgerm agglutinin, a nuclear import inhibitor. The mechanism of p53 transcription-independent apoptosis involves p53 activation of Bax in the cytoplasm and subsequent release of cytochrome *c* and caspase activation (Chipuk *et al.*, 2004). Evidence also supports the role of p53 in releasing pro-apoptotic proteins (e.g. Bid) from sequestration by anti-apoptotic proteins (e.g. Bcl-x_L), altering the net functional balance of the Bcl-2 family of proteins. The protein PUMA has provided a link between the transcriptional and cytoplasmic functions of p53 (Chipuk *et al.*, 2005). Data suggest that p53 activates transcription of *PUMA*; PUMA protein then acts as an enabler to release p53 from Bcl-x_L in the cytoplasm, so that p53 can directly activate Bax (Figure 7.7). In summary, p53 functions in both the nucleus and the cytoplasm by transcription-dependent and transcription-independent means, and these functions are linked by PUMA.

 HOW DO WE KNOW THAT?

Immunoprecipitation

(See Chipuk *et al.*, 2005.)

UV treatment of cells results in DNA damage and induction of apoptosis. One question investigated by Chipuk *et al.* (2005) was whether there are any regulators of the Bcl-x_L–p53 complex. They analyzed the extracts of cells that were treated with UV by a technique called immunoprecipitation. Here, an antibody against Bcl-x_L was used to isolate proteins complexed to Bcl-x_L, and these proteins were analyzed by SDS–polyacrylamide gel electrophoresis and visualized by silver staining and also by western blot. Both p53 and PUMA were identified

(look carefully at the data in Figures 1 and S1A in Chipuk *et al.*, 2005).

Another experiment was designed to investigate the kinetics of the formation of the p53–Bcl-x_L complex after treatment with UV. Bcl-x_L was immunoprecipitated at specific time points after UV treatment and analyzed. The data show that the amount of p53 complexed with Bcl-x_L decreased over time after UV treatment, and this correlated with the induction of apoptosis (see Figure S2C in Chipuk *et al.*, 2005). What do the asterisks above the fifth band in the row labeled p53 represent? Read the methods, and see how they determined the time of induction of apoptosis.

7.2 Apoptosis and cancer

Evasion of apoptosis is one of the hallmarks of cancer (Figure 1.1). Tumor cells produce many signals, such as those in response to DNA damage and oncogene activation, that normally induce apoptosis. Through tumor suppression pathways, most cells that acquire carcinogenic characteristics are eliminated by apoptosis. However, tumor cells that acquire mutations that allow them to escape from the apoptotic response survive and proliferate. The avoidance of apoptosis permits further accumulation of mutations. This draws our attention to a difference that develops between tumor cells and normal cells—as tumor cells receive many apoptosis-inducing signals (such as oxidative stress and oncogene activation), compared with normal cells, tumor cells are "closer" to triggering an apoptotic response. However, the apoptotic pathway is often defective in tumor cells. This has stimulated the concept that repair of the apoptotic pathway in tumor cells by targeting p53 may be a valuable strategy for new cancer therapeutics (see Chapter 6).

Evidence suggests that there is a fundamental difference in the state of caspase activation between cancer cells and normal cells because of the stresses characteristic of cancer cells mentioned earlier—cancer cells contain activated caspases that are inhibited by upregulated IAPs, but normal cells contain inactive procaspases that require proteolytic cleavage. Therefore, cancer cells are "closer" to triggering an apoptotic response, compared with a normal cell. In support of this, processed caspase-3 has been identified in tumor cells. Tumor cells share this characteristic with the normal *Drosophila* apoptotic process whereby activated caspases are inhibited by IAPs in normal cells and induction of apoptosis requires release from IAP inhibition. In summary, apoptotic signals stimulate procaspase processing in normal cells, while apoptotic signals stimulate the cessation of IAP inhibition of processed caspases in cancer cells.

TRAIL receptors

A subfamily of TNF receptors, called TRAIL receptors (TRAILR1 and TRAILR2, also known as death receptor 4 and 5—DR4, DR5), has been found to elicit a differential sensitivity to apoptosis between normal cells and cancer cells. Its ligand TRAIL (TNF-related apoptosis-inducing ligand) induces apoptosis in many cancer cells (regardless of the *p53* gene profile), but not in most normal cells. This subfamily signals apoptosis in a similar manner to TNF receptors recruiting adaptors (e.g. FADD) and an initiator caspase, caspase-8, to the membrane (Figure 7.3). As TRAIL and its receptors are expressed in most organs, it is hypothesized that the addition or loss of regulatory molecules determines whether apoptosis will be induced in particular cell types.

The differences identified between normal and tumor cells create an opportunity for designing drugs that target the apoptotic pathways and suggest that attempts to restore apoptotic activity will not affect normal cells. Such drugs are likely to have few side effects. Before discussing apoptotic therapies, we will first examine a sample of mutations in the apoptosis pathways that have been identified in cancers.

Mutations that affect the extrinsic pathway

Sunburn represents skin cells that are undergoing UV-induced apoptosis and is an important tumor-suppressive response to get rid of cells with damaged DNA. Both the intrinsic and extrinsic pathways are activated during UV exposure. UV causes the clustering of Fas death receptors and activation of the caspase cascade (Raj *et al.*, 2006). Mutations in the Fas pathway may reduce the apoptotic response and increase the risk of skin cancer. Somatic mutations in Fas receptors have been reported in melanomas and SCCs.

Suppression of caspase gene expression in the development of particular cancers, such as small-cell lung carcinoma and neuroblastomas, has also been demonstrated. Remember that caspase-8 is the first caspase activated by the death receptors (Figure 7.3), and it maintains a top position in the initiation of the caspase cascade. Loss of caspase-8 expression observed in cancers is a result of both epigenetic and mutational alterations; hypermethylation of the caspase-8 promoter, deletions, and missense mutations have been identified. Caspase-8 deficiency is particularly characteristic of neuroblastomas and small-cell lung cancer. Methylation of the caspase gene promoter was found to be the mechanism of loss of expression in a majority of cell lines examined. Also, a deletion of Leu62 was identified in human vulval SCCs, and this mutation blocks the interaction of caspase-8 with the adaptor FADD, thus abolishing its link with the death signal and receptor.

Mutations that affect regulators of the intrinsic pathway

Alterations of the intrinsic pathway of apoptosis are much more common than alterations in the extrinsic pathway during carcinogenesis. Mutations that bypass the transduction of apoptotic signals that are triggered by DNA damage, a dominant characteristic of cancer cells, are favored by natural selection. A major contribution to intrinsic pathway alterations occurs through mutations that affect the p53 pathway. Mutations in the *p53* gene itself are the most frequently found mutations in cancer cells, and current estimations may be underestimated as many earlier studies restricted their analysis to exons 5–9 only, rather than the full length of the gene. The *p53* mutations provide the cancer cells with a survival advantage by disrupting apoptosis. Abnormal methylation and loss of heterozygosity for *p73*, a *p53* family member capable of inducing apoptosis, are frequent in lymphomas. Also, the *p73* gene undergoes alternative splicing to generate several RNAs, including amino-terminally deleted variants which can act as inhibitors of *p53*. These variants are overexpressed in many cancers. In addition, mutations in genes involved in the upstream regulation of *p53* (e.g. *ATM* and *Chk2*; see Chapter 6 and Figure 6.6) and also in the downstream targets of p53 have also been identified in human tumors. Mutations in molecular components that affect MDM2, the major regulator of p53 activity, are common in tumors that maintain wild-type p53 alleles.

Bcl-2, the first member of the Bcl-2 family of proteins to be discovered, was initially identified from a chromosomal translocation, t(14;18) in B-cell lymphomas, hence the name *bcl*. In t(14;18), the *Bcl-2* gene is translocated to a position juxtaposed to the immunoglobulin heavy chain enhancer. As a consequence of its relocation next to a strong promoter, oncogenic activation of the *Bcl-2* gene occurs. Overexpression of the anti-apoptotic protein Bcl-2 leads to insufficient apoptotic turnover and accumulation of B-cells. This translocation is found not only in most cases of follicular B-cell lymphomas, but also in other types of cancer such as gastric, lung, and prostate. Overexpression of the *Bcl-2* gene due to a reduction in miR-15a and miR-16-1 is common in some chronic leukemias. This demonstrates that post-transcriptional control by miRNAs plays a role in the deregulation of apoptosis. In this case, miR-15a and miR-16-1 are tumor suppressors.

Aberrant expression of most of the genes in the Bcl-2 family is linked to carcinogenesis. Anti-apoptotic members of the Bcl-2 family may function as oncogenes, and pro-apoptotic members may act as tumor suppressor genes. Mutations in genes that code for pro-apoptotic proteins, such as deletions in the *bak* and *bid* genes, are characteristic of some tumors. *Bax* is mutated in over 50% of a specific class of colon tumors. Remember that p53 regulates many genes of the Bcl-2 family. Thus, many mutations in *p53* that are common in tumors also affect the transcriptional regulation of its target genes, including those of the Bcl-2 family.

Molecules involved in events downstream of the release of mitochondrial apoptotic factors also play a role in tumorigenesis. The gene encoding Apaf-1, the co-activator of caspase-9 upon its release from the mitochondria, is mutated and transcriptionally repressed in metastatic melanoma. Note that epigenetic inactivation, in addition to mutation, plays a role in the inactivation of the apoptotic pathway.

The induction of inhibitors of apoptosis also plays a role in carcinogenesis. XIAP is induced in many types of cancer, including leukemias, lung cancer, and prostate cancer. As XIAP acts to suppress caspase-9, -3, and -7, it affects downstream caspases that are common to both the extrinsic and intrinsic pathways.

Alternative death pathways

The observation that many apoptotic stimuli do not require caspases has led to studies of alternative death pathways. In addition to necrosis, autophagy and mitotic catastrophe are other non-apoptotic mechanisms of cell death. **Autophagy** (meaning "eating oneself") acts as a recycling system for the cell whereby proteins and components of damaged organelles that require degradation are targeted to the lysosomes. Cells can recycle the resulting products of degradation. This process is important under starvation conditions and for ridding the cell of defective organelles. Targeted proteins and organelles are surrounded by a double-membraned structure called an autophagosome. The contents of the autophagosome are degraded upon fusion of these vesicles with lysosomes. Excessive autophagy triggers non-apoptotic cell death. Mitotic catastrophe, another type of cell death, is caused by aberrant mitosis. Defects in genes (e.g. *BECN1*) required for these processes can contribute to tumorigenesis (see "Activity" on p. 175).

The details of the molecular events involved in caspase-independent cell death are not known fully. However, it is known that they also utilize proteases and facilitate permeabilization of the mitochondrial outer membrane. Alternative proteases, such as calpains, cathepsins, and serine proteases, cleave target proteins to bring about morphological changes characteristic of programmed cell death. Calpains, like caspases, are found as inactive zymogens in the cytoplasm. Cathepsins become activated in lysosomes before being translocated into the cytoplasm and/or nucleus. Apoptosis-inducing factor (AIF) is one molecular player released from the mitochondrial intermembrane space that induces caspase-independent DNA degradation.

The abnormal expression of molecules involved in alternative death pathways is observed in tumor cells. For example, mutations in genes that encode the tumor suppressor proteins, Bin1 and promyelocytic leukemia (PML) protein, which induce alternative death pathways are found in human cancers. Bin activates a caspase-independent pathway that is blocked by a serine protease inhibitor. As more is learned about the molecular players of alternative death pathways, new potential drug targets will be uncovered. In fact, several drugs in clinical trials (e.g. EB1089/seocalcitol, a vitamin D analog) induce calpain-dependent and caspase-independent cell death.

7.3 Apoptosis and chemotherapy

Disruption of the apoptotic pathway has important effects on the clinical outcome of chemotherapy. In order for chemotherapy to be successful, cells must be capable of undergoing apoptosis. Chemotherapeutic agents act primarily by inducing DNA damage. This damage consequently triggers the intrinsic apoptotic pathway. Some types of chemotherapy induce particular cells of the immune system to produce TNF, thus triggering the extrinsic pathway. Drugs with varying structures indirectly elicit the same morphological changes typical of apoptosis. However, remember that one of the hallmarks of cancer cells is that they evade apoptosis. Many tumors have defective apoptotic pathways and are inherently resistant to chemotherapies, regardless of whether or not they have been previously exposed to the drugs. This type of resistance contrasts the classical acquired mechanisms that are associated with drug accumulation and drug stability such as the use of the P-glycoprotein pump (see Chapter 2). As resistance to chemotherapy is a major clinical problem, elucidating the role of apoptosis in drug responses is important for future therapeutic strategies.

Drug resistance can arise through mutations in genes that code for molecular regulators of apoptosis. These mutations serve to uncouple drug-induced damage from the activation of apoptosis. As mentioned earlier, mutations in the p53 pathway are common in cancer cells and greatly contribute to the inherent drug resistance observed for many cancers. Cells engineered to have a *p53* knock-out are resistant to drug-induced apoptosis. Yet, even some "gain-of-function" *p53* mutations may confer resistance to specific chemotherapies. One p53 mutant induces the expression of the *dUTPase* gene and results in resistance to 5-FU. Therefore, on one hand, both "loss-of-function" and some "gain-of-function" mutations can give rise to resistance. On the other hand, one mutant p53 sensitizes some cells to taxanes.

The upregulation of the anti-apoptotic members of the Bcl-2 family and the downregulation of the pro-apoptotic members of the Bcl-2 family in tumors are associated with an increased resistance to chemotherapies. For example, loss of Bax, a pro-apoptotic protein, increases drug resistance in human colorectal cancer cells to the antimetabolite 5-FU and non-steroidal anti-inflammatory drugs (NSAIDs) used as chemopreventative agents (Zhang *et al.*, 2000). Overexpression of Bcl-2 in metastatic tumors may contribute to the fact that they are notoriously chemoresistant. Since Bcl-2 is overexpressed in about 75% of breast cancer, the use of Bcl-2-targeting drugs (such as BH3 mimetics, see "Regulation of the Bcl-2 family of proteins" on p. 172) is being investigated to improve clinical responses to chemotherapy (Merino *et al.*, 2015).

Overall, these observations point to an important clinical implication—the genotype of a tumor, especially with respect to the *p53* and *Bcl-2* gene families, is an important factor that influences the effectiveness of therapy.

There is another important implication of treating cells that have non-functional apoptotic pathways with chemotherapy. The lack of an apoptotic effect in response to extensive DNA damage caused by these genotoxic drugs provides an opportunity for the accumulation of mutations. Consequently, the risk of carcinogenesis increases. Indeed, therapy-related leukemia, whereby a new cancer arises after the administration of chemotherapy, is a clinical problem. Therapy-related leukemias have relatively short latency times. Specific cytogenetic aberrations are associated with different chemotherapeutic agents; chromosomal deletions of chromosome 5 and/or chromosome 7 are characteristic of alkylating drugs.

 Therapeutic strategies

7.4 Apoptotic drugs

The ability to trigger apoptosis in tumor cells is an important strategic design for cancer therapeutics. This is supported by the fact that many successful conventional chemotherapies work by triggering apoptosis, albeit indirectly. Using the knowledge of the molecular players in the apoptotic pathways enables us to design *direct* apoptotic inducers or, alternatively, inhibitors of anti-apoptotic proteins. These approaches crucially bypass the need for a drug to be mutagenic and avoid therapy-related leukemias. In addition, the induction of apoptotic factors in normal cells should have little effect on these cells, as they are not poised to trigger apoptosis to the same degree as tumor cells (see "Pause and think").

Next is a description of strategies targeted against the TRAIL and its receptor, the Bcl-2 family, and caspases. Note that many potential drugs developed using these strategies have been tested in the clinic, but none have yet been approved.

Targeting TRAIL and its receptor

The differential activity of TRAIL and its receptor in normal cells versus cancer cells suggests that they are good molecular targets for apoptotic therapies. Approximately 80% of cancer cell lines are sensitive to TRAIL ligand and can be induced to undergo apoptosis. In addition, apoptosis induced via a death receptor is thought to be independent of p53, and so the many cancers with inactivated p53 mutations may still be vulnerable to such an approach.

The administration of recombinant human TRAIL ligand (dulanermin) has exhibited promising anti-tumor activity in animal models, and Phase I trials have been initiated for the treatment of solid tumors (de Bruyn *et al.*, 2013). Caution is being exercised because of the hepatic toxicity known to result from related Fas and TNF-α ligand administration and also because of one study that reported a TRAIL-induced apoptotic response in human hepatocytes in culture, even though toxicity was not observed in mice and non-human primates. The variability of the effects

PAUSE AND THINK

What molecules would you target if asked to design an apoptotic drug? One method not discussed in this section is to target the apoptosome. Think of the molecules involved, and ask yourself how you can induce apoptosome formation. One way is to create cytochrome c mimetics that can trigger a change in Apaf-1 for apoptosome formation and caspase-9 activation. Another way is to find compounds that trigger the release of cytochrome c and procaspase 9 from the mitochondrial intermembrane space.

among different species reminds us that animal studies have limitations for predicting the toxicity of cancer therapies in humans. Also, the critical eye will notice upon reading of the literature that different structural forms of recombinant TRAIL were used in different studies, because of variation in the preparation of the recombinant protein, and thus may be a source of the different toxicities. TRAIL without extraneous amino acid residues showed no toxicity in non-human primates (see references within Fesik, 2005). Although partial responses were seen in two sarcoma patients, effects were disappointing as part of combination therapy in Phase II studies.

Using TRAIL receptor agonistic monoclonal antibodies (e.g. mapatumumab) that recognize the extracellular domain of the receptor is another strategy that has been tested in clinical trials. Although development and testing are ongoing, preliminary results have been disappointing (see references within Bates and Lewis, 2013; Bremer *et al.*, 2006).

Regulation of the Bcl-2 family of proteins

The Bcl-2 family is another target for the design of apoptotic drugs (Kang and Reynolds, 2009). Three main strategies have been used, and these are illustrated in Figure 7.8: (a) antisense RNA, (b) small molecules to inhibit protein function and protein–protein interactions of anti-apoptotic molecules, and (c) drugs that induce the activity of pro-apoptotic molecules. As Bcl-2 is overexpressed in a broad range of tumors, inhibition of its expression by antisense RNA is one strategy that has been employed to create a new cancer drug. G-3139 (oblimersen sodium, Genasense™) is an 18-mer modified antisense oligonucleotide that is complementary to the first six codons downstream of the translational start site on the Bcl-2 mRNA.

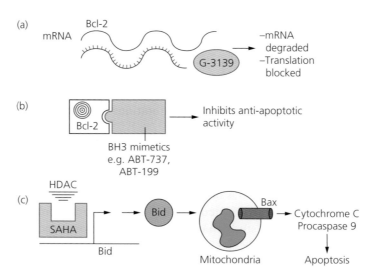

Figure 7.8 Drug strategies that target the Bcl-2 family of proteins. Therapeutic agents are shown in red.

Upon hybridization, translation is inhibited, and the mRNA is degraded. This alters the balance of pro- and anti-apoptotic factors in favor of apoptosis. G-3139 was tested alone and in combination with chemotherapeutic drugs in Phase III clinical trials but failed to show an effect. It is **not** approved by the FDA.

Structural studies of Bcl-2 have revealed that a hydrophobic groove forms the binding site for the BH3 region of pro-apoptotic Bcl-2 family members. Small-molecule inhibitors that bind to Bcl-2/Bcl-x_L and interfere with protein–protein interactions with pro-apoptotic molecules, in order to induce apoptosis, have been identified (e.g. ABT-737, Abbott Laboratories). These small-molecule inhibitors are also called BH3 mimetics, as they mimic the action of BH3 proteins binding to the BH3 hydrophobic groove of anti-apoptotic proteins. This displaces pro-apoptotic molecules and induces apoptosis. ABT-737 and an oral version navitoclax (ABT-263) target Bcl-2, Bcl-x_L, and Bcl-w. The drug was developed by structure-based design, using nuclear magnetic resonance-based screening, and binds to the anti-apoptotic proteins with high affinity. Several other small molecules, including AT-101 (Ascenta) and GX15-070 (Gemin X), also bind anti-apoptotic Bcl-2 molecules and are in Phase I or II clinical trials. One of the newest BH3 mimetics, ABT-199, binds more selectively to Bcl-2 and has shown promising anti-tumor effects in follicular lymphoma.

Suberoylanilide hydroxamic acid (SAHA) works by an opposing, non-specific mechanism; instead of inhibiting an anti-apoptotic factor, it promotes the activity of a pro-apoptotic factor. SAHA is a HDAC inhibitor and therefore acts to induce the expression of epigenetically suppressed genes. SAHA induces the expression of the pro-apoptotic proteins, such as Bid, which activates Bax. Bax translocates into the mitochondria and results in the release of apoptotic factors such as cytochrome *c*. SAHA (vorinostat) has received approval for treatment of non-Hodgkin's lymphoma.

Direct and indirect activation of caspases

Selective activation of caspases is the most obvious apoptotic target. However, because they comprise a large family of over 13 members and are in every cell type, selectivity has not been easy. Yet, screening for small-molecule caspase activators is ongoing. Procaspase-3, the pro-enzyme of a key effector caspase, is inhibited by an intramolecular interaction facilitated by three aspartate residues, called the "safety-catch." Screens are being pursued for small molecules that are able to interfere with this intramolecular inhibitory conformation of procaspase-3. More recently, a caspase-8-selective small molecule has been identified through an *in silico* chemical screen (Bucur *et al.*, 2015). This molecule directly binds to the dimerization interface of caspase-8 and triggers apoptosis when combined with TRAIL. However, the activation of caspases by *indirect* methods is promising. Targeting XIAP represents the last step against cell

> **PAUSE AND THINK**
>
> Do you think that the *pro-apoptotic* Bcl-2 proteins, such as Bim and Bid, can influence the sensitivity of cells to BH3 mimetics? Why? See Renault *et al.* (2014) for preliminary data that investigate this hypothesis.

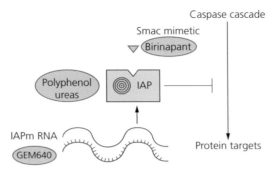

Figure 7.9 Drug strategies that target IAPs and indirectly activate caspases. Therapeutic agents are shown in red.

death by caspases. This may be done by using direct XIAP inhibitors or Smac mimetics that relieve XIAP inhibition (Figure 7.9). As endogenous caspase inhibitor XIAP is overexpressed in many cancers, it is a good molecular target for new cancer therapeutics. A synthetic chemical screen aimed at inhibiting XIAP activity identified a class of polyphenylureas that directly relieves the inhibition of caspase-3 and caspase-7, but not caspase-9 (Schimmer *et al.*, 2004). This is feasible, because one domain of XIAP directly blocks the active site of caspase-3 and caspase-7, while another distinct domain inhibits caspase-9. The small-molecule inhibitors bind to the XIAP domain known to block the active site of caspase-3 and caspase-7, the downstream caspases. Furthermore, these compounds induced apoptosis in a range of tumor cell lines and showed anti-tumor activity in animal tumor models. Little toxicity was observed for normal cells. This was the first demonstration that relief of caspase inhibition can induce tumor cell apoptosis. IAP antisense oligonucleotides, such as GEM640 (AEG35156), have been designed to reduce XIAP mRNA and protein but have not proved successful thus far. Smac mimetics, such as birinapant (TL32711), are another class of drugs that target IAPs and are being tested in pre-clinical models and in the clinic.

Time is needed to see whether apoptotic drugs can be successful cancer therapeutic targets.

■ CHAPTER HIGHLIGHTS—REFRESH YOUR MEMORY

- Apoptosis is an important tumor suppression mechanism.

- Caspases, aspartate proteases, are the main molecular players during apoptosis.

- Apoptosis can be triggered by extracellular death signals or internal stimuli that act via an extrinsic and intrinsic pathway, respectively.

- TNF/TNF receptor and Fas/Fas receptor signaling are paradigms of the extrinsic pathway.

- The mitochondria store apoptotic molecules involved in the intrinsic pathway.

- The Bcl-2 family regulates the permeability of the outer mitochondrial membrane.

- p53 induces apoptosis by both transcription-dependent and transcription-independent means.

- Evasion of apoptosis is a hallmark of cancer cells.

- A tumor cell is "closer" to eliciting an apoptotic response than a normal cell if the apoptotic pathway were to be functional.

- Caspase activity is regulated differently in normal cells and tumor cells: normal cells require procaspase processing, while tumor cells require release of processed caspases from IAPs.

- Alterations in the p53 and Bcl-2-related pathways play a major role during carcinogenesis.

- Chemotherapies act indirectly via DNA damage to induce apoptosis.

- Tumors with defective apoptotic pathways are resistant to chemotherapies.

- Mutations in apoptotic proteins enable cancer cells to both survive and become drug-resistant.

- Treatment with chemotherapies can cause therapy-related leukemia.

- Apoptotic drugs aim to trigger apoptosis directly and do not require genotoxic activity.

ACTIVITY

1. Think about a therapeutic strategy that targets Smac/DIABLO, and write a proposal for the development of such a drug (hint: begin with Martinez-Ruiz *et al.*, 2008).

2. Beclin-1 is involved in the induction of autophagy in response to starvation. Studies of the gene encoding Beclin, *BECN1*, have demonstrated that impaired autophagy is linked to tumorigenesis. Critically discuss the experimental evidence that supports this link, beginning with the following—White, E., 2015.

FURTHER READING

Constantinou, C., Papas, K.A., and Constantinou, A.I. (2009) Caspase-independent pathways of programmed cell death: the unraveling of new targets of cancer therapy? *Curr. Cancer Drug Targets* **9**: 717–728.

Danial, N.N. and Korsmeyer, S.J. (2004) Cell death: critical control points. *Cell* **116**: 205–219.

Delbridge, A.R.D. and Strasser, A. (2015) The BCL-2 protein family, BH3-mimetics and cancer therapy. *Cell Death Differ.* **22**: 1071–1080.

Elkholi, R., Renault, T.T., Serasinghe, M.N., and Chipuk, J.E. (2014) Putting the pieces together: how is the mitochondrial pathway of apoptosis regulated in cancer and chemotherapy? *Cancer Metab.* **2**: 16–31.

Er, E., Oliver, L., Cartron, P.-F., Juin, P., Manon, S., and Vallette, F.M. (2006) Mitochondria as the target of the pro-apoptotic protein Bax. *Biochim. Biophys. Acta* **1757**: 1301–1311.

Fulda, S. and Debatin, K.M. (2006) Extrinsic versus intrinsic apoptosis pathways in anticancer chemotherapy. *Oncogene* **25**: 4798–4811.

Indran, I.R., Tufo, G., Pervaiz, S., and Brenner, C. (2011) Recent advances in apoptosis, mitochondria and drug resistance in cancer cells. *Biochim. Biophys. Acta* **1807**: 735–745.

Johnstone, R.W., Ruefli, A.A., and Lowe, S.W. (2002) Apoptosis: a link between cancer genetics and chemotherapy. *Cell* **108**: 153–164.

Kirkin, V., Joos, S., and Zornig, M. (2004) The role of Bcl-2 family members in tumorigenesis. *Biochim. Biophys. Acta* **164**: 229–249.

Ledgerwood, E.C. and Morison, I.M. (2009) Targeting the apoptosome for cancer therapy. *Clin. Cancer Res.* **15**: 420–424.

Okada, H. and Mak, T.W. (2004) Pathways of apoptotic and non-apoptotic death in tumor cells. *Nat. Rev. Cancer* **4**: 592–603.

Ow, Y.P., Green, D.R., Hao, Z., and Mak, T.W. (2008) Cytochrome c: functions beyond respiration. *Nat. Rev. Mol. Cell Biol.* **9**: 532–542.

Reed, J.C. (2006) Proapoptotic multidomain Bcl-2/Bax-family proteins: mechanisms, physiological roles, and therapeutic opportunities. *Cell Death Differ.* **13**: 1378–1386.

Shiozaki, E.N. and Shi, Y. (2004) Caspases, IAPs and Smac/DIABLO: mechanisms from structural biology. *Trends Biochem. Sci.* **39**: 486–494.

Vousden, K.H. and Lu, X. (2002) Live or let die: the cell's response to p53. *Nat. Rev. Cancer* **2**: 594–604.

Walensky, L.D. and Gavathiotis, E. (2011) Bax unleashed: the biochemical transformation of an inactive cytosolic monomer into a toxic mitochondrial pore. *Trends Biochem. Sci.* **36**: 642–652.

Wong, R.S.Y. (2011) Apoptosis in cancer: from pathogenesis to treatment. *J. Exp. Clin. Cancer Res.* **30**: 87.

Zhivotovsky, B. and Orrenius, S. (2003) Defects in the apoptotic machinery of cancer cells: role in drug resistance. *Semin. Cancer Biol.* **13**: 125–134.

▓ WEB SITE

Apoptotic Pathways (Genentech)—animation of both the intrinsic and extrinsic apoptotic pathways. https://m.youtube.com/watch?v=SyvOPXeg4ig

▓ SELECTED SPECIAL TOPICS

Bates, D.J.P. and Lewis, L.D. (2013) Manipulating the apoptotic pathway: potential therapeutics for cancer patients. *Br. J. Clin. Pharmacol.* **76**: 381–395.

Bremer, E., van Dam, G., Kroesen, B.J., de Leij, L., and Helfrich, W. (2006) Targeted induction of apoptosis for cancer therapy: current progress and prospects. *Trends Mol. Med.* **12**: 382–393.

Bucur, O., Gaidos, G., Yatawara, A., Pennarun, B., Rupasinghe, C., Roux, J., *et al.* (2015) A novel caspase 8 selective small molecule potentiates TRAIL-induced cell death. *Sci. Rep.* **5**: 9893.

Chen, D.J. and Huerta, S. (2009) Smac mimetics as new cancer therapy. *Anticancer Drugs* **20**: 646–658.

Chipuk, J.E., Bouchier-Hayes, L., Kuwana, T., Newmeyer, D.D., and Green, D.R. (2005) PUMA couples the nuclear and cytoplasmic proapoptotic function of p53. *Science* **309**: 1732–1735(supporting online material: http://www.sciencemag.org/cgi/content/full/309/5741/1732/DC1).

Chipuk, J.E., Kuwana, T., Bouchier-Hayes, L., Droin, N.M., Newmeyer, D.D., Schuler, M., *et al.* (2004) Direct activation of Bax by p53 mediates mitochondrial membrane permeabilization and apoptosis. *Science* **303**: 1010–1014.

De Bruyn, M., Bremer, E., and Helfrich, W. (2013) Antibody-based fusion proteins to target death receptors in cancer. *Cancer Lett*. **332**: 175–183.

Fesik, S.W. (2005) Promoting apoptosis as a strategy for cancer drug discovery. *Nat. Rev. Cancer* **5**: 876–885.

Frantz, S. (2004) Lessons learnt from Genasense's failure. *Nat. Rev. Drug Discov*. **3**: 542–543.

Kamer, I., Sarig, R., Zaltsman, Y., Niv, H., Oberkovitz, G., Regev, L., *et al*. (2005) Proapoptotic BID is an ATM effector in the DNA-damage response. *Cell* **122**: 593–603.

Kang, M.H. and Reynolds, C.P. (2009) Bcl-2 inhibitors: targeting mitochondrial apoptotic pathways in cancer therapy. *Clin. Cancer Res*. **15**: 1126–1132.

Lui, X.-H., Yu, E.Z., Li, Y.-Y., Rollwagen, F.M., and Kagan, E. (2006) RNA interference targeting Akt promotes apoptosis in hypoxia-exposed human neuroblastoma cells. *Brain Res*. **1070**: 24–30.

Martinez-Ruiz, G., Maldonado, V., Ceballos-Cancino, G., Grajeda, J.P., and Melendez-Zajgla, J. (2008) Role of Smac/DIABLO in cancer progression. *J. Exp. Clin. Cancer Res*. **27**: 48.

Merino, D., Lok, S.W., Visvader, J.E., and Lindeman, G.J. (2015). Targeting BCL-2 to enhance vulnerability to therapy in estrogen receptor-positive breast cancer. *Oncogene* doi:10.1038/onc.2015.287 [Epub ahead of print]

Otsuki, Y., Li, Z., and Shibata, M.A. (2003) Apoptotic detection methods—from morphology to gene. *Prog. Histochem. Cytochem*. **38**: 275–339.

Raj, D., Brash, D.E., and Grossman, D. (2006) Keratinocyte apoptosis in epidermal development and disease. *J. Invest. Dermatol*. **126**: 243–257.

Renault, T.T., Elkholi, R., Bharti, A., and Chipuk, J.E. (2014) B cell lymphoma-2 (BCL-2) homology domain 3 (BH3) mimetics demonstrate differential activities dependent upon the functional repertoire of pro- and anti-apoptotic BCL-2 family proteins. *J. Biol. Chem*. **289**: 26481–26491.

Schimmer, A.D., Welsh, K., Pinilla, C., Wang, Z., Krajewska, M., Bonneau, M.-J., *et al*. (2004) Small-molecule antagonists of apoptosis suppressor XIAP exhibit broad anti-tumor activity. *Cancer Cell* **5**: 25–35.

Subburaj, Y., Cosentino, K., Axmann, M., Pedrueza-Villalmanzo, E., Hermann, E., Bleicken, S., *et al*. (2015) Bax monomers form dimer units in the membrane that further self-assemble into multiple oligomeric species. *Nat. Commun*. **6**: 8042.

White, E. (2015) The role of autophagy in cancer. *J. Clin. Invest*. **125**: 42–46.

Yu, X., Achehan, D., Menetret, J.-F., Booth, C.R., Ludtke, S.J., Riedl, S., *et al*. (2005) A structure of the human apoptosome at 12.8 Å resolution provides insights into this cell death platform. *Structure* **13**: 1725–1735.

Zhang, L., Yu, J., Park, B.H., Kinzler, K.W., and Vogelstein, B. (2000) Role of BAX in the apoptotic response to anticancer agents. *Science* **290**: 989–992.

Zhivotovsky, B. and Orrenius, S. (2003) Defects in the apoptotic machinery of cancer cells: role in drug resistance. *Semin. Cancer Biol*. **13**: 125–134.

Zinkel, S.S., Hurov, K.E., Ong, C., Abtahi, F.M., Gross, A., and Korsmeyer, S.J. (2005) A role for proapoptotic BID in the DNA-damage response. *Cell* **122**: 579–591.

Chapter 8

Cancer stem cells and the regulation of self-renewal and differentiation pathways: focus on colon cancer and leukemias

Introduction

As described in Chapter 1, the balance between cell growth, differentiation, and apoptosis affects the net number of cells in the body, and aberrant regulation of these processes can give rise to tumors. In this chapter, we will describe the characteristics of cells at different degrees of differentiation and discuss their relationship to cancer. This will include a review of the "cancer stem cell model" that states that subpopulations of cells with stem cell properties initiate and maintain the cancer phenotype. We will also investigate the molecular mechanisms that underlie the regulation of self-renewal and differentiation and examine specific mutations in these pathways that can lead to cancer. Lastly, new cancer therapeutics designed to target aspects of self-renewal and differentiation pathways are presented. Let us begin with an overview of the process of differentiation during development and in the adult.

We seldom reflect upon our own **ontogeny** or individual development. The processes involved in the development of a complete person from a formless fertilized egg are almost magical. Hundreds of specialized cell types must form from the fertilized egg and its unspecialized progeny cells called **embryonic stem cells** that reside in the inner cell mass. The process whereby cells become specialized to perform a particular function is called **differentiation** and relies on the regulation of a particular subset of genes that define a certain cell type. All cells in the body (except red blood cells and germ cells) contain a full complement of genes of the human genome, but it is the expression of a subset of genes that makes one cell type different from another—for example, a brain cell expresses different genes from a liver cell. Lineage-specific transcription factors responsible for turning on cell type-specific genes, as well as epigenetic mechanisms, are important in this process. During our development, different cell types are organized into varying tissues by pattern formation; although the same cell types are present in an arm and a leg, the morphology, or form of the structures, differs. Regulated gene expression is also important for patterning during development.

In addition to *embryonic* stem cells, there are also stem cells in the adult that are involved in the regeneration of tissues during the lifetime of the individual. In fact, stem cells are believed to be present in all tissues. Some stem cells are continually active to replace cells, as they mature and die off. For example, adult **hematopoietic** stem cells (HSCs), stem cells that give rise to blood, self-renew and differentiate to sustain the different types of blood cells over the lifetime of the individual. The success of bone marrow transplantations illustrates the remarkable regenerative capabilities of HSCs. Other stem cells remain dormant until a physiological signal is received. Hair follicle stem cells respond to a wound. Breast stem cells strongly respond to pregnancy hormones and, to a lesser extent, to hormones within the female monthly cycle. It has recently been demonstrated that stem cells show differentiation plasticity—for example, HSCs can give rise to non-hematopoietic cells.

But perhaps this should not be too surprising. Recent cloning experiments have demonstrated that a nucleus from a differentiated cell can be reprogrammed to direct the development of another individual. The cloning of Dolly the sheep from a mammary cell nucleus is a notable demonstration that the pattern of gene expression of a differentiated cell is not permanently fixed.

The process of differentiation is fueled by a source of stem cells in both the embryo and the adult. Stem cells self-renew while at the same time give rise to cells that are more committed to differentiate along a particular cell lineage. Withdrawal from the cell cycle is characteristic of differentiated cells. A block in differentiation results in a higher net number of cells and therefore is a mechanism for tumor formation in some cancers such as leukemia.

This chapter will focus on two features of differentiation pathways that have implications for carcinogenesis: first, the characteristics of stem cells in normal tissue and in tumors, and second, lineage-specific transcription factors and epigenetics that act as master switches for sets of genes during the differentiation process.

8.1 Cancer stem cells

Two defining features of stem cells are their ability to self-renew and their ability to give rise to committed progenitors of differentiated cell types of one or more cell lineages. Upon cell division, one daughter cell maintains the characteristics of a stem cell, including the ability to self-renew, and the other daughter cell shows characteristics of commitment towards differentiation (Figure 8.1). These features are shared with cancer stem cells (CSCs), rare cells within a tumor that have the ability to self-renew and to give rise to phenotypically diverse cancer cells with limited proliferative potential that make up the rest of the tumor.

CSCs have been functionally defined by having the ability to initiate new tumors when transplanted into host animals, and their existence supports the CSC model (see "Introduction" on p. 178).

It has been shown in several types of cancer that tumors are maintained in a growing cancerous state by only a small fraction of particular tumor cells. For example, it has been established that only about one in a million acute myeloid leukemia (AML) cells can develop into new leukemias when transferred *in vivo* and that these cells expressed the same markers (CD34$^+$, CD38$^-$) as normal HSCs. In general, CSCs have surface proteins called markers, which are characteristic of the stem cell normally present in the tissue, and they have been used to isolate enriched populations of CSCs. Brain CSCs display normal neural stem cell markers. Interestingly, the proportion of brain CSCs identified in a variety of brain cancers correlates with the course of the disease, or prognosis. Fast-growing tumors, such as glioblastomas, had more brain CSCs than slow-growing tumors like astrocytomas. Evidence for the existence of breast CSCs was obtained by testing whether human breast cancer cells could give rise to new tumors when grown in immunocompromised mice (Al-Hajj *et al.*, 2003).

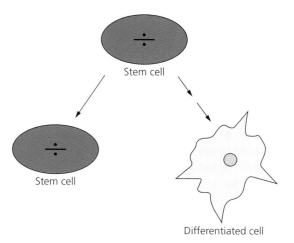

Figure 8.1 Features of cancer stem cells shared with stem cells. Cancer stem cells have the ability to self-renew and can also give rise to daughter cells that show a commitment towards differentiation and contribute to the bulk of the tumor.

A minority subpopulation of breast CSCs was isolated, based on the expression of cell surface markers (CD44$^+$, CD24$^{-/low}$), and these cells showed a 10- to 50-fold increase in ability to form tumors in animals, compared with the bulk of breast tumor cells. Furthermore, these cells not only were able to demonstrate the ability to self-renew but were also able to give rise to cells with different characteristics or phenotypes that made up the bulk of the tumor. These observations support the concept that these cells are breast CSCs. Subsequently, CSCs were identified in a range of solid tumors and include colon and pancreatic CSCs (reviewed in Visvader and Lindeman, 2008). Colon CSCs, along with prostate and brain CSCs, overexpress the cell surface antigen CD133. Pancreatic and breast CSCs both express CD44. More recently, there has been a debate over exact definitions of CSCs and conflicting results in different mouse/human model systems. Rosen and Jordan (2009) suggest that this is to be expected in such "inherently unstable conditions as cancer."

In summary, the CSC model is supported by the identification of a small minority of CSCs that drive tumorigenesis in some cancers, similar to the small number of adult stem cells that drive the growth of normal tissues.

Relevance of stem cell biology to carcinogenesis

Two proposals for the relevance of stem cell biology to carcinogenesis have been suggested. The first proposal is that self-renewal provides increased opportunities for carcinogenic changes to occur. The second proposal suggests that altered regulation of self-renewal directly underlies carcinogenesis. We will explore each of these next where we will find that the concepts from both proposals will intertwine.

Self-renewal provides an extended window of time for mutation

Stem cells are long-lived targets for chance mutations, compared with many differentiated cells that die within days or months. The accumulation of mutations necessary for carcinogenesis is more likely to occur in stem cells that self-renew over the lifetime of an individual, rather than in mature cells that exit the cell cycle and/or undergo apoptosis after a brief period. This concept supports the proposal that tumors are likely to arise from stem cells. Because restricted progenitor cells self-renew for shorter periods of time, they are less likely than stem cells to become oncogenic. Let us look at the process of skin carcinogenesis. Skin is a tissue with a clearly defined hierarchical organization of differentiation. The differentiation pathway begins in the basal layers and leads to the formation of the dead, cornified outer layer of the skin. Epithelial cells of the skin have a turnover rate of 60 days in humans. However, malignant transformation involving the accumulation of specific mutations takes 18 months or more, so the 60-day lifespan of a differentiating cell is not long enough to allow a sufficient number of mutations to accumulate. Self-renewal is a quality of stem cells that allows for the accumulation of transforming mutations, as an individual mutation may be passed on to daughter cells which are themselves susceptible to additional mutations and can pass on accumulated mutations to future progeny cells. This rationale suggests that the accumulation of mutations required for the initiation of skin cancer is likely to occur in the normal stem cell or early progenitor cell compartment.

Deregulation of self-renewal

Stem cells must maintain a balance between self-renewal and differentiation. One proposal for the relevance of stem cell biology to carcinogenesis is that the loss of this balance by stem cells can lead to unregulated self-renewal, a hallmark of cancer. Therefore, tumor cells may arise from stem cells. Alternatively, differentiated cells may acquire a mutation that reactivates a self-renewal program. This concept is supported by experiments that showed that minor, but specific, genetic alterations of differentiated cells can produce induced pluripotent stem cells (iPSCs) (Takahashi *et al.*, 2007, and others). That is, differentiated cells can be modified to create stem cell-like cells.

This discussion leads to the question of the nature of the cell that initiates carcinogenesis. It may be hypothesized that tumors arise from stem cells within a tissue or alternatively from more differentiated cells that acquire the stem cell quality of self-renewal. There is supporting evidence for both. Alternatively, a continuum of target cells relative to different states of differentiation may exist: stem cells, progenitor cells, and terminally differentiated cells may all be targets for transformation. Further, the stage of differentiation of the target cell may affect the malignant potential and severity of the cancer.

Both of these proposals, that cancer can initiate either in a stem cell that has lost regulation of self-renewal or in a differentiated cell that has obtained the ability to self-renew, are supported by the identification of CSCs.

PAUSE AND THINK

Mammary stem cells react to physiological cues, such as hormones, to provide a source of proliferation and differentiation during pregnancy for the creation of a milk-generating breast. A major factor that protects women from breast cancer is an early first full-term pregnancy (see Chapter 11). It is suggested that the depletion of stem cells as a result of the burst of differentiation that occurs during pregnancy is the reason why pregnancy is protective against breast cancer. There may be fewer breast stem cells that have the potential of becoming breast cancer stem cells over time in women who have had children in early adulthood.

Molecular mechanisms of self-renewal

Let us examine the molecular mechanisms of self-renewal. The molecular mechanisms that regulate self-renewal of stem cells are beginning to be understood. There is evidence that the Wnt signaling pathway, which is important for regulating pattern formation during development, is involved in the self-renewal process of stem cells during development, in the adult, and also in cancer. When the Wnt-regulated transcription factor Tcf (see "The Wnt signaling pathway" below) is deleted in mice by gene knock-out procedures, the resulting phenotype is a lack of stem cells in the intestines. In addition, HSCs respond to Wnt signaling *in vivo* and require Wnt signaling for self-renewal. Data from DNA arrays show that the gene expression pattern in response to Wnt signaling is similar between colon stem cells and colon cancer cells, but differs in differentiated colon cells, suggesting that Wnt signaling plays a role in stem (cancer) cell self-renewal. The Hedgehog signaling pathway, which is also important for regulating pattern formation in the embryo, has also been implicated in the process of stem cell self-renewal. Both the Wnt and Hedgehog signaling pathways will be discussed next.

The Wnt signaling pathway

Wnt proteins (of which there are 19 members) are secreted intercellular signaling molecules that act as a ligand to trigger a specific signal transduction pathway (Figure 8.2). Lipid modifications of Wnt protein by the protein Porcupine play a role in its secretion from the cell. It is easiest to examine the cell in the absence of Wnt ligand first (Figure 8.2a). In this state, several proteins associate together in the cytoplasm to form a degradation complex. The degradation complex consists of axin, adenomatous polyposis coli (APC) protein, glycogen synthase kinase 3β (GSK3β), and casein kinase I (CKI). Axin and APC form a structural scaffold for GSK3β and CKI which are serine/threonine kinases. An important transcriptional co-activator, called β-catenin (Figure 8.2a, yellow triangular shape), is modified by this complex via sequential phosphorylation by CKI and GSK3β. β-catenin is further modified by ubiquitination via newly recruited ubiquitin ligase. Ubiquitin acts as molecular flag that targets β-catenin for degradation by **proteasomes**. As β-catenin is not available to associate with the Tcf/LEF (T-cell factor/lymphoid enhancing factor) family of transcription factors in unstimulated cells, target genes under the regulation of β-catenin–Tcf are repressed (shown by a red "x"). In the absence of β-catenin, Tcf associates with the transcriptional repressor Groucho.

Wnt signaling occurs through the binding to transmembrane Wnt receptors, in which there are ten members (including G-protein-coupled and tyrosine kinase receptors). Upon binding of Wnt ligand to its seven-pass transmembrane receptor Frizzled and co-receptor LRP (low-density

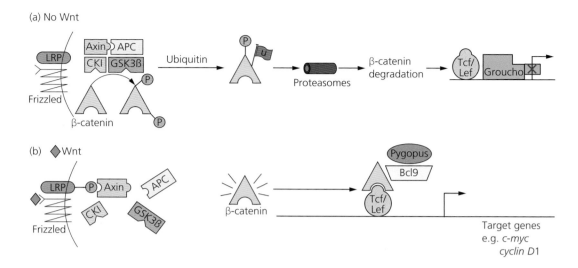

Figure 8.2 The Wnt signaling pathway. See text for details. (P) and (u) mark events of phosphorylation and ubiquitination.

lipoprotein receptor-related protein), a conformational change is induced, and the cytoplasmic tail of LRP is phosphorylated by GSK3β and CKI. Axin is recruited to the phosphorylated co-receptor LRP. This disrupts the assembly of the degradation complex (Figure 8.2b). In addition, an inhibitor of GSK3β, dishevelled protein, is activated via phosphorylation (not shown). These events allow β-catenin to escape degradation and move into the nucleus where it can act as a co-activator of the Tcf/LEF family of transcription factors to regulate specific target genes (e.g. *c-myc*, *cyclin D*, and genes that code for adhesion molecules from the ephrin (Eph) receptor family). Activation of target genes also depends on nuclear proteins Bcl9 (also known as legless) and Pygopus (see "Pause and think").

PAUSE AND THINK

There are several molecular similarities and differences between this signaling transduction pathway and the others previously described. What are they? Phosphorylation is an important regulatory process in many signaling transduction pathways (e.g. EGF), but, in the case of the Wnt pathway, phosphorylation leads to the recruitment of a ubiquitin ligase and is tied closely to protein degradation. Also, similar to the transcription factor E2f, β-catenin is prevented from acting until a specific signal is received: phosphorylation of RB is the signal in the case of E2F, and Wnt pathway activation is the signal in the case of β-catenin.

Self-test Close this book, and try to redraw Figure 8.2. Check your answer. Correct your work. Close the book once more, and try again. Note that this signaling pathway is dependent on β-catenin and is referred to as the "canonical" pathway, but Wnt can also modulate β-catenin-independent pathways, known as "non-canonical" pathways.

Wnt signaling and cancer

Wnt1 was one of the first proto-oncogenes discovered. Viral integration induced oncogene activation and subsequent cancer of the mammary gland in mice. Mutations that constitutively activate the Wnt signaling pathway have been identified in several types of cancer.

Mutations that result in the constitutive activation of the Wnt pathway are responsible for 90% of colorectal cancer. This translates in human terms to 50,000 lives per year in the USA alone. Most of the mutations inactivate the function of APC or activate β-catenin, but rarely alter the ligand Wnt. Colorectal cancer can be classified into two forms: familial forms and sporadic forms. Patients with the inherited cancer predisposition syndrome familial adenomatous polyposis coli (FAP) carry a germline mutation in the *APC* gene and develop high numbers of polyps in the colon (polyposis) in early adulthood. As a result of having many polyps, these patients have an increased risk of colorectal cancer. The *APC* gene acts as a true tumor suppressor gene in that both copies of the *APC* gene are inactivated in colorectal tumors. Most mutations are truncating mutations and occur in the coding sequences for the central region of the APC protein (codons 1250–1500), referred to as the mutation cluster region, in both germline and somatic cases (see "Pause and Think").

The small intestine/colon is a well-studied model system for examining the link between stem cells, the Wnt pathway, and cancer. Intestinal tissue is highly regenerative; stem cells and epithelial progenitors, also called transit-amplifying cells, that reside in the crypts give rise to more differentiated cells that migrate up along the villi (Figure 8.3a and Figure 8.4). The stem cells

> **PAUSE AND THINK**
>
> What is the ultimate molecular consequence of having inactivating mutations in *APC*? Constitutive activation of Tcf transcription factors.

(a)　　　　　　　　(b)

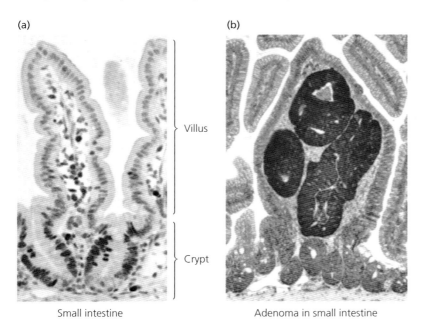

Small intestine　　　　Adenoma in small intestine

Figure 8.3 Comparison of normal epithelium and adenomas in the small intestine of a mouse. (a) Normal epithelial in the small intestine. Proliferative cells are stained for a cell cycle marker (brown nuclei). (b) An adenoma residing inside the villus in mouse small intestine. Tissue was stained for β-catenin (brown). An accumulation of β-catenin is seen throughout cells in the adenoma and aberrant crypt. From Radtke, F. and Clevers, H. (2005) Self-renewal and cancer of the gut: two sides of a coin. *Science* **307**: 1904–1909. Reprinted with permission from the AAAS.

renew over the lifespan of the individual, while progenitor cells have a limited self-renewal capacity (about four divisions). Upon reaching the top of the villi, fully differentiated cells undergo apoptosis. The intestinal epithelium is renewed within a few days. Normally, Wnt signaling is required to maintain the stem cells and progenitors of the crypt. *Lgr5* is one target gene of the Wnt pathway that is restricted to the stem cells of the intestine. Its protein product Lgr5 is a transmembrane receptor that binds to the ligand R-spondin and physically interacts with Frizzled/LRP complexes to enhance Wnt/β-catenin signaling. It is an important marker of adult stem cells.

Colorectal cancer seems to follow a sequence of progression from benign polyps or adenoma (Figure 8.3b) to carcinoma *in situ*, and finally invasive carcinoma. The sequence is paralleled by the accumulation of mutations.

The location of the initial event, the cell of origin of adenomas, induced by constitutively active Wnt signaling, has been shown to be the stem cell of the small intestine (Barker *et al.*, 2009). Barker and colleagues deleted the *Apc* gene in long-lived intestinal stem cells using genetically engineered mice that allow examination of oncogenes/tumor suppressor genes in different cellular compartments (see "Activity" at the end of the chapter on p. 201). Adenomas were induced within several weeks. They also showed that deletion of the *Apc* gene in progenitor transit-amplifying cells or differentiated cells does not trigger tumor formation (Figure 8.4). Therefore, their data support the cancer stem cell concept that a small subset of cells is able to self-renew and maintain a tumor.

Loss of Apc is considered the initiating event in most colorectal cancers and is sufficient to induce adenomas in mice.

There is now evidence that Apc is crucial for maintenance of colorectal cancer and more surprisingly that restoring Apc levels in colorectal cancer cells can revert them to functioning normal cells *in vivo*, even when the cancer cells contain other oncogenic mutations such as K-ras and p53 mutations (Dow *et al.*, 2015; see Box "How do we know

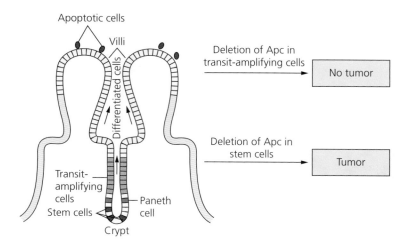

Figure 8.4 Schematic representation of the cells of the intestinal crypts and villi. Tumors initiate from stem cells, and not transit-amplifying cells.

that?"). Earlier in the text, we have mentioned oncogene addiction. This study demonstrates addiction to the inactivation of a tumor suppressor gene. That is, tumors are dependent upon the inactivation of the tumor suppressor Apc for their maintenance. Remarkably, tumor cells were shown to contribute to normal crypt stem cell function when Apc was restored.

? HOW DO WE KNOW THAT?

Conditional and reversible control of gene expression in transgenic mice

(See Dow *et al.*, 2015.)

The initiating event in colorectal cancer is the loss of Apc. Subsequently, additional mutations in genes, such as K-ras and p53, accumulate. It is well established that Apc inactivation in mice can induce adenomas in the small and large intestine. In order to investigate if Apc is also required for maintenance and cancer progression, transgenic mice containing K-ras and p53 mutations and capable of conditional and reversible control of Apc expression were created. Conditional and reversible control of Apc expression was accomplished by using inducible short hairpin RNAs (shRNAs). Remember specific shRNAs can hybridize to complementary mRNAs and block their translation. In brief, DNA constructs coding for the shRNA linked to green fluorescent protein (GFP) were designed such that treatment with doxycycline drives shRNA expression and Apc

silencing. GFP acts as a marker for shRNA expression. Withdrawal of doxycycline halts shRNA expression and restores endogenous Apc expression. What would you expect to see on a western blot (a technique that detects protein) during doxycycline treatment and withdrawal for Apc, GFP, a Wnt target, and a control? Western blot analysis showed expected protein profiles for Apc, GFP, Wnt target c-myc, and control protein β-actin, during doxycycline treatment and withdrawal (Figure 8.5b). Immunohistochemical and immunofluorescent stains of intestinal tissue during doxycycline treatment show expansion of alkaline phosphatase and keratin 20 negative progenitor cells and hyperproliferation (Ki67), and Lgr5 positive stem cells outside of their normal position at the base of the crypt, whereas withdrawal of doxycycline illustrates restoration of phenotype (Figure 8.5a). See Dow *et al.* (2015) for details. These results validate the Wnt pathway as an effective drug target for colorectal cancer.

Mutations in the Wnt signaling cascade also promote other types of cancers. Activating mutations of β-catenin that affect the regulatory sequences essential for its targeted degradation can lead to skin tumors. Mutations in the axin gene are found in hepatocellular carcinoma. Many of the axin gene mutations lead to protein truncations that delete the axin–β-catenin binding sites. These observations suggest that some transforming mutations may function to reactivate the self-renewal pathway. The cells carrying these mutations can be thought of as *de novo* stem cells, that is cells that have acquired stem cell characteristics as a result of mutation, and were not produced from self-renewal of other stem cells.

The Hedgehog signaling pathway

The Hedgehog (Hh) signaling pathway also plays important roles in embryonic development, tissue self-renewal, and carcinogenesis. Except for its function in tissue repair and maintenance, it is mainly inactive in adults. It is essential for pattern formation in many tissues, including the neural tube, skin, and gut. Similar to Wnt proteins, Hh proteins (three

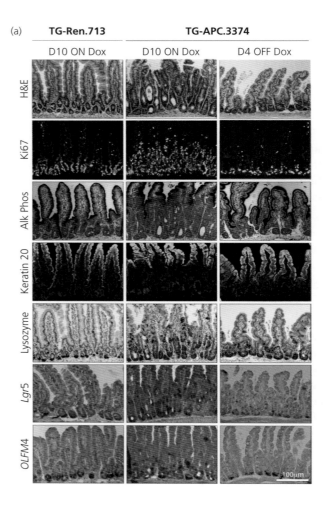

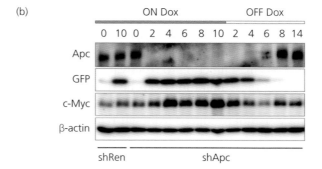

Figure 8.5 Apc restoration promotes cellular differentiation and re-establishes crypt homeostasis in colorectal cancer. (a) Immunohistochemical and immunofluorescent stains and in situ hybridizations (Lgr5) from TG-Ren.713 (shRen; control) and TG-Apc.3374 (shApc) intestine following dox treatment (left two panels) and withdrawn from dox (right panel). (b) Western blot analysis from intestinal villi as indicated. See "How do we know that?" on p. 187. Reprinted from Dow, L.E. *et al.* (2015) Apc restoration promotes cellular differentiation and re-establishes crypt homeostasis in colorectal cancer. *Cell* **161**: 1539–1552. Copyright (2015) with permission from Elsevier.

members: Sonic, Desert, and Indian) are secreted intercellular signaling molecules that act as a ligand to trigger a specific signal transduction pathway (Figure 8.6). Two transmembrane proteins, Patched and Smoothened (related to Frizzled described earlier), are responsible for signal

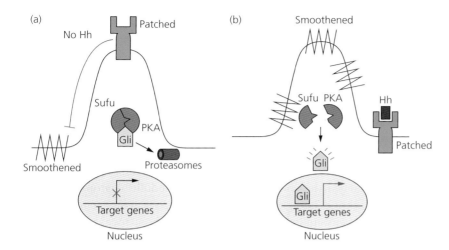

Figure 8.6 The Hedgehog signaling pathway. See text for details.

transduction by Hh; the interaction of these two molecules are regulated within the cilium. In the absence of Hh (Figure 8.6a), Patched is localized in the cilia and inhibits Smoothened by preventing its localization to the cilia and thus suppresses the pathway. The important zinc finger transcription factor, Gli, is sequestered by a protein complex in the cytoplasm that induces cleavage of Gli by the proteasomes. The resulting product is a repressor that is transported to the nucleus and inhibits Hh target gene expression. Upon binding of Hh to Patched (Figure 8.6b), Patched is translocated from the cilia, allowing Smoothened to relocate to the cilia, and inhibition of Smoothened is relieved. Smoothened transduces a signal into the cell and causes the large protein complex to dissociate and release the transcription factor Gli (shaded yellow), so that it can be translocated to the nucleus and regulate the expression of Hh target genes. Hh target genes include *Gli*, which amplifies the initial Hh signal at the transcriptional level. Although Hh target genes may be cell type-specific, they may include: *Cyclin Ds*, *Bcl2*, *VEGF*, and *Snail* (the last two are important for angiogenesis and metastasis, respectively; see Chapters 9 and 10). The signaling events downstream of Smoothened are not well defined in mammals, but the **su**ppressor of **fu**sed (Sufu) and protein kinase A (PKA) have been identified as negative regulators that reside in the protein complex.

Self-test Close this book, and try to redraw Figure 8.6. Check your answer. Correct your work. Close the book once more, and try again.

Hedgehog signaling and cancer

Patched is defined as a tumor suppressor gene; patients with Gorlin syndrome carry a germline mutation in one copy of *Patched* and have a predisposition to develop skin, cerebellar, and muscle tumors (BCC, medulloblastomas, and rhabdomyosarcomas, respectively). Inactivating mutations in *Patched* and activating mutations of *Smoothened* were identified in sporadic (opposed to familial) human BCC tumors, and *Gli-1* expression was found in nearly all BCCs. In fact, all sporadic BCCs possess an activated Hh signaling pathway.

It has been suggested that, as BCCs are tumors that include hair follicle differentiation, they may originate from a source of stem cells that reside in a structure of the hair follicle, called the hair follicle bulge, which has acquired an aberrant Hh signaling pathway. These CSCs could self-renew and give rise to the differentiated hair follicle cells observed in BCC tumors. Similarly, it has been proposed that medulloblastoma, the most common childhood malignant brain tumor, arises from neuron precursors that possess an inappropriately activated Hh pathway. Activation of this pathway by mutation is observed in 30% of sporadic medulloblastomas. Molecular evidence of an activated Hh pathway was also reported for gliomas. *Gli* was originally identified as an amplified gene in cultured glioma cells. Unlike Wnt, where mutations involved in carcinogenesis rarely affect this ligand, Hh is overexpressed in upper gastrointestinal tumors. Pre-clinical studies also point to paracrine signaling, such that secreted, tumor-derived Hh stimulates *Gli* expression in the tumor-associated stroma, but not in the tumor itself. Stromal feedback to the tumor is suggested to promote tumor growth.

The role of Hh signaling has been extended to chronic myeloid leukemia (CML). Zhao *et al.* (2009) have demonstrated that loss of Smoothened inhibits HSC renewal and decreases the induction of CML by the oncogenic BCR–ABL chromosomal translocation. They also show that inhibition of the Hh pathway inhibits the growth of Gleevec™-resistant mice and human CML. This has direct implications for a new approach to the development of drugs for Gleevec™-resistant cancers. Taken together, it is apparent that the Hh signal pathway is relevant for several types of human cancer and is implicated in the maintenance of CSCs.

Additional properties of tumors that may be due to cancer stem cells

Tumors display heterogeneity, plasticity, and an ability to migrate; these properties may be due to CSCs.

CSCs, in addition to being able to self-renew, also have the ability to give rise to more differentiated cell types with limited proliferative capacity. A tumor contains a mix of cells that have different genetic and physical characteristics. It is a mass of genotypically and phenotypically heterogeneous cells despite all the cells being of clonal origin (see Section 1.2, "Evidence suggests that cancer is a disease of the genome at the cellular level" on p. 5). This heterogeneity may reflect aberrant differentiation of CSCs, in addition to accumulation of different mutations. Teratocarcinoma is

PAUSE AND THINK

Does *Smoothened* act as a tumor suppressor gene or an oncogene? It is an oncogene, as activating mutations have been isolated. Would you expect to see gain- or loss-of-function mutations in SuFu in cancer cells? Yes, like Patched, loss-of-function mutations have been reported in medulloblastoma.

an obvious example whereby the tumor contains undifferentiated stem cells and non-proliferative differentiated cells such as bone and cartilage. Although the degree of differentiation for other cancers is less obvious, CSCs from other cancers have been shown to generate more cells with different phenotypes. Note that, although CSCs possess some characteristics of normal stem cells, they are nevertheless transformed cells containing complex genetic and epigenetics alterations. Thus, they may be unable to undergo normal differentiation. The normal pattern formation of cells that underlies the morphology of a normal organ is abandoned.

In addition, the degree of differentiation of a cell may also affect the outcome of oncogene activation. The use of selective gene promoters to drive the expression of a *Ras* oncogene in different cell populations with different degrees of differentiation (e.g. stem cells or committed progeny cells) resulted in tumors with different malignant potential (i.e. malignant carcinomas versus benign papillomas). Although additional studies are needed, this suggests that the activation of an oncogene may be carcinogenic in some states of differentiation, and not in others, within a particular cell lineage.

Progeny of both normal and cancer stem cells also exhibit developmental plasticity whereby cells may switch between two developmental pathways. The development program, called the epithelial–mesenchymal transition (EMT), describes cell type changes that involve cells losing epithelial cell traits and gaining mesenchymal cell traits. Interestingly, EMT is observed in some tumor cells, is linked to metastasis, and gives rise to stem-like characteristics (discussed in further detail in Chapter 9). Reversible phenotype switching in tumor cells may also account for the development of drug resistance (reviewed in Kemper *et al.*, 2014). In addition, it has been demonstrated that non-stem tumor cells may convert into a stem-like state (Chaffer *et al.*, 2011). This concept can be supported by the observation that, in particular normal tissues such as the mouse testis, progenitor cells can dedifferentiate into stem cells at low frequencies. A striking demonstration has suggested that tumor cells can be reprogrammed to become normal and totipotent during experimental manipulation (like fully differentiated cells during cloning experiments). When placed in early mouse embryos, teratocarcinoma cells can mimic stem cells and can contribute to normal development.

Another feature of CSCs is the ability to migrate to other tissues of the body. The inherent ability of a stem cell to migrate may cause these cells to be aggressively metastatic, if transformed. The myeloid leukemias support this view: transformed stem cells are likely to be malignant, while transformed committed progenitor cells are likely to be benign. Metastasis of tumor cells from a primary site to secondary sites is a characteristic that makes cancer lethal (details are discussed in Chapter 9). The success of metastasis may be based on the number of CSCs in the primary tumor; non-cancer stem cells in a primary tumor may not have the ability to form new tumors at distant sites.

8.2 The regulation of differentiation by gene expression

The process of differentiation is dependent upon the expression of a specific subset of genes that defines a particular type of cell. Regulation of gene expression can include both inhibitory and inductive mechanisms. The polycomb group of protein repressors and hematopoietic lineage-specific transcription factors are two examples of important regulatory mechanisms involved in a stem cell's ability to form differentiated progeny.

Polycomb proteins silence gene expression in stem cells and cancer

The polycomb group (PcG) of proteins represses the transcription of specific sets of genes by epigenetic modifications (see Chapter 3). PcG proteins do not have the ability to bind to specific DNA motifs, but they are thought to be recruited to genes by transcription factors and/or lncRNAs (e.g. HOTAIR) which can bind to genomic DNA in a sequence-specific manner. As p53 has been nicknamed the "guardian of the genome," the PcG proteins have been nicknamed the "guardians of stemness." This is because the target genes that they repress include a large number of developmental regulators that promote differentiation (Figure 8.7). Transcription factors, such as the homeobox proteins of the Dlx and Pax family and the Fox and Sox families that are crucial during development, are among the hundreds of target genes identified by mapping of PcG proteins to human DNA. Thus, PcG proteins are implicated in stem cell maintenance. In addition, they also repress key tumor suppressor pathways. This can be seen by repression of the genetic locus *INK4a/CDKN2A* that encodes the cdk inhibitor INK4a (p16), and ARF (p14), an inducer of p53, and suggests that the PcG proteins have oncogenic potential.

The mechanism of epigenetic regulation involves the formation of two PcG repressive complexes, PRC2 and PRC1. PRC2 consists of proteins EED, EZH1, EZH2, and SUZ12. This complex contains histone methyltransferase activity and targets lysine 27 (and lysine 9) of histone H3. The

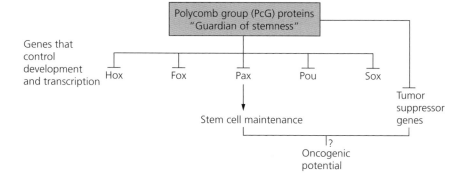

Figure 8.7 Polycomb group proteins repress the expression of many developmental regulators (only a sample is shown). The role of PcG proteins in stem cell maintenance and in repressing tumor suppressor genes suggests that they may have oncogenic potential.

trimethylated histone H3 product may serve as an anchor for PRC1. PRC1, which contains Bmi-1 and chromobox protein subunits, causes the addition of a ubiquitin molecule to histone H2A at lysine 119. This molecular flag is found on 10% of H2A proteins and represents one of the most abundant epigenetic modifications. It is proposed that repression involves direct inhibition of the transcriptional machinery, recruitment of methyltransferases, and chromatin compaction.

Upon differentiation, there is de-repression of PcG target genes. Although the mechanism for the de-repression is unknown, some evidence suggests that PcGs are removed from the promoter regions of their target genes and/or a specific histone lysine demethylase is involved in reversing the epigenetic modifications initiated by the PcG proteins.

PcG proteins and cancer

The ability of PcG proteins to suppress the differentiation and promote self-renewal of stem cells implicates them in oncogenesis, as abnormal regulation of differentiation can lead to cancer. Bmi-1 and EZH2 are over-expressed in a variety of different cancers. In AML, most leukemic cells have a limited capacity for proliferation but are replenished by rare leukemic stem cells (see Section 8.1, "Cancer stem cells" on p. 180). Therefore, the ability of stem cells to self-renew is important in the maintenance of this disease. A PcG protein repressor Bmi-1 has been demonstrated to be essential for the control of self-renewal in HSCs and in leukemic stem cells (Lessard and Sauvageau, 2003). *In vitro*, leukemic stem cells that lack Bmi-1 show growth arrest in the G_1 phase of the cell cycle and begin to differentiate. *In vivo*, mice with a *Bmi-1* gene knock-out show a progressive depletion of all blood cells, indicating the essential role of Bmi-1 in HSCs. In addition, in mouse models using leukemic stem cells lacking Bmi-1, smaller numbers of leukemic cells are detected in the peripheral blood, compared with controls, indicating that these CSCs have proliferative defects. This is an example of how a common gene can regulate self-renewal in both normal and cancer stem cells. Bmi-1 exerts its effects partially by repressing the expression of two cdk inhibitors p16 and p14 via chromatin remodeling. The role of *Bmi-1* as a human oncogene is supported by the identification of *Bmi-1* gene amplification in some lymphomas. Several other PcG proteins are linked to oncogenesis: SUZ12 is overexpressed in breast and colon cancers, and EZH2 is overexpressed in lymphoma and breast and prostate tumors. This evidence supports the theory that cells with stem cell properties drive tumor formation and progression. PcG proteins may contribute to carcinogenesis both by the silencing of tumor-suppressing pathways and by inducing and maintaining the stem cell state (Figure 8.7).

Role of lineage-specific transcription factors in differentiation and cancer

The induction of lineage-specific gene transcription is dependent on lineage-specific transcription factors. AML serves as an important paradigm for examining the mechanisms of differentiation. It illustrates how disruption of the function of a transcription factor can interfere with differentiation and lead to cancer.

AML is a disease characterized by a block in the differentiation of the granulocyte or monocyte lineage (Figure 8.8). There are several subtypes of AML. The classification system of this disease is still evolving but should eventually reflect molecular features at the point of the differentiation block. The lineage is organized as a hierarchy that begins with pluripotent HSCs. These cells self-renew and form progenitor cells (common lymphoid progenitors (CLPs) and common myeloid progenitors (CMPs)). These progenitor cells differentiate into several types of more specific progenitor cells, including granulocyte/monocyte progenitors (GMPs) and megakaryocyte/erythroid progenitors (MEPs). GMPs are common to both the monocyte and granulocyte lineages. Several transcription factors have been identified to be important in the development of hematopoietic lineages. One factor, Runx-1 (also known as AML1; Figure 8.8), is involved in almost all lineages, as it is an important factor for stem cells. Others are lineage-specific factors (differentiation factors) such as Pu.1 and CCAAT/enhancer-binding protein α (C/EBPα). Lineage-specific transcription factors activate a particular set of lineage-specific genes and/or inhibit the cell cycle for terminally differentiated cells. Focus on the myeloid pathway. Pu.1 is involved in the differentiation of the CMP cell and, later on, in the differentiation of monocytes/macrophages. Most myeloid-specific genes have Pu.1 sites in their promoters. C/EBPα, a zinc finger transcription factor, functions in the differentiation of granulocytes.

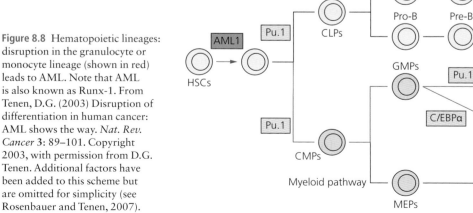

Figure 8.8 Hematopoietic lineages: disruption in the granulocyte or monocyte lineage (shown in red) leads to AML. Note that AML is also known as Runx-1. From Tenen, D.G. (2003) Disruption of differentiation in human cancer: AML shows the way. *Nat. Rev. Cancer* **3**: 89–101. Copyright 2003, with permission from D.G. Tenen. Additional factors have been added to this scheme but are omitted for simplicity (see Rosenbauer and Tenen, 2007).

Many mutations that are typically found in AML affect specific transcription factors; both chromosomal translocations (e.g. t(8;21)) and coding region mutations are common. The gene for the AML1 transcription factor is disrupted in the t(8;21) translocation, and this translocation leads to AML. The chromosomal translocation t(8;21) is identified in both HSCs and more differentiated cells in patients, thus providing additional evidence that the transforming mutations of AML occur in HSCs.

Mutations in lineage-specific transcription factors are found in patients with AML subtypes that are consistent with their role in normal hematopoiesis. Pu.1 mutations are found in the earliest stage (M0: very immature leukemia) and in monocytic leukemias, reflecting the early role of Pu.1 in myeloid precursor cells and in the development of monocytes/macrophages. Approximately 10% of patients with AML carry a mutation in c/EBPα, and most of these cases are associated with the granulocytic subtype, reflecting the role of c/EBPα in granulocyte differentiation. Thus, mutation of transcription factors involved in differentiation is an important mechanism behind oncogenesis.

Acute promyelocytic leukemia (APL), a subtype of AML, is most often characterized by the chromosomal translocation t(15;17) that results in the fusion of the *PML* gene with the retinoic acid receptor alpha (RARα) gene to create a hybrid protein, PML–RAR, with altered functions. As described in Chapter 3, RARs (α, β, and γ) are members of the steroid hormone receptor superfamily and act as ligand-dependent transcription factors that are important effectors of the essential role of RA in cell differentiation. The wild-type receptors bind to the RARE in target genes as RAR–RXR heterodimers. In the absence of RA, the receptors associate with HDAC–co-repressor complexes that silence target genes by histone deacetylation and subsequent chromatin compaction (Figure 8.9a). Upon binding of RA, the receptor undergoes a change in shape that causes the receptor to dissociate from the HDAC–co-repressor complex and allows the receptor to act as an activator in order to transcriptionally induce its target genes and promote differentiation (Figure 8.9b). The oncogenic fusion protein (shown in red) maintains both the DNA-binding domain and the ligand-binding domain of the RARα receptor. It has a higher affinity for HDAC and does not dissociate in the presence of physiological concentrations of RA (Figure 8.9c). In addition, the ability of the fusion protein to form homodimers is essential for the development of the disease, suggesting that they act in a dominant negative manner by blocking wild-type RAR–RXR heterodimers or by recruiting novel co-repressors. The normal role of the PML protein may also be disrupted in the fusion protein. PML protein is normally found in nuclear organelles called nuclear bodies and regulates p53, which acts as a pro-apoptotic protein. Thus, the oncogenic effect of PML–RAR is thought to include blocking differentiation and inducing self-renewal of APL cells.

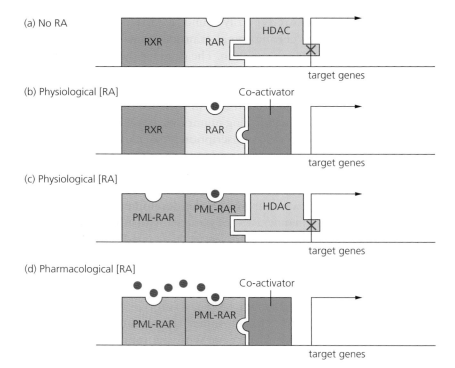

Figure 8.9 The activity of the RAR and the PML–RAR fusion protein in varying conditions of retinoic acid concentration. Expression of target genes leads to differentiation.

◎ Therapeutic strategies

The concept of CSCs has important implications for the design and testing of new cancer drugs. First, as CSCs support the growth and migration of the tumor, drugs need to target this small subset of cells within the tumor. Many existing conventional drugs give hopeful initial responses that are followed by disappointing latter reoccurrences, perhaps because of resistant CSCs. Drugs targeted at CSCs, in addition to those that target non-stem cells, bearing in mind the concept of tumor cell plasticity, may prevent reoccurrence and actually cure metastatic cancer.

The best scenario would be to find a drug that would target CSCs without affecting normal stem cells of the same tissue. In fact, there is evidence that suggests this may be possible (Yilmaz *et al.*, 2006). It has been demonstrated in mice that deletion of PTEN (a tumor suppressor phosphatase protein) resulted in leukemia-initiating cells that could transfer disease upon transplantation to irradiated mice, but also caused initial proliferation and later depletion of normal HSCs. Thus, a distinction was made between the CSCs and normal stem cells of the hematopoietic system: PTEN deletion promotes the generation of leukemia-initiating cells but the depletion of normal stem cells. A drug that is able to reverse the downstream effects of PTEN deletion would lead to a depletion of leukemia-initiating cells and prevent the depletion of normal stem cells.

The efficacy of such new drugs should be determined by their effect on the CSC population, and not on overall tumor regression. Drugs may be successful at killing all of the cells of a tumor, except CSCs, so that measuring tumor regression would not reflect the fact that the most dangerous tumor cell types remained unaffected. Differences in drug resistance between CSCs and other tumor cells is a possible explanation for such a scenario. Stem cells express high levels of ATP-binding cassette (ABC) transporters (e.g. P-glycoprotein), members of the multi-drug resistance gene family (see Chapter 2). The ability of the ABC transporters in stem cells to inhibit the accumulation of the fluorescent dyes Hoechst 33342 and rhodamine 123 provides a means of helping to sort stem cells. The populations of tumor cells that do not accumulate these dyes are referred to as "side population" cells. Stem cells are predominantly found in this population. This is a property that protects these long-living cells from foreign toxins and is not usually maintained upon differentiation. A side population of cells with high efflux capacity has also been identified in tumor cells. This suggests that CSCs have inherent drug resistance and may be an alternative mechanism to the acquired drug resistance discussed in Chapter 2. Therefore, therapeutic strategies involving the administration of ABC inhibitors, along with chemotherapies, are being investigated.

The rarity of CSCs in tumor cell populations and the difficulty of propagating them in culture made it difficult to use high-throughput screening approaches to search for new drugs. However, the knowledge that stem-like cells can be produced by induction of the EMT (see Chapter 9) in epithelial cells has provided a new resource that can be used for high-throughput screening for CSC-specific drugs. Gupta *et al.* (2009) reported the success of such an approach.

Next are several examples of drug strategies that target self-renewal or differentiation pathways.

8.3 Inhibitors of the Wnt pathway

The importance of the Wnt pathway in several cancers, particularly colorectal cancer, suggests that the molecular components of this pathway are good targets for new therapeutics (see "Pause and Think").

PAUSE AND THINK

What molecules would you target in this pathway?

Note that, since many cancers have mutations in the *Apc* and *Axin* genes, molecular targets upstream of these proteins in the Wnt signaling pathway would not be useful, as they are not needed for activation in these cases. Recall the previous example given in Chapter 4 where drugs that target EGFR do not work in the presence of Ras mutations. Strategies, such as the use of antibodies that block Wnt binding to its receptors, may be used in cases where Wnt or its receptors are overexpressed in the tumor.

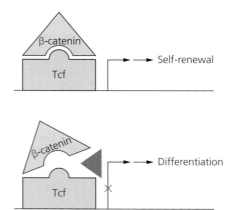

Figure 8.10 Drug strategy to inhibit the β-catenin–Tcf interaction.

Disruption of the protein–protein interaction between β-catenin and the Tcf transcription factors (Figure 8.10) is one strategy that has been investigated. This interaction occurs downstream of the APC degradation complex and is the endpoint effect of Wnt signaling. Drugs acting at this stage would counter both inactivating mutations in APC, axin, and GSK3β, and activating mutations in β-catenin that cause inappropriate formation of β-catenin–Tcf complexes. The observation that Tcf inhibition induces the differentiation of colorectal cancer cells into epithelial villi is evidence that supports this approach. From a high-throughput screen, Lepourcelet *et al.* (2004) identified three natural compounds that acted as inhibitors of the β-catenin–Tcf interaction. Also, these compounds that share a core chemical structure inhibited the expression of two Tcf target genes and inhibited proliferation of colorectal cancer cells. Additional promise has been seen in liver cancer and leukemia pre-clinical models (for references, see Polakis, 2012). Although these compounds have not yet been further developed and tested in the clinic, this strategy holds promise for the development of new cancer therapeutics. As these drugs target a molecular pathway that is important in self-renewal, they have a greater chance of tumor eradication, rather than just tumor regression.

8.4 Inhibitors of the Hh pathway

Inhibitors of the Hh signaling pathway are being investigated as cancer therapeutics (Figure 8.11). The first Hh pathway inhibitor cyclopamine, a steroidal alkaloid, was isolated from wild corn lilies. (Its name comes from the cyclopic effects—formation of a single eye—observed from its action as a teratogen. Pregnant sheep that ingested high quantities of wild corn lilies gave birth to cyclopic lambs.) Cyclopamine suppresses the Hh pathway by inhibiting the activity of the transmembrane protein Smoothened. As a result of the inhibition of Smoothened, transcription of target genes is repressed. In models of medulloblastoma (see Box "How do we know that?", p. 199), cyclopamine

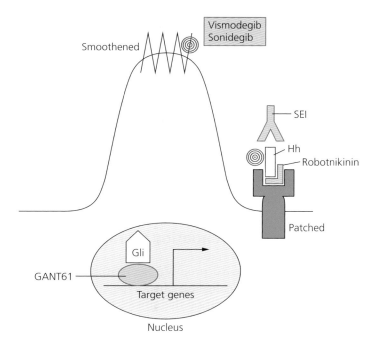

Figure 8.11 Some examples of drug strategies that inhibit the Hh pathway. Potential therapeutic agents are shown in red. Vismodegib (Erivedge™; GDC-0449) was approved for the clinic. Information from Amakye, Jagani, and Dorsch, 2013.

treatment blocked the growth of medulloblastoma cells and affected the regulation of genes involved in differentiation. Expression of a neuronal stem cell marker, neurofilament nestin, decreased, while expression of a marker of neuronal differentiation, Neuro D, increased. Cyclopamine does not possess appropriate drug characteristics (e.g. bioavailability, optimal pharmacokinetics). However, small-molecule inhibitors have been identified, some with a 100-fold increase in potency, compared to cyclopamine.

Small-molecule antagonists and Hh-blocking antibodies are being developed as means of inhibiting the Hh pathway for new cancer treatments. An orally active small-molecule inhibitor of Smoothened called vismodegib (Erivedge™; GDC-0449) was approved to treat BCC. Several other Smoothened inhibitors, such as sonidegib (LDE225), LY2940680, BMS-833923, and PF 04449913, are being tested in clinical trials, while some, such as saridegib and IPI-926, have been discontinued (Amakye *et al.*,

PAUSE AND THINK

What other approaches can you propose to inhibit the Hh signaling pathway? Start by thinking about molecular targets that can be accessed from the outside of the cell. What type of drugs can be used?

Yes, monoclonal antibodies that bind to Hh ligands are being developed. What about molecular targets that lie downstream of Smoothened? Molecules that target Gli are being investigated. One small molecule called GANT61 blocks the binding of Gli to DNA. See Figure 8.11.

 HOW DO WE KNOW THAT?

Pre-clinical testing in medullablastoma models

Transgenic mouse models that mimic the molecular defect of human tumors can be created such as mice that are **heterozygous** for the *Ptc1* gene and null for the *p53* gene. These mice have a 100% incidence of developing medulloblastoma. Growth inhibition by an inhibitor of the Hh pathway can be examined in **allografts** of these tumors and cell lines originating from the allografts. Allografts involve the transplantation of tissue from one individual to another. Tumor cells from the medulloblastomas can be taken from the transgenic mice and transplanted subcutaneously in nude mice. The effect on tumor growth by a drug that can enter the brain can also be examined *in vivo*. See references in Lau et al. (2012).

PAUSE AND THINK

The choice of experimental model for developing an anti-cancer drug is vital for predicting the outcome in patients. Which do you suppose produces the strongest evidence for proceeding to clinical trials?

Ex vivo analysis (xenografts, allografts) removes the tumor from its normal environment, and cell culture lines may not truly represent the disease in humans. Testing the effects of a drug on a tumor located in its normal anatomical location may be better.

2013; Ng and Curran, 2011). Some of these inhibitors block translocation of Smoothened to the cilia, and some do not.

8.5 Inhibitors of PcG proteins

As PcG proteins are "guardians of stemness," inhibitors of these proteins could target self-renewal. Proof of concept has been demonstrated in a study that investigated the effects of a small molecule inhibitor of Bmi-1 on colorectal CSCs. The data showed that the Bmi-1 inhibitor PTC-209 decreased CSCs and irreversibly inhibited tumor growth in colorectal cancer xenografts (Kreso *et al.*, 2014).

8.6 Leukemia and differentiation therapies

Differentiation therapy aims to promote the maturation and differentiation of cells, such that a malignant phenotype changes into a benign phenotype. The use of differentiation therapy for the treatment of APL, a subtype of AML, has been one of the great success stories of the last few decades, though the scenario is more complicated than originally thought. Retinoid treatment, using all-trans retinoic acid (ATRA), demonstrated complete hematological remission. However, most patients retained a small number of APL cells (stem cells, also called leukemia-initiating cells) and relapsed within a few months. This mismatch between initial response and long-term outcome was puzzling. The remaining cells had a high capability for self-renewal and were resistant to differentiation. Later, it was found that ATRA treatment, combined with arsenic trioxide, had synergenic effects that eliminated the resistant cells, and this combination became a cure for the disease. Pharmacological doses (high concentrations) of retinoids convert PML–RARα from a repressor to a transcriptional activator (Figure 8.9d) and induces its proteolysis. Normally, ligand-dependent activation of nuclear hormone receptors is often followed by degradation as a means of negative feedback for turning off the signal. Retinoid treatment in this context leads to the degradation of the oncogene. Arsenic trioxide induces the degradation of PML and PML–RARα by directly binding to PML and flagging it for degradation by the proteasome. PML regulates stem cell maintenance. To summarize, retinoid plus arsenic trioxide treatment enables the target genes of RARα to be expressed, and the block of differentiation is overcome. The gene encoding the differentiation-specific transcription factor C/EBP is one of the targets of retinoid treatment. In addition, the degradation of the PML–RARα oncoprotein by the proteasome is achieved. This may be another important mechanism of drug action, as cancers are often dependent on their oncogenes. This combination therapy is one of the most successful for the treatment of leukemia.

CHAPTER HIGHLIGHTS—REFRESH YOUR MEMORY

- Stem cells are characterized by their ability to self-renew and to form more differentiated progeny simultaneously. These properties are shared with CSCs.

- The CSC model states that subpopulations of cells with stem cell properties initiate and maintain the cancer phenotype.

- Stem cells are more likely to accumulate mutations, compared with other cells, and may often be the cell of origin for cancer.

- Deregulation of self-renewal pathways may contribute to cause initiation.

- Evidence suggests that the Wnt and Hh signaling pathways are involved in self-renewal.

- The transcriptional co-activator β-catenin is stabilized in the presence of Wnt.

- Mutations that inappropriately activate the Wnt signaling pathway promote carcinogenesis, particularly colon cancer.

- A germline mutation in the *APC* gene causes FAP.

- Hh signaling exerts its effects via the Gli zinc finger transcription factors.

- Inappropriate activation of the Hh pathway is linked to many cancers, particularly BCC.

- A germline mutation in the *Patched* gene causes Gorlin's syndrome.

- Both stem cells and malignant tumor cells migrate to other tissues in the body.

- The degree of differentiation of the transformed founder cell may determine metastatic potential.

- The PcG of proteins are nicknamed the "guardians of stemness."

- PcG proteins act as epigenetic gene silencers and maintain the stem cell state.

- AML is an important paradigm for the role of differentiation in cancer.

- Lineage-specific transcription factors are commonly mutated in AML.

- Drugs that target the self-renewal pathways of CSCs are more likely to achieve a cure.

- Interference with both the Wnt and Hh signaling pathways is a recent therapeutic strategy that is being explored.

- Retinoid therapy, with arsenic trioxide, is successful as a differentiation therapy for APL.

ACTIVITY

1. Stem cells of the intestinal crypt have been identified as the cells of origin for cancer of the intestine (Barker *et al.*, 2009). They used a genetically engineered inducible *cre* mouse *in vivo* system to examine Apc deletions in different cell compartments of the intestine. Learn and describe the details of the *cre* mouse system (start with the website: http://cre.jax.org/introduction.html). Draw a diagram to summarize your findings. Refer back to the paper by Barker *et al.*, 2009, and try to understand their experimental approach.

FURTHER READING

Anastas, J.N. and Moon, R.T. (2013) WNT signalling pathways as therapeutic targets in cancer. *Nat. Rev. Cancer* **13**: 11–26.

Bracken, A.P. and Helin, K. (2009) Polycomb group proteins: navigators of lineage pathways led astray in cancer. *Nat. Rev. Cancer* **9**: 773–784.

Clevers, H. and Nusse, R. (2012) Wnt/b-catenin signaling and disease. *Cell* **149**: 1192–1205.

Dean, M., Fojo, T., and Bates, S. (2005) Tumour stem cells and drug resistance. *Nat. Rev. Cancer* **5**: 275–284.

Dos Santos, G.A., Kats, L., and Pandolfi, P.P. (2013) Synergy against PML-RARa: targeting transcription, proteolysis, differentiation, and self-renewal in acute promyelocytic leukemia. *J. Exp. Med.* **210**: 2793–2802.

Holland, J.D., Klaus, A., Garratt, A.N., and Birchmeier, W. (2013) Wnt signaling in stem and cancer stem cells. *Curr. Opin. Cell Biol.* **25**: 254–264.

Lee, T., Jenner, R., Boyer, L., Guenther, M., Levine, S., Kumar, R., *et al.* (2006) Control of developmental regulators by polycomb in human embryonic stem cells. *Cell* **125**: 301–313.

Magee, J.A., Piskounova, E., and Morrison, S.J. (2012) Cancer stem cells: impact, heterogeneity, and uncertainty. *Cancer Cell* **21**: 283–296.

Morceau, F., Chateauvieux, S., Orsini, M., Trécul, A., Dicato, M., and Diederich, M. (2015) Natural compounds and pharmaceuticals reprogram leukemia cell differentiation pathways. *Biotech. Adv.* **33**: 785–797.

Orkin, S.H. and Zon, L.I. (2008) Hematopoiesis: an evolving paradigm for stem cell biology. *Cell* **132**: 631–644.

Pardal, R., Clarke, M.F., and Morrison, S.J. (2003) Applying the principles of stem-cell biology to cancer. *Nat. Rev. Cancer* **3**: 895–902.

Pattabiraman, D.R. and Weinberg, R.A. (2014) Tackling the cancer stem cells—what challenges do they pose? *Nat. Rev. Drug Discov.* **13**: 497–512.

Perez-Losada, J. and Balmain, A. (2002) Stem-cell hierarchy in skin cancer. *Nat. Rev. Cancer* **3**: 434–443.

Reya, T. and Clevers, H. (2005) Wnt signaling in stem cells and cancer. *Nature* **434**: 843–850.

Richly, H., Aloia, L., and Di Croce, L. (2011) Roles of the Polycomb group proteins in stem cells and cancer. *Cell Death Dis.* **2**: e 204.

Scales, S.J. and de Sauvage, F.J. (2009) Mechanisms of Hedgehog pathway activation in cancer and implications for therapy. *Trends Pharm. Sci.* **30**: 303–312.

Schepers, A. and Clevers, H. (2012) Wnt signaling, stem cells, and cancer of the gastrointestinal tract. *Cold Spring Harb. Perspect. Biol.* **4**: a 007989.

Smalley, M. and Ashworth, A. (2003) Stem cells and breast cancer: a field in transit. *Nat. Rev. Cancer* **3**: 832–844.

Sparmann, A. and van Lohuizen, M. (2006) Polycomb silencers control cell fate, development and cancer. *Nat. Rev. Cancer* **6**: 846–856.

Tang, D.G. (2012) Understanding cancer stem cell heterogeneity and plasticity. *Cell Res.* **22**: 457–472.

Vescovi, A.L., Galli, R., and Reynolds, B.A. (2006) Brain tumor stem cells. *Nat. Rev. Cancer* **6**: 425–436.

Visvader, J.E. and Lindeman, G.J. (2012) Cancer stem cells: current status and evolving complexities. *Cell Stem Cell* **10**: 717–728.

Vries, R.G.J., Huch, M., and Clevers, H. (2010) Stem cells and cancer of the stomach and intestine. *Mol. Oncol.* **4**: 373–384.

Werner, E. (2011) Cancer Networks: A general theoretical and computational framework for understanding cancer. http://arxiv.org/abs/1110.5865: 1–121.

■ **WEB SITE**

The Wnt homepage http://www.stanford.edu/group/nusselab/cgi-bin/wnt/

■ **SELECTED SPECIAL TOPICS**

Al-Hajj, M., Wicha, M.S., Benito-Hernandez, A., Morrison, S.J., and Clarke, M.F. (2003) Prospective identification of tumorigenic breast cancer cells. *Proc. Natl. Acad. Sci. USA* **100**: 3983–3988.

Amakye, D., Jagani, Z., and Dorsch, M. (2013) Unraveling the therapeutic potential of the Hedgehog pathway in cancer. *Nat. Med.* **19**: 1410–1422.

Barker, N., Ridgway, R.A., van Es, J.H., van de Wetering, M., Begthel, H., van den Born, M., *et al.* (2009) Crypt stem cells as the cells-of-origin of intestinal cancer. *Nature* **457**: 608–611.

Bushue, N. and Wan, Y.J. (2010) Retinoid pathway and cancer therapeutics. *Adv. Drug Deliv. Rev.* **62**: 1285–1298.

Chaffer, C.L., Brueckmann, I., Scheel, C., Kaestil, A.J., Wiggins, P.A., Rodrigues, L.O., *et al.* (2011) Normal and neoplastic nonstem cells can spontaneously convert to a stem-like state. *Proc. Natl. Acad. Sci. U.S.A.* **108**: 7950–7955.

Dow, L.E., O'Rourke, K.P., Simon, J., Tschaharganeh, D.F., van Es, J.H., Clevers, H., *et al.* (2015) Apc restoration promotes cellular differentiation and re-establishes crypt homeostasis in colorectal cancer. *Cell* **161**: 1539–1552.

Gupta, P.B., Onder, T.T., Jiang, G., Tao, K., Kuperwasser, C., Weinberg, R.A., *et al.* (2009) Identification of selective inhibitors of cancer stem cells by high-throughput screening. *Cell* **138**: 645–659.

Kemper, K., de Goeje, P.L., Peeper, D.S., and van Amerongen, R. (2014) Phenotype switching: tumor cell plasticity as a resistance mechanism and target for therapy. *Cancer Res.* **74**: 5937–5941.

Kreso, A., van Galen, P., Pedley, N.M., Lima-Fernandes, E., Frelin, C., Davis, T., *et al.* (2014) Self-renewal as a therapeutic target in human colorectal cancer. *Nat. Med.* **20**: 29–36.

Kwok, C., Zeisig, B.B., and So, C.W. (2006) Forced homo-oligomerization of RARalpha leads to transformation of primary hematopoietic cells. *Cancer Cell* **9**: 73–74.

Lau, J., Schmidt,C., Markant, S.L., Taylor, M.D., Wechsler-Reya, R.J. and Weiss, W.A. (2012) Matching mice to malignancy: molecular subgroups and models of medulloblastoma. *Childs Nerv. Syst.* **28**: 521–532.

Lepourcelet, M., Chen, Y.-N.P., France, D.S., Wang, H., Crews, P., Petersen, F., *et al.* (2004) Small-molecule antagonists of the oncogenic Tcf/beta-catenin protein complex. *Cancer Cell* **5**: 91–102.

Lessard, J. and Sauvageau, G. (2003) Bmi-1 determines the proliferative capacity of normal and leukemic stem cells. *Nature* **423**: 255–260.

Ng, J.M.Y. and Curran, T. (2011) The Hedgehog's tale: developing strategies for targeting cancer. *Nat. Rev. Cancer* **11**: 493–501.

Polakis, P. (2012) Drugging Wnt signaling in cancer. *EMBO J.* **31**: 2737–2746.

Radtke, F. and Clevers, H. (2005) Self-renewal and cancer of the gut: two sides of a coin. *Science* **307**: 1904–1909.

Rosen, J.M. and Jordan, C.T. (2009) The increasing complexity of the cancer stem cell paradigm. *Science* **324**: 1670–1673.

Rosenbauer, F. and Tenen, D.G. (2007) Transcription factors in myeloid development: balancing differentiation with transformation. *Nat. Rev. Immunol.* **7**: 105–117.

Schuijers, J. and Clevers, H. (2012) Adult mammalian stem cells: the role of Wnt, Lgr5 and R-spondins. *EMBO J.* **31**: 2685–2696.

Takahashi, K., Tanabe, K., Ohnuki, M., Narita, M., Ichisaka, T., Tomoda, K., *et al.* (2007) Induction of pluripotent stem cells from adult human fibroblasts by defined factors. *Cell* **131**: 861–872.

Tenen, D.G. (2003) Disruption of differentiation in human cancer: AML shows the way. *Nat. Rev. Cancer* **3**: 89–101.

Visvader, J.E. and Lindeman, G.J. (2008) Cancer stem cells in solid tumors: accumulating evidence and unresolved questions. *Nat. Rev. Cancer* **8**: 755–768.

Von Hoff, D.D., Lo Russo, P.M., Rudin, C.M., Reddy, J.C., Yauch, R.L., Tibes, R., *et al.* (2009) Inhibition of the Hedgehog pathway in advanced basal-cell carcinoma. *N. Engl. J. Med.* **361**: 1164–1172.

Yilmaz, O.H., Valdez, R., Theisen, B.K., Guo, W., Ferguson, D.O., Wu, H., *et al.* (2006) *Pten* dependence distinguishes haematopoietic stem cells from leukaemia-initiating cells. *Nature* **441**: 475–482.

Zhao, C., Chen, A., Jamieson, C.H., Fereshteh, M., Abrahamsson, A., Blum J., *et al.* (2009). Hedgehog signaling is essential for maintenance of cancer stem cells in myeloid leukemia. *Nature* **458**: 776–779.

Chapter 9

Metastasis

Introduction

Most cells of the body normally remain resident within a particular tissue or organ (though hematopoietic cells are a notable exception). Liver cells remain in the liver and cannot be found in the lung, and vice versa. Organs have well-demarcated boundaries defined by surrounding **basement membranes**. Basement membranes are acellular structures made up of a fabric of extracellular matrix (ECM) proteins: predominantly laminins, type IV collagen, and proteoglycans. Cancer is distinctly characterized by the spreading of tumor cells throughout the body. The process by which tumor cells from a primary site invade and migrate to other parts of the body is called **metastasis**. Metastasis is the fundamental difference between a benign and malignant growth, and represents the major clinical problem of cancer. A primary tumor can be surgically removed relatively easily, whereas, once hundreds or more metastases have been established throughout the body, they are practically impossible to remove. Sadly, many solid tumors have initiated metastasis at the time of diagnosis.

The spread of cells throughout the body results in physical obstruction, competition with normal cells for nutrients and oxygen, and invasion and interference with organ function. Interestingly, specific cancers metastasize to particular sites, and this is called **organotropism**. Many of the preferences observed for the spread of specific cancers to specific metastatic locations can be explained by the directionality of blood flow, as first proposed by James Ewing in the 1930s. As the bloodstream is the predominant means of long-distance transport, organs in close proximity "*en route*" are likely to be main sites of metastasis for a particular primary tumor. However, about one-third of the locations of frequent metastases is puzzling in this regard. For example, a particular kidney cancer often metastasizes to the thyroid, a relationship that cannot be explained simply by anatomy. One explanation of this observation was described over 100 years ago in the "seed and soil" theory proposed by Stephen Paget. It described cancer cells as "seeds" requiring a match with optimal environments or "soils" to succeed. The ability of cancer cells to metastasize is dependent on the interactions of their cell surface molecules with the microenvironment, including neighboring cells and the ECM. Molecular evidence suggest that receptors lining the capillaries in the organs to which cancer spreads influence the destination of metastasized cells, and these findings support the "seed and soil" theory. This theory is also supported by the more recent concept, proposed by David Lyden and colleagues, of the establishment of a **pre-metastatic niche**, a site of future metastasis that is altered in preparation for the arrival of tumor cells. However, unlike Paget's theory where the "soil" is already established, in the pre-metastatic niche theory, the primary tumor itself prepares the distant "soil" or microenvironment via tumor-secreted factors. In addition, germline mutations may play a role in the susceptibility of an individual to develop metastases. This is supported by studies using mice with different genetic backgrounds. Further research is needed to discover the factors required for a tumor cell to be successful in metastasis. Although cancers are largely successful in metastasizing in the long run, on the cellular level, only 1 in 10,000 metastasizing cells can establish a recognized metastatic lesion.

PAUSE AND THINK

Why is the spread of cells throughout the body lethal?

9.1 How do tumors spread?

Metastasis is the spread of tumor cells from a primary tumor that, as we have learned from next-generation sequencing, is not clonal. A primary tumor is composed of subpopulations of genetically identical cells, called subclones, that differ by mutations obtained through an ongoing evolutionary process (discussed in Chapter 1). This leads us to try to understand how clonal evolution affects metastasis. Spreading can be either monoclonal (being seeded by one cell or subclone) or polyclonal (being seeded by two or more subclones). The pattern of spreading can be either linear (from primary tumor to metastasis) or branched (with one primary tumor seeding two or more other metastases). There is evidence from whole-genome sequencing of metastases and primary tumors that all possible combinations occur (Gundem *et al.*, 2015; Figure 9.1). Interestingly, the data demonstrated that metastases can be seeded not only from cells of the primary tumor, but also from subclones from other metastases. This is referred to as cross-seeding, and again the pattern could be either linear or branched. Some metastases within a patient are more closely related to each other than to the primary tumor.

9.2 The process of metastasis

Tumor cells undergo several major steps during metastasis: invasion, intravasation, transport, extravasation, and metastatic colonization (Figure 9.2). An additional layer of complexity must be added to this

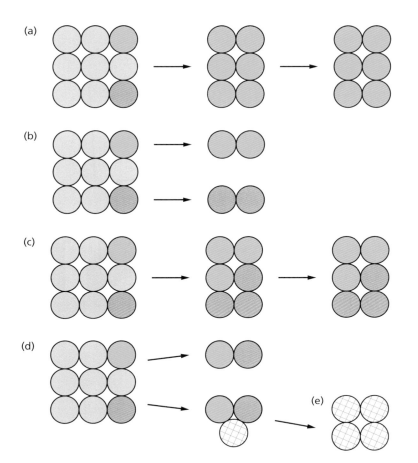

Figure 9.1 Ways that metastases spread: (a) monoclonal linear, (b) monoclonal branched, (c) polyclonal linear, (d) polyclonal branched, and (e) cross-seeding. New metastases may arise from both the primary tumor and other metastases.

model, because a tumor cannot be viewed simply as an isolated mass of cancer cells, but instead as a tissue that recruits and receives signals from surrounding normal cells, called the tumor microenvironment or tumor-associated stroma. The interaction of tumor cells with the tumor microenvironment affects the metastatic ability of individual cancer cells: not all cells of a primary tumor contain the same ability to metastasize. The subpopulation of CSCs in a primary tumor also affects metastatic potential. The interaction of the primary tumor with distant locations via signaling molecules to create a future site of metastasis has also been demonstrated (see Section 9.7). Thus, both individual tumor cells and the tumor microenvironment play a role in the process of metastasis. The interactions between cancer cells and their microenvironment are complex, but they are beginning to be understood. As we examine each of the major steps that a tumor cell takes during metastasis, take note of the contributions made by the tumor and host microenvironments, as these too play an important role in metastasis. Also, try to evaluate the molecular

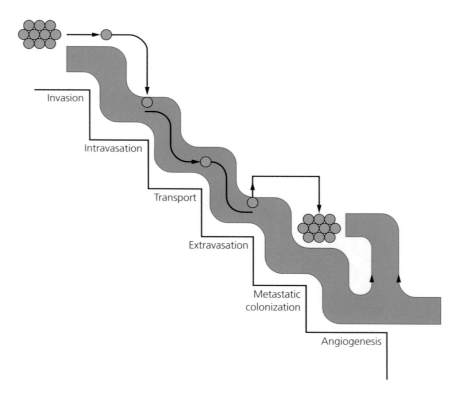

Figure 9.2 Steps of metastasis.
Note: Angiogenesis is required for
metastatic colonization and is fully
discussed in Chapter 10.

components as possible therapeutic targets, and think about strategies
that may be used to develop new drugs. Examples of therapeutic strategies will be described at the end of the chapter.

9.3 Invasion and the epithelial–mesenchymal transition

In order for cells of the primary tumor to invade into their local environment (Figure 9.2, top left), cells must break free from the normal molecular
constraints that link adjacent cells to each other and acquire migratory
characteristics. The acquisition of the characteristics needed for invasion
may be orchestrated by a mechanism called the epithelial–mesenchymal
transition (EMT) which involves the conversion of a sheet of closely connected epithelial cells into highly mobile mesenchymal cells. (Note that
the term "transition" reflects its transient nature, as EMT is reversible by
the process termed the mesenchymal–epithelial transition.) EMT is common in early embryogenesis. The early embryo requires cell invasion and
motility during early development to orchestrate pattern formation. For
example, gastrulation requires coordinated cell movement and detach-
ment, and invasion from/into adjacent tissue. Many of the cell processes
that are active in early embryogenesis are reactivated in cancer cells. EMT
is characterized by loss in cell polarity, deconstruction of epithelial cell–cell

junctions, changes in cell shape, downregulation of epithelial markers such as E-cadherin, upregulation of mesenchymal proteins such as N-cadherin, secretion of specific proteases, and increased cell protrusions and motility. These molecular characteristics are observed in migrating tumor cells.

Induction of EMT

Signals from the tumor stroma (e.g. growth factors hepatocyte growth factor (HGF), EGF, PDGF, and transforming growth factor-beta (TGF-β)) induce EMT in neighboring tumor cells via their specific receptors (MET receptor, EGFR, PDGF-R, and TGFR, respectively). Note that these receptors are kinase receptors. Ligand-bound receptors activate signal transduction pathways (such as the MAPK or PI3K pathway) and ultimately activate specific transcription factors (e.g. Twist, Snail, Slug, ZEB1, Goosecoid, FOXC2). These transcription factors control sets of genes needed to trigger EMT (Figure 9.3). For example, Snail binds to E-box sequences in epithelial genes, such as those in the promoter region of the E-cadherin promoter, and recruits the Polycomb repressor complex. This leads to histone modifications and epigenetic regulation to repress gene expression. In a comparison of tumors derived from cell lines with varying capabilities for metastasis, gene expression profiling revealed the transcription factor Twist as one of the most differentially expressed genes. Inhibition of Twist resulted in the loss of several steps of metastasis, including intravasation. Twist expression has been documented for several human tumors (e.g. metastatic melanoma). These findings suggest that Twist is a key regulator of metastasis.

The relationships between cancer cells and stem cells were discussed in Chapter 8. Evidence, such as the induced expression of stem cell markers, suggests that EMT produces stem cell-like cells (Mani *et al.*, 2008). Thus,

Figure 9.3 The epithelial–mesenchymal transition (EMT). Cells of the stroma release EMT-inducing signals such as TGFβ and HGF. These signals activate their corresponding receptors TGFβR and MET receptor on tumor cells, and the signal is transduced to the nucleus. This leads to the activation of EMT transcription factors, such as Twist, which, in turn, regulate a set of genes whose products direct EMT. Mesenchymal cells express N-cadherin (N-cad) and secrete proteases such as matrix metalloproteinases (MMPs).

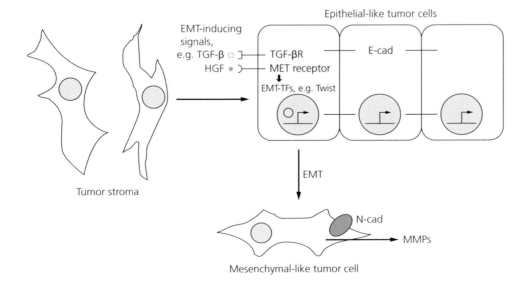

EMT not only enables tumor cells to acquire a migratory phenotype, but also promotes their self-renewal capability-qualities that are essential for metastasis, including successful colonization. Thus, it is proposed that metastasis-initiating cells are CSCs. Gene signatures of early-stage human metastatic breast cancer cells support this view in that they show increased expression of stem cell- and EMT-associated genes (Lawson *et al.*, 2015).

Tools of invasion: cell adhesion molecules, integrins, and proteases

Let us examine a few groups of molecules that are important tools of invasion: cell adhesion molecules (CAMs), integrins, and proteases.

Cell adhesion molecules

CAMs and cadherins are two families of proteins that mediate homotypic (same cell type) and heterotypic (different cell types) cell recognition. They "hook" cells into place extracellularly (Figure 9.4). Cadherins (Figure 9.4a) are calcium-dependent transmembrane glycoproteins that interact, via catenins (Figure 9.4b), with the cytoskeleton (Figure 9.4c).

Catenins also bind to transcription factors and induce gene expression in the nucleus. Thus, intercellular interactions are networked to mediators of intracellular functions.

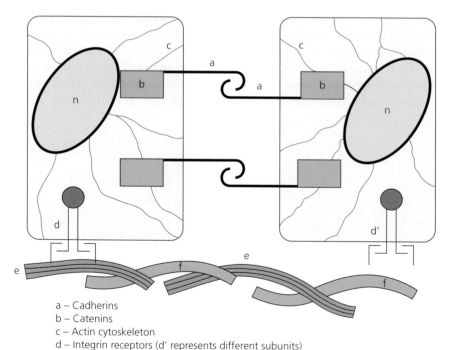

a – Cadherins
b – Catenins
c – Actin cytoskeleton
d – Integrin receptors (d' represents different subunits)
e – Collagen
f – Fibronectin
n – Nucleus

Figure 9.4 Cell adhesion molecules and associated components.

Several lines of evidence support the role of cadherins in metastasis. E-cadherin is the predominant CAM in epithelial cells that plays a role in cell-cell junctions. E-cadherin acts as a tumor suppressor that normally functions to secure cell–cell adhesion and suppresses metastasis of tumor cells to distant sites, and, as noted in the discussion above, it is downregulated during EMT. Cells treated with antibodies that block the function of E-cadherin became invasive in collagen gels, indicating an increased metastatic potential. Transfection of the *E-cadherin* gene into metastatic epithelial cells can render them non-invasive. Mutations in the extracellular domain and methylation in the promoter region of the *E-cadherin* gene have been identified in gastric and prostate carcinomas, as expected for a tumor suppressor gene (see "Pause and think").

Cell behaviors are influenced by specific cadherin family members. Cadherin switching plays a role in gaining a migratory phenotype characteristic of the EMT.

Integrins

Cells must also break free from the normal molecular constraints by the ECM. Integrin receptors (Figure 9.4d) are a family of more than 24 heterodimers made up of a range of α and β subunits (Figure 9.5) that mediate cell–ECM interactions and intracellular signal transduction. The recognition of the different components of the ECM, for example collagen (Figure 9.4e), fibronectin (Figure 9.4f), or laminin, depends on the α and β subunit composition. Many ligands for integrin receptors contain the three-amino acid sequence Arg(R)–Gly(G)–Asp(D) (where the letter in parentheses represents the single letter abbreviation for the corresponding amino acid) that is involved in integrin binding. Upon ligand binding, the integrins cluster in the membrane and affect the cytoskeleton through interaction with actin-binding proteins and specific kinases such as focal adhesion kinase (FAK). In contrast to most transmembrane receptors, the cytoplasmic tail of integrins does not exhibit any catalytic activity (e.g. kinase activity) itself. Data suggest that FAK mediates cell motility through recruitment of Src and activation of the RAS pathway (discussed in Chapter 4). In addition to this typical outside-of-the-cell to inside-of-the-cell signaling, integrins also mediate "inside–outside" signaling. Intracellular signals mediated at the cytoplasmic domain of integrins induce a conformational change in the extracellular domain and thus regulate the affinity of the integrins for their ECM ligands. Integrins also have a role in anoikis—apoptosis triggered in response to lack of ECM ligand binding and loss of cell adhesion. Integrins without suitable ECM ligands recruit caspase-8 to the membrane and trigger apoptosis. Thus, integrin-dependent cell anchorage is crucial for survival of the cell.

Altered integrin receptor expression in tumor cells is observed during EMT and can enable the mobility and invasion of metastasizing cells by modifying membrane distribution and/or allowing adherence to different ECMs. Regulation by the cancer cell must result in precise intermediate

PAUSE AND THINK

Classify the types of evidence presented here that support the idea that E-cadherin is a tumor suppressor gene, using the description of types of evidence presented in Box "How do we know that?" in Chapter 1, on p. 9.

strengths of adhesion to produce the maximum rate of cell migration, allowing cells to advance their leading edge and to release their lagging edge. The role of integrins in motility is obvious in melanoma cells in which their invasive front edge shows a strong pattern of expression of integrin $\alpha v\beta 3$ that is absent in preneoplastic melanomas. An altered expression of a specific integrin heterodimer may be permissive for invasion. For example, an increase of $\alpha 6\beta 4$, a laminin-binding integrin, promotes invasion through the basement membrane and the laminin matrix often secreted by epithelial tumors. Further still, altered integrin expression may facilitate invading cells to overcome anoikis.

Proteases

Invasion of tumor cells into the surrounding tissue requires the action of specific proteases that degrade a path through the ECM and stroma. Increased expression of specific proteases is a characteristic of EMT.

Serine proteases and matrix metalloproteinases (MMPs) are two families that are important. MMPs can cleave the extracellular domain of E-cadherin and so can contribute to the loss of epithelial cell–cell junctions seen during EMT. Although some tumor cells can synthesize MMPs, more often tumor cells induce surrounding stromal cells to produce MMPs. One appropriately named protein, called extracellular matrix metalloproteinase inducer (EMMPRIN), is upregulated on the membrane of tumor cells and induces production of MMPs in adjacent stromal cells. The family of MMPs can not only degrade structural components of the ECM, but also cleave other proteins residing on the outside of cells (e.g. endothelial cell growth factors), and thus are likely to play an important role in metastasis, including intravasation (see Section 9.4 "Intravasation"). Normally, these zinc-dependent proteinases are tightly regulated at several levels, in addition to gene expression. First, they are synthesized as latent enzymes and require proteolytic cleavage to be activated. Also, endogenous tissue inhibitors (TIMPs) regulate their function. A tip in the balance between MMPs and TIMPs can signal invasion. MMPs are upregulated in almost all tumors, and their expression profile can indicate the degree of tumor progression in some cancers.

9.4 Intravasation

Intravasation is the entry of a tumor cell into a blood or lymphatic vessel. The process requires several steps: the tumor cell must attach to the stromal face of the vessel, degrade the basement membrane (absent in lymphatic vessels) using MMPs and serine proteases, and pass between the endothelial cells (transendothelial migration) into the bloodstream. The structural features of tumor-associated blood vessels influence the mechanics of

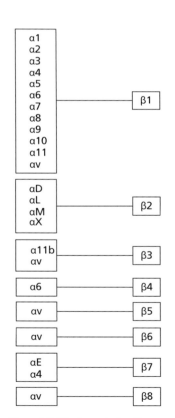

Figure 9.5 The integrin family: α and β subunit heterodimers.

intravasation. New blood vessels stimulated by the tumor are leaky and tortuous, allowing easy access, compared with blood vessels in normal tissues. Direct visualization in animal models by multiphoton microscopy has established that the process of intravasation is assisted by tumor-associated macrophages which "guide" tumor cells to vessels (Wyckoff *et al.*, 2007— videos are freely available in supplementary data online). This process involves the colony-stimulating factor 1 (CSF1) receptor on macrophages and EGFR on tumor cells. Macrophages associate with blood vessels and are a source of EGF that bind EGFR on tumor cells; tumor cells produce CSF1 that can interact with CSF1 receptors on macrophages, leading to chemotaxis-mediated co-migration. Antibodies that block CSF1 receptor function decreased intravasation, compared to controls. Further research to understand additional molecular events involved in this process will be needed to develop therapeutic strategies that target intravasation.

9.5 Transport

Tumor cells within the bloodstream are called circulating tumor cells (CTCs). They give us the potential to "see" metastasis. Transport through the bloodstream is "one-way." Tumor cells travel singly or as clumps with platelets, called emboli, in the direction of blood flow. Emboli may protect tumor cells from sheer forces and immune cells inside the bloodstream. Specific cancers have favored sites of metastasis, and, for some cancers, this is explained by the concept of the first-pass organ. The first-pass organ is the first organ *en route* via the bloodstream that lies downstream from the primary tumor site. The large size of tumor cells (20–30 µm) relative to the diameter of capillaries (about 8 µm) suggests that a large fraction of tumor cells get physically trapped in the first capillary beds that they encounter. The lung is the first-pass organ for cells of the breast via the superior vena cava that also receives drainage from the lymphatic vessels. Thus, the lung is a common site of metastasis from breast cancer. The liver is the first-pass organ for cells via the hepatic portal vein and is particularly vulnerable because of sinusoids, areas where blood is in direct contact with hepatocytes. Therefore, the liver is a common site of metastasis from pancreatic or colorectal cancer. Overall, only a very small proportion of tumor cells that enter the bloodstream are successful in setting up a metastatic colony.

9.6 Extravasation

Extravasation is the exit of a tumor cell from a blood or lymphatic vessel. It is important to note that the location of extravasation, along with the microenvironment of the stroma, influences the site of metastases.

The steps of extravasation are similar to intravasation, but in reverse: the tumor cell must attach to the endothelial side of the blood vessel, pass through the endothelial cells and basement membrane, and migrate into the surrounding stroma. Although intravasation and extravasation are processes that require crossing the endothelium (transendothelial migration), the mechanisms involved differ, because they approach the endothelium from opposite sides. Note that the structural features of a blood vessel at a distant site from the primary tumor differ from the new vasculature induced by the tumor where intravasation occurs, and thus extravasating cells require different mechanisms, compared with intravasating cells.

Tumor cells are arrested in the circulation as a result of mechanical entrapment, as mentioned earlier, and form molecular links to blood vessel cells through ligands and receptors on cancer cells and endothelial cells. Members of the selectin family of adhesion molecules are important for the attachment of cancer cells to the endothelium, particularly E-selectin that is specifically expressed on endothelial cells in response to inflammatory cytokines produced by cancer cells or associated white blood cells. They are calcium-dependent transmembrane receptors that mediate interactions with cancer cells by binding to various glycoprotein ligands presented on adhering cells. Endothelial selectins are differentially expressed on the vasculature of different organs and may support the "seed and soil" theory discussed previously. Signaling between the selectins and their ligands appears to be bidirectional, in that signal transduction in both participating cells has been demonstrated. That is, signaling initiates from both the selectin cytoplasmic tails and from their activated ligands. For example, cross-linking of E-selectin induces tyrosine phosphorylation in *endothelial cells* and also modifies endothelial cell shape. In contrast, stress-activated protein kinase-2 (SAPK2/p38), an isoform of MAPK, is induced in *cancer cells* upon binding of E-selectin (on endothelial cells) and is necessary for transendothelial migration. This suggests that binding to E-selectin on the endothelium by cancer cells not only mediates adhesion to the endothelium, but also triggers a signal transduction cascade that is important for transendothelial migration by the cancer cells. (CD44 and MUC1 are examples of molecules expressed on cancer cells that bind selectins.)

Transendothelial migration of cancer cells involves moving in between endothelial cells through endothelial cell junctions (paracellular transendothelial migration). A number of molecules have been proposed from *in vitro* studies to be involved in this process, including vascular endothelial growth factor (VEGF) (see Chapter 10) and molecules involved in the structure of the junction (Rho and Rac GTPases).

(Watch this space: Though controversial, there is some limited evidence that some cancer cells can pass directly through endothelial cells, a process called transcellular transendothelial migration.)

The final step of extravasation is invading the basement membrane. Although proteases are known to be important for this step in intravasation, there is a lack of evidence for their role in extravasation, and further studies are needed to reveal the molecular events of this process.

9.7 Metastatic colonization

Sites of metastasis not only are influenced by the location of extravasation, but also depend upon the ability of tumor cells to adapt to specific organ microenvironments to form a *successful* metastatic colony. The words in the term "metastatic colonization" have been precisely chosen to describe the last stage of metastasis. Let us examine the "familiar" concept of colonization. The British established distant colonies in the New World, the growth of which was dependent on the environment, including the surrounding waterways and harbors. As the settlers moved west, some of the new environmental conditions were unfavorable for growth. Other locations encouraged the development of a new means of water access and resulted in a flourishing settlement. Metastatic colonization is the establishment of a progressively growing tumor at a distant site, involving the formation of new blood vessels as an essential process to provide nutrients and oxygen. Evidence suggests that tumor cells that have already reached a secondary organ, but have not yet grown to become a metastasis, can occur early during carcinogenesis, sometimes from the earliest signs of the formation of a primary tumor. These cells that have spread but have not yet colonized are called disseminated tumor cells (DTCs). Since many tumor cells in a metastasis show an epithelial morphology, it is hypothesized that such tumor cells need to undergo a mesenchymal-to-epithelial transition (a reversal of EMT) to be successful.

It is important to contrast successful colonization with the situation of DTCs that do not expand and remain dormant for years as micrometastases. Micrometastases maintain an overall balance between proliferation and apoptosis, or enter a state of quiescence (non-proliferation that is reversible) and do not demonstrate progressive growth. In some cases, this may be due to a period of time before angiogenesis or by action of the immune system. Molecular factors that may induce quiescence in DTCs include stress signaling, genes involved in quiescence of normal cells, and dormancy-permissive microenvironments. Data suggest that factors from stable blood vessels induce DTC quiescence (as opposed to new blood vessels that promote outgrowth) (Ghajar *et al.*, 2013). Alternatively, DTCs stimulated by interactions with their new environment may grow into secondary tumors called metastases. Although a few molecular players in this process have been identified (e.g. fibronectin), further studies into the regulation of metastatic colonization are greatly needed.

The pre-metastatic niche

The "seed and soil" theory states that metastases selectively colonize specific organs because of a "match" between the migrating tumor cell and a "suitable" environment. The concept of a pre-metastatic niche, a site of future metastasis that is altered as a result of factors released by the primary tumor, in preparation for the arrival of tumor cells supports the "seed and soil" theory. Evidence for the pre-metastatic niche was presented in a ground-breaking paper (reviewed by Kaplan *et al.*, 2006 and Oppenheimer, 2006). The migration of green fluorescently labeled bone marrow-derived cells in mice was examined after the inoculation of specific red fluorescently labeled mouse tumor cells. The data demonstrated that green-labeled cells arrive at common sites of metastasis and form clusters before the arrival of red-labeled tumor cells (Figure 9.6). The bone marrow-derived cells express VEGF receptor 1 (VEGFR-1) (discussed in Chapter 10) and integrin VLA-4 that binds to fibronectin-rich environments induced by tumor-specific growth factors, facilitating cell clustering.

Importantly, further experiments showed that factors released by the primary tumor orchestrate the formation of the pre-metastatic niche in selective locations. Conditioned media from one type of tumor were able to redirect the location of metastatic colonization of

Figure 9.6 Evidence of the pre-metastatic niche. Top row: immunochemistry to examine the arrival of bone marrow-derived cells (green) and tumor cells (red) in the lung. Bottom row: a schematic representation of the data shown in the top row. R1, bone marrow-derived cells (green); R2, endothelial progenitor cells (blue); fibronectin (FN) (yellow); fibroblasts (FB) (orange); TB, terminal bronchiole; BV, bronchial vein. Reproduced from Kaplan, R.N., *et al.* (2006) Preparing the soil: The premetastatic niche. *Cancer Res.* **66**: 11089–11093, with permission from the American Association for Cancer Research.

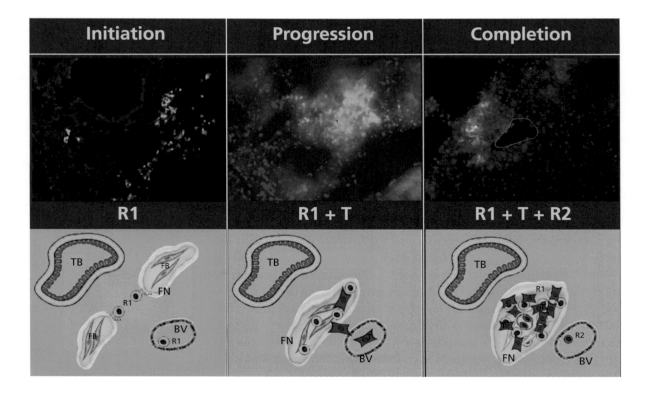

another tumor type. For example, melanoma-conditioned media repro-grammed the metastatic profile of mouse lung cancer cells to locations normally colonized by melanoma cells. Similar results were obtained when human cancer cells were tested. This suggests that the establishment of the pre-metastatic niche is a prime mediator of organotropism. As discussed in Kaplan *et al.* (2006) and Oppenheimer (2006), these data suggest that tumor type-specific factors released from the primary tumor facilitate changes to the microenvironment of a distant and *future* colonization site before tumor cells arrive. Bone marrow cells respond to these systemic factors by migrat-ing to the pre-metastatic niche and are involved in preparing a favorable environment for the cancer cells to colonize (for more on migrating bone marrow cells and cancer, see also p. 315). There is now evidence that these systemic factors include tumor-secreted exosomes (see "Exosomes").

Exosomes

Exosomes are small vesicles (30–100 nm) that carry protein and nucleic acids. They are packaged in multivesicular bodies that fuse with the cell membrane and are released into the circulation. Exosomes are an impor-tant means of intercellular communication between cancer cells and non-cancer cells in their microenvironment and further away. They are partly responsible for the systemic effects that primary tumors exert to influence the metastatic process. They can carry and transfer DNA, RNA, and pro-tein to cells to which they can fuse, and this is referred to as horizontal transfer. The result of cancer exosome transfer can provide "education" to, and change the behavior of, their receiving cells. This was described in a melanoma mouse model system by David Lyden and colleagues who provided data that tumor-derived exosomes act as systemic factors that deliver biomolecules, such as tyrosine kinase receptor MET, and "educate" bone marrow-derived cells to mobilize to a pre-metastatic niche (Peinado *et al.*, 2012). They showed that, if mice were treated with concentrated tumor exosomes and then used as a bone marrow donor, this "educated" bone marrow would increase the number of metastasis-associated bone marrow-derived cells and the metastatic burden. They also tried to block the function of exosomes by using RNA interference to reduce exosome production, and this showed reduced bone marrow-derived cell mobiliza-tion and metastasis. This research group have also identified a role for exosomes in pancreatic cancer. Pancreatic exosomes promote the estab-lishment of a pre-metastatic niche for liver metastasis in a stepwise fashion (Costa-Silva *et al.*, 2015). First, pancreatic exosomes selectively fuse with Kupffer cells in the liver. Kupffer cells upregulate genes involved in fibrosis, including TGF-β, in response to macrophage migration inhibitory factor enriched in the exosomes. TGF-β then induces the expression of fibronec-tin in another liver cell type, stellate cells. Fibronectin deposits recruit bone marrow-derived cells, an important component of the pre-metastatic niche, as also described earlier. Thus, the microenvironment is altered and supports the survival and growth of pancreatic tumor cells.

These studies using injections of concentrated exosomes or preconditioned cells strongly suggest that exosomes are important in metastasis, but direct evidence for horizontal transfer was lacking until recently. *In vivo* imaging and the Cre–LoxP system have been used to visualize extracellular vesicle transfer between tumor cells in living mice (Zomer *et al.*, 2015). A DNA construct that would express GFP only if the cell received the regulator CRE-recombinase was used. Upon exposure to CRE-containing exosomes, receiving cells became fluorescent green, indicating transfer from exosomes had occurred. Similar to previous studies, uptake of cancer exosomes by less malignant cells enhanced metastatic capacity.

The transfer of miRNAs is an efficient means of significantly altering gene expression and reprogramming receiving cells. The miRNAs carried by exosomes and other extracellular vesicles, such as ectosomes that bud from the cell membrane, have been shown to promote metastasis (Le *et al.*, 2014) and even stimulate non-tumorigenic cells *in vitro* to form tumors (Melo *et al.*, 2014). New data regarding exosomes are contributing to our understanding of the molecular mechanisms that underlie the "pre-metastatic niche" theory. Membrane-bound molecules present on tumor-derived exosomes have been shown to act as an "address" for organotropic delivery of exosomes and organotropic metastasis (Hoshino *et al.*, 2015). Integrin subunit composition was identified as the molecular mechanism responsible for organ-specific uptake and for initiating the formation of the pre-metastatic niche at these sites. Exosomes of one tumor type were shown to redirect the pattern of metastasis of a different tumor type, and integrin-blocking peptides could block exosomal delivery to specific organs. These findings provide hope for future therapeutic interventions.

Metastasis suppressor genes

A new class of genes, called **metastasis suppressor genes**, have been identified by their low expression in metastatic cells, compared with non-metastatic tumor cells. They are defined by their ability to inhibit overt metastasis without affecting the growth of the primary tumor. Many products of these genes are involved in the regulation of growth of metastatic cells at secondary sites and may maintain these cells as stable, non-proliferating dormant cells. Thus, loss of function (analogous to the mechanism of tumor suppressor genes) increases the metastatic propensity of a cancer cell. Approximately 23 metastasis suppressor genes have been identified since the discovery of the first, *NM23*. The protein product of the *NM23* gene functions as a nucleoside diphosphate kinase and a histidine kinase, and it is suggested that the latter may contribute to its metastasis inhibitory effects. *MKK4* (mitogen-activated protein kinase kinase 4) is a metastasis suppressor gene whose protein product affects metastatic colonization. It is hypothesized that MKK4 protein induces apoptosis in response to the stress of a new microenvironment and thus suppresses metastatic colonization. Both NM23 and MKK4 appear to promote dormancy of micrometastatic colonies. The mechanism of action of metastasis suppressor proteins includes regulation of common signal transduction pathways (e.g. MAPK) and gap junction communication, although these are only beginning to be elucidated. In addition, a new class of miRNAs that suppresses metastasis, called metastasis suppressor miRNAs, has been identified and includes miR-335 and miR-126 (see Section 3.6, "MicroRNAs (miRNAs) and regulation of mRNA expression" on p. 68).

◎ Therapeutic strategies

Unlike our success in studying abnormal cell growth, the study of metastasis has not been easy. Our incomplete understanding of the complexities of metastasis is the reason that targeted drugs for prevention and treatment of metastasis are lacking. Please note that this section discusses many more *potential* strategies than it does approved therapeutics.

Perhaps, in theory, one could envisage a therapy targeted at each of the major steps of metastasis. Protease and integrin inhibitors are obvious molecular targets to block invasion, but the development of such drugs has been met with mixed success. More recently, inhibiting EMT pathways and inducing metastasis suppressors are providing new strategies for drug development. Perhaps the best target would be to block the final step of metastatic colonization, instead of merely attempting to block escape.

Some examples of therapeutic strategies that target the processes of metastasis are discussed in the following sections (Figure 9.7).

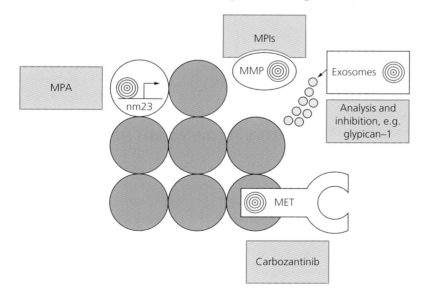

Figure 9.7 Therapeutic strategies that target metastasis. Carbozantinib is approved for clinical use.

9.8 Metalloproteinase inhibitors (MPIs)

There was a rapid response by pharmaceutical companies to develop MPIs (such as Marimastat and Neovastat) because of the evidence of the role of MMPs in metastasis. These molecules appear to function in several steps of metastasis, including invasion and intravasation. The initial wave of clinical trials proved to be disappointing, although informative for future trials. The development of drugs exceeded the rate at which basic research was able to uncover the details of the MMP family. First, about 24 different family members have been identified. The knowledge of temporal and spatial expression patterns, functional roles, and roles in different cancers of the individual family members lagged behind the initial development of small molecule inhibitors and natural product drugs. The trials were hindered by unexpected side effects (musculoskeletal pain) and poor design. There was difficulty in measuring efficacy, and the drugs were administered only to patients with advanced disease, even though pre-clinical evidence suggested that administration at early stages of disease was crucial. No MPIs have received approval as a cancer therapy thus far. Targeting membrane-bound MMPs by selective therapeutic antibodies is a more recent strategy for the development of the next generation of MPIs (Devy and Dransfield, 2011) and promises to improve upon the severe toxicities associated with the broad-spectrum inhibitors developed earlier.

c-MET inhibitors for blocking EMT

The tyrosine kinase receptor MET (**m**esenchymal-**e**pithelial **t**ransition factor; also known as the hepatocyte growth factor receptor) and its ligand hepatocyte growth factor (HGF) play a role in EMT that is important for motility and invasion of cancer cells, in addition to other processes important in carcinogenesis. A small molecule tyrosine kinase inhibitor that targets c-MET and VEGFR, called cabozantinib (Cometriq™), was approved for the treatment of thyroid cancer in 2012.

9.9 Strategies for restoring metastasis suppressors

Similar to drug development strategies that aim to reactivate tumor suppressor genes, drug development strategies that reactivate metastasis suppressor genes are also being developed. Gene analysis has revealed that mutations in the coding region of the NM23 gene are rare, and this suggests that gene expression is inhibited in cancer cells. Therefore, agents that could reactivate gene expression may be drug candidates. The promoter of the NM23 gene is regulated by the glucocorticoid response pathway.

Medroxyprogesterone acetate (MPA), a progesterone and glucocorticoid agonist, was shown to induce expression of the *NM23* promoter. A Phase II trial (NCT00577122) is currently testing MPA, with and without chemotherapy, in metastatic breast cancer. Other strategies for reactivating metastasis suppressor genes that are being tested in pre-clinical trials include administration of recombinant metastasis suppressor proteins and gene therapy approaches (Smith and Theodorescu, 2009).

Targeting metastatic colonization

Targeting metastatic colonization might be one of the best weapons (in addition to immunotherapy) against metastatic disease. Data from the National Cancer Institute (Surveillance, Epidemiology, and End results) indicate the following distribution for cancer patients at the time of diagnosis (Sun and Ma, 2015): localized, 61%; regional, 32%; metastatic, 5%; and unstaged, 2%.

Since signs of invasion or **micro**metastatic tumor cells are evident in many cases of localized and regional cancers (total = 93%), inhibiting metastatic colonization is key to a successful therapeutic strategy, as it may be too late to stop early steps of metastasis. Targets include inhibiting mesenchymal-to-epithelial transition, interfering with the metastatic niche, or targeting dormant cells directly. Reactivation of dormant DTCs may be the reason why some cancers return after many years. If therapies could prevent the reactivation of dormant disseminated cells, then we would have hit an aspect of the Achilles heel of cancer—but our knowledge is weak. We are just beginning to understand necessary components. For example, fibrosis can reactivate dormant DTCs. One anti-fibrotic drug, an antibody against connective tissue growth factor (FG-3019), is in Phase II clinical trials and may prove to be an anti-metastatic agent. There is some evidence that DNA methylation inhibitor 5-azaC can be used to reprogramme tumor cells into dormancy. Also, pre-clinical evidence is unveiling the "education" of host stromal cells by incoming DTCs and identifying factors (e.g. POSTN) that are induced and necessary for successful colonization. Such data support a strategy that prevents metastatic niche formation to block metastatic colonization (Malanchi *et al.*, 2012).

Exosomes for non-invasive diagnostic and prognostic applications

Since cancer cells are known to secrete exosomes, identifying cancer-specific exosomes in the bloodstream would be a welcomed non-invasive method for early detection of cancer cells. Data suggest that analysis of circulating exosomes in melanoma patients can help predict stage, prognosis, and survival: high total protein concentration indicates stage 4 disease and low survival (Peinado *et al.*, 2012). Melo *et al.* (2015) have identified a cell surface proteoglycan, glypican-1 (GPC-1), that is enriched on cancer exosomes from patients with early- and late-stage pancreatic cancer. The method was sensitive and specific, and did not identify healthy people or those with benign pancreatic disease. GPC-1 is a likely candidate for further development as a pancreatic cancer diagnostic tool.

9.10 Targeting several steps of metastasis at once

As we have seen in this chapter, there are several steps involved in the metastasis of a cell from a primary tumor to a secondary site. Recently, sets of genes or "gene signatures" that are associated with primary tumor growth and risk of metastasis have been identified. Gupta *et al.* (2007) analyzed the function of four genes of a lung metastasis gene signature (the EGFR ligand *epiregulin*, *COX2*, and *MMP1* and *MMP2*) in human cancer cells. The experimental approach involved using the technique of shRNA interference to generate cells that simultaneously targeted four genes for reduced expression. Biological parameters of tumor progression and metastasis were then examined in mouse models. Their results demonstrated that, when all four genes were inactivated, tumor growth and lung metastasis were inhibited. Thus, the four genes mediate the processes of tumor growth, angiogenesis, migration, intravasation, and extravasation in mice (Gupta *et al.*, 2007). More interestingly from a therapeutic context, a combination of existing drugs (cetuximab, an anti-EGFR antibody; celecoxib, a cyclo-oxygenase (COX) inhibitor; and GM6001, an MMP inhibitor) was also able to target the protein products of the four genes and demonstrated inhibition of growth and metastasis. These results hold great potential for the future, because, if we can catch cancer early and block metastasis, we have a hope of managing cancer.

 HOW DO WE KNOW THAT?

In vivo tumor growth and metastasis assays to analyze the effects of drugs that target four specific gene products

(See Gupta *et al.*, 2007.)

The experimental system (Figure 9.8a) was as follows: human cancer cells (originally obtained from cells that had metastasized to the lung in a breast cancer patient) were injected into the mammary fat pad of immunosuppressed mice. After allowing 24 days for primary tumor growth, mice were treated with the named drugs, and tissues were harvested from mice at specified times.

In vivo assays: to assess the effects on primary tumor growth, the tumor volume was measured and compared with controls. Look at the data in Figure 9.8b. Was primary tumor growth inhibited with a combination of treatments? Results are attributed to tumor cell apoptosis. How do they know that? You will need to refer to the online supplementary information associated with the article. To assess the effects of these drugs on intravasation, the presence of human cells in the blood of mice was examined by the detection of human-specific GAPDH expression using real-time PCR. Look at Figure 9.8c. Did a combination of treatments cause a decrease in the number of circulating tumor cells? To assess extravasation and colonization ➡

→ of the lung, cryosections of lung tissue were analyzed by immunohistochemistry using a human-specific fluorescent antibody to detect tumor cells, and 4,6-diamidino-2-phenylindole (DAPI) stain to detect all nuclei (Figure 9.8d and e). The metastatic burden (number and size distribution of metastases) was calculated as the area of fluorescence normalized to the area of DAPI staining. Look at Figure 9.8f. Did the results show that a combination of drugs inhibited lung metastasis? In summary, primary tumor growth and metastasis were inhibited by specific drugs in the mouse model described earlier.

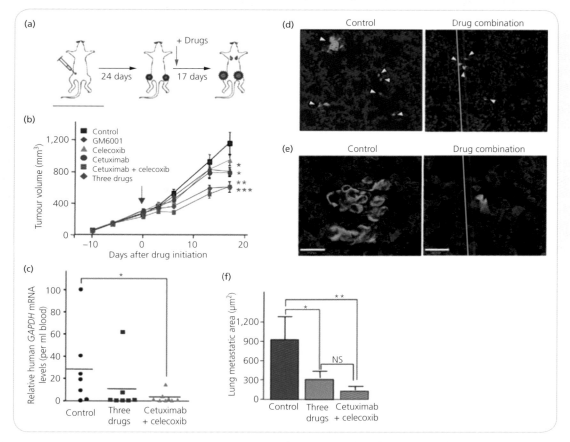

Figure 9.8 (a) A schematic representation of the time course for tumor cell implantation, primary tumor growth, and therapeutic treatment to assess effects on primary tumor and lung metastatic progression. Animals were treated with vehicle control, cetuximab, celecoxib, GM6001, or the indicated combinations. (b) Tumor volume measurements of mice treated with vehicle control or targeted therapies, either individually or in combination. $n = 6$; error bars indicate standard error of the mean (s.e.m.); *, $P < 0.05$; **, $P < 0.01$; ***, $P < 0.001$; based on a two-sided Student's t-test, compared to control-treated animals. (c) Blood from mice was isolated, and red blood cells lysed. RNA from the remaining cells was extracted for qRT-PCR. The presence of circulating tumor cells was assessed as a function of human-specific *GAPDH* expression, relative to murine *B2m*, in 3 ml of mouse blood perfusate. $n = 7–8$; bars indicate median *GAPDH* expression; *, $P < 0.05$; based on a two-sided Student's t-test. (d) Staining for tumor cells in lung cryosections using a human-specific vimentin antibody. Nuclei were visualized using 4,6-diamidino-2-phenylindole (DAPI). Arrowheads indicate tumor cell clusters in representative ×10 images of lungs from control (left) and animals treated with all three drugs (right). (e) Confocal imaging at ×63 magnification of tumor cells (vimentin, green) and lung vasculature (lectin, red) from control (left) and treated animals (three drugs, right). Scale bars, 20 µm. (f) Digital quantification of lung metastatic burden. Fluorescein isothiocyanate (FITC)-stained cancer cells in the lungs were quantified as metastatic lesions, and normalized to the area of nuclear DAPI staining. Shown on the graph is the average lung metastatic area per 5×10^5 µm^2 DAPI area. $n = 10$; error bars indicate s.e.m. Adapted by permission from Macmillan Publishers Ltd: *Nature*, from Gupta, G.P. *et al.* (2007) Mediators of vascular remodeling co-opted for sequential steps in lung metastasis. *Nature* **446**: 765–770, copyright (2007).

▤ CHAPTER HIGHLIGHTS—REFRESH YOUR MEMORY

- The major steps involved in metastasis of a tumor cell are: invasion, intravasation, transport, extravasation, and metastatic colonization.

- Complex interactions between primary tumor cells and the tumor-associated stroma affect the ability of individual cells to metastasize.

- Integrins are receptors that mediate cell–ECM interactions and, with respect to the exterior and interior of a cell, mediate bidirectional signaling.

- The steps involved in intravasation and extravasation are reversed but similar (with some exceptions), but the mechanisms for each are different.

- Metastatic colonization is characterized by progressive growth of a tumor at a distant site and requires the growth of new blood vessels (see Chapter 10 "Angiogenesis" on p. 228).

- Different cancers metastasize to specific locations. This may be explained either by: anatomy with respect to the direction of blood flow from the primary tumor; matching with "tumor-receptive" environments, as described by the "seed and soil" hypothesis; or by the establishment of the pre-metastatic niche by signals from the primary tumor.

- The concept of a pre-metastatic niche describes a site of future metastasis that is altered as a result of factors released by the primary tumor, in preparation for the arrival of tumor cells.

- Tumor exosomes are an important means of intercellular communication and carry important biological molecules that can be transferred to recipient cells.

- Evidence suggests that tumor exosomes are involved in metastasis and establishing the pre-metastatic niche.

- Micrometastases do not show net growth and may stay dormant for years.

- Loss of function of metastasis suppressor genes results in an increase in metastatic capability. Many metastasis suppressor genes play a role in dormancy.

- Many steps of the metastatic cascade are candidate targets for new therapies.

▤ ACTIVITY

1. Discuss the clinical and experimental evidence that supports a role of MET/HGF in cancer. Suggest additional strategies that could target MET, using your knowledge of similar molecular targets (Cecchi *et al.*, 2010; Liu *et al.*, 2009).

▤ FURTHER READING

Bourboulia, D. and Stetler-Stevenson, W.G. (2010) Matrix metalloproteinases (MMPs) and tissue inhibitors of metalloproteinases (TIMPs): positive and negative regulators in tumor cell adhesion. *Semin. Cancer Biol.* **20**: 161–168.

Chaffer, C.L. and Weinberg, R.A. (2011) A perspective on cancer cell metastasis. *Science* **331**: 1559–1564.

Cook, L.M., Hurst, D.R., and Welch, D.R. (2011) Metastasis suppressors and the tumor microenvironment. *Semin. Cancer Biol.* **21**: 113–122.

Desgrosellier, J.S. and Cheresh, D.A. (2010) Integrins in cancer: biological implications and therapeutic opportunities. *Nat. Rev. Cancer* **10**: 9–22.

Geiger, T.R. and Peeper, D.S. (2009) Metastasis mechanisms. *Biochem. Biophys. Acta* **1796**: 293–308.

Ghajar, C.M., Peinado, H., Mori, H., Matei, I.R., Evason, K.J., Brazier, H., *et al.* (2013) The perivascular niche regulates breast tumour dormancy. *Nat. Cell Biol.* **15**: 807–817.

Giancotti, F.G. (2013) Mechanisms governing metastatic dormancy and reactivation. *Cell* **155**: 750–764.

Horak, C.E., Lee, J.H., Marshall, J.-C., Shreeve, S.M., and Steeg, P.S. (2008) The role of metastasis suppressor genes in metastatic dormancy. *APMIS* **116**: 586–601.

Kalluri, R. and Weinberg, R.A. (2009). The basics of epithelial-mesenchymal transition. *J. Clin. Oncol.* **119**: 1420–1428.

Kang, Y. and Pantel, K. (2013) Tumor cell dissemination: emerging biological insights from animal models and cancer patients. *Cancer Cell* **23**: 573–579.

Lamouille, S., Xu, J., and Derynck, R. (2014) Molecular mechanisms of epithelial-mesenchymal transition. *Nat. Rev. Mol. Cell Biol.* **15**: 178–196.

Langley, R.R. and Fidler, I.J. (2007) Tumor cell-organ microenvironment interactions in the pathogenesis of cancer metastasis. *Endocr. Rev.* **28**: 297–321.

Lyden, D., Welch, D.R., and Psaila, B. (2011) *Cancer Metastasis, Biologic Basis and Therapeutics*. Cambridge University Press, Cambridge.

Peinado, H., Zhang, H., Matei, I.R., Costa-Silva, B., Hoshino, A., Rodrigues, G., *et al.* (2016) Pre-metastatic niches: organ-specific homes for metastases. *Nat. Rev. Cancer* **in press**.

Reymond, N., Bordad'Agua, B., and Ridley, A.J. (2013) Crossing the endothelial barrier during metastasis. *Nat. Rev. Cancer* **13**: 858–869.

Roy, R., Yang, J., and Moses, M.A. (2009) Matrix metalloproteinases as novel biomarkers and potential therapeutic targets in human cancer. *J. Clin. Oncol.* **27**: 5287–5297.

Sosa, M.S., Bragado, P., and Aguirre-Ghiso, J.A. (2014) Mechanisms of disseminated cancer cell dormancy: an awakening field. *Nat. Rev. Cancer* **14**: 611–622.

Valastyan, S. and Weinberg, R.A. (2011) Tumor metastasis: molecular insights and evolving paradigms. *Cell* **147**: 275–292.

Yang, J., Mani, S.A., and Weinberg, R.A. (2006) Exploring a new twist on tumor metastasis. *Cancer Res.* **66**: 4549–4552.

Yilmaz, M., Christofori, G., and Lehembre, F. (2007) Distinct mechanisms of tumor invasion and metastasis. *Trends Mol. Med.* **13**: 535–541.

WEB SITE

Clinical trials http://www.cancer.gov/clinicaltrials

SELECTED SPECIAL TOPICS

Cecchi, F., Rabe, D.C., and Bottaro, D.P. (2010) Targeting the HGF/Met signaling pathway in cancer. *Eur. J. Cancer* **46**: 1260–1270.

Costa-Silva, B., Aiello, N.M., Ocean, A.J., Sing, S., Zhang, H., Thakur, B.K., *et al.* (2015) Pancreatic cancer exosomes initiate pre-metastatic niche formation in the liver. *Nat. Cell Biol.* **17**: 816–826.

Devy, L. and Dransfield, D.T. (2011) New strategies for the next generation of matrix metalloproteinase inhibitors: selectively targeting membrane-anchored MMPs with therapeutic antibodies. *Biochem. Res. Int.* **2011**: 191670.

Gundem, G., Van Loo, P., Kremeyer, B., Alexandrov, L.B., Tubio, J.M.C, Papaemmanuil, E., *et al.* (2015) The evolutionary history of lethal metastatic prostate cancer. *Nature* **520**: 353–357.

Gupta, G.P., Nguyen, D.X., Chiang, A.C., Bos, P.D., Kim, J.Y., Nadal, C., *et al.* (2007) Mediators of vascular remodeling co-opted for sequential steps in lung metastasis. *Nature* **446**: 765–770.

Hoshino, A., Costa-Silva, B., Shen, T.-L., Rodrigues, G., Hashimoto, A., Mark, M.T., *et al.* (2015) Tumour exosome integrins determine organotropic metastasis. *Nature* **527**: 329–335.

Kaplan, R.N., Rafii, S., and Lyden, D. (2006) Preparing the "soil": the premetastatic niche. *Cancer Res.* **66**: 11089–11093.

Lawson, D.A., Bhakta, N.R., Kessenbrock, K., Prummel, K.D., Yu, Y., Yakai, K., *et al.* (2015) Single-cell analysis reveals a stem-cell program in human metastatic breast cancer cells. *Nature* **526**: 131–135.

Le, M.T.N., Hamar, P., Guo, C., Basar, E., Perdigao-Henrigues, R., Balaj, L., *et al.* (2014) miR-200-containing extracellular vesicles promote breast cancer metastasis. *J. Clin. Invest.* **124**: 5109–5128.

Liu, X., Newton, R.C., and Scherle, P.A. (2009) Developing c-MET pathway inhibitors for cancer therapy: progress and challenges. *Trends Mol. Med.* **16**: 37–45.

Malanchi, I., Santamaria-Martinez, A., Susanto, E., Peng, H., Lehr, H.-A., Delaloye, J.-F., *et al.* (2012) Interactions between cancer stem cells and their niche govern metastatic colonization. *Nature* **481**: 85–89.

Mani, S.A., Guo, W., Liao, M.-J., Eaton, E.N., Ayyanan, A., Zhou, A., *et al.* (2008) The epithelial-mesenchymal transition generates cells with properties of stem cells. *Cell* **133**: 704–715.

Melo, S.A., Luecke, L.B., Kahlert, C., Fernandez, A.F., Gammon, S.T., Kaye, J., *et al.* (2015) Glypican-1 identifies cancer exosomes and detects early pancreatic cancer. *Nature* **523**: 177–182.

Melo, S.A., Sugimoto, H., O'Connell, J.T., Kato, N., Villanueva, A., Vidal, A., *et al.* (2014) Cancer exosomes perform cell-independent microRNA biogenesis and promote tumorigenesis. *Cancer Cell* **26**: 1–15.

Oppenheimer, S.B. (2006) Cellular basis of cancer metastasis: a review of fundamentals and new advances. *Acta Histochem.* **108**: 327–334.

Peinado, H., Alečković, M., Lavotshkin, S., Matei, I., Costa-Silva, B., Moreno-Bueno, G., *et al.* (2012) Melanoma exosomes educate bone marrow progenitor cells toward a pro-metastatic phenotype though MET. *Nat. Med.* **18**: 883–891.

Peinado, H., Lavotshkin, S., and Lyden, D. (2011) The secreted factors responsible for pre-metastatic niche formation: old sayings and new thoughts. *Semin. Cancer Biol.* **21**: 139–146.

Reardon, D.A., Nabors, L.B., Stupp, R., and Mikkelsen, T. (2008) Cilengitide: an integrin-targeting arginine-glycine-aspartic acid peptide with promising activity for glioblastoma multiforme. *Expert Opin. Investig. Drugs* **17**: 1225–1235.

Smith, S.C. and Theodorescu, D. (2009) Learning therapeutic lessons from metastasis suppressor proteins. *Nat. Rev. Cancer* **9**: 253–264.

Sun, Y. and Ma, L. (2015) The emerging molecular machinery and therapeutic targets of metastasis. *Trends Pharmacol. Sci.* **36**: 349–359.

Tucker, G.C. (2006) Integrins: molecular targets in cancer therapy. *Curr. Oncol. Rep.* **8**: 96–103.

Wyckoff, J.B., Wang, Y., Lin, E.Y., Li, J., Goswami, S., Stanley, E.R., *et al.* (2007) Direct visualization of macrophage-assisted tumor cell intravasation in mammary tumors. *Cancer Res.* **67**: 2649–2656.

Zomer, A., Maynard, C., Verweij, F.J., Kamermans, A., Schäfer, R., Beerling, E., *et al.* (2015) In vivo imaging reveals extracellular vesicle-mediated phenocopying or metastatic behaviour. *Cell* **161**: 1046–1057.

Chapter 10

Angiogenesis

Introduction

Cancer cannot be successful without the formation of new blood vessels—it is needed by both the primary tumor and metastatic colonies. **Angiogenesis** is the process of forming new blood vessels from pre-existing ones by the growth and migration of endothelial cells (the building blocks of vessels) in a process called "sprouting." Although this process is common during embryogenesis, it rarely occurs in the adult, being reserved for wound healing and the female reproductive cycle. With respect to cancer, angiogenesis is essential for tumors. All cells must be within 100–200 μm of a blood vessel (the diffusion limit of oxygen) in order to receive essential oxygen and nutrients. Cells within the core of a tumor that do not receive sufficient oxygen and nutrients die by necrosis (Figure 10.1). Sprouting of pre-existing vessels requires major reorganization involving destabilization of the mature vessel, proliferation and migration of endothelial cells, and maturation. It is regulated by the interaction of soluble mediators and their cognate receptors. Malignant cells in culture and host stromal cells induced by a tumor *in vivo* have been shown to be sources of these soluble mediators. The neovasculature that is formed in cancer is unlike that formed in wound healing. It is leaky and tortuous and provides direct entry, allowing cells easy access to the circulation. The neovasculature is also different at the molecular level from resting endothelium. For example, the integrins αvβ3 and αvβ5 are upregulated in angiogenic vessels, compared with mature vessels. The proliferating endothelial cells of the sprouting vessel need to interact with components of the ECM that it is invading. Factors that promote angiogenesis induce the expression of these integrins. Molecular differences are not limited to the endothelium, as the supporting pericytes and ECM show specific angiogenic markers (e.g. NG2 and oncofetal fibronectin, respectively). Therefore, many components of angiogenic vasculature are molecularly distinct from normal vessels.

Figure 10.1 Necrotic core of a tumor. Histopathological section from paraffin-embedded tumor tissue of a SHH-type medulloblastoma highlighting geographic necrosis within the tumor (haematoxylin and eosin (HE) stain). Courtesy of Jensflorian, image licensed under the Creative Commons Attribution Share Alike 4.0 International license.

10.1 The angiogenic switch

The regulation of angiogenesis is dependent upon the dynamic balance of angiogenic inducers and inhibitors. Increasing the activity of the inducers or decreasing the activity of the inhibitors tips the balance of the "angiogenic switch" to the "on" position, and vice versa. We will focus on only a few of these (Figure 10.2).

A PIONEER IN THE FIELD . . . of angiogenesis: in memory of Judah Folkman

Folkman's pioneering discoveries into the mechanism of angiogenesis opened up a new field of cancer research and supported his ground-breaking idea that tumors are dependent on angiogenesis. His laboratory identified the first angiogenic inhibitor and carried out clinical trials of anti-angiogenic therapies. He also investigated the observation that some tumors remain dormant, sometimes indefinitely, because of the production by the tumor of an angiogenic inhibitor, but can become angiogenic when production of the inhibitor decreases.

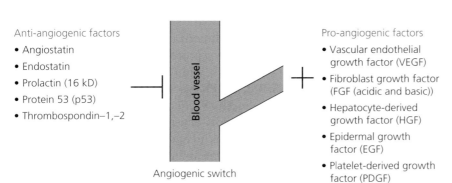

Anti-angiogenic factors
- Angiostatin
- Endostatin
- Prolactin (16 kD)
- Protein 53 (p53)
- Thrombospondin–1,–2

Blood vessel

Angiogenic switch

Pro-angiogenic factors
- Vascular endothelial growth factor (VEGF)
- Fibroblast growth factor (FGF (acidic and basic))
- Hepatocyte-derived growth factor (HGF)
- Epidermal growth factor (EGF)
- Platelet-derived growth factor (PDGF)

Figure 10.2 The angiogenic switch: examples of anti-angiogenic factors and pro-angiogenic factors discussed in the text.

PAUSE AND THINK

Could VEGFRs, Tie receptors, and ephrin receptors be additional targets for kinase inhibitor therapeutics?

Angiogenic inducers

Growth factors, both non-specific (e.g. EGF, FGF, HGF, PDGF) and endothelial-specific (e.g. VEGF), are examples of angiogenic inducers. Although the non-specific growth factors (e.g. FGF and PDGF) affect many cell types, they are still important for angiogenesis. Three families of vascular endothelium-specific growth factors and their tyrosine kinase transmembrane receptors have been identified: vascular endothelial growth factors (VEGFs), and VEGF receptors (VEGFRs), angiopoietins and Tie receptors, and ephrins and ephrin receptors.

VEGF is the star player involved in the initiation of angiogenesis, while angiopoietins and ephrins are important for subsequent maturation. The VEGF family currently consists of five family members (VEGF-A to D and placental growth factor PlGF) which transmit their signal via three VEGFR tyrosine kinases (VEGFR-1, VEGFR-2, and VEGFR-3). The interaction of VEGF-A with its receptor VEGFR-2 is responsible for the majority of angiogenic effects. VEGFR-1 acts as a decoy by regulating the amount of VEGF-A available to VEGFR-2, having only weak kinase activity. By binding VEGF-A, it restricts the angiogenic response induced by VEGFR-2. VEGFR-3 and its ligand VEGF-C play a role in the development of the lymphatic vascular system. VEGF-A is secreted by a range of tumor cells. The tumor cells also affect the surrounding stromal cells, and induction of the VEGF promoter in surrounding non-transformed cells has been demonstrated, suggesting a collaboration between normal and transformed cells. For example, a subset of tumor-infiltrating immune cells have been shown to adopt a pro-angiogenic phenotype and produce high levels of VEGF (Bruno *et al.*, 2014). Also, reserves of VEGF are found in the ECM and are released by MMPs. Not only does VEGF induce endothelial cell proliferation, but it can also induce permeability and leakage. This feature may be important for the initiation of angiogenesis, as it has been suggested that the existing mature vessels must be destabilized before sprouting begins. To summarize, VEGFR-2 mediates the endothelial effects of VEGF, while VEGFR-1 is inhibitory and VEGFR-3 is vital for lymphatic vessels. Although many of the details of the signal transduction pathway of VEGFs have yet to be elucidated, the VEGF-A signal transduction pathway (Figure 10.3) appears to be very similar to the signal transduction pathways for EGFs (compare with Figure 4.2): dimerization, autophosphorylation, creation of high-affinity binding sites for proteins with SH2 domains (e.g. VEGFR-associated protein (VRAP), Sck, and phospholipase Cγ), and subsequent activation of the RAS, Raf, MAPK cascade. VEGF-responsive genes include the EGFR ligand, epiregulin, *COX2*, and matrix metalloproteinases *MMP1* and *MMP2*. In addition, a PI3K-dependent pathway is triggered. This activates AKT and leads to a block in apoptosis and, via nitric oxide production, to increased vascular permeability. Several other important intracellular molecules, such as Src, are also implicated.

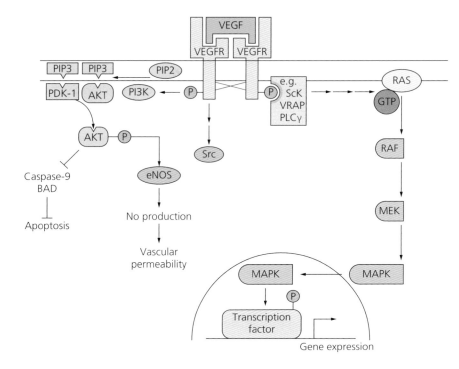

Figure 10.3 The VEGF-A signal transduction pathway. One molecule of VEGF-A binds to two VEGFR-2 receptors, facilitating dimerization and autophosphorylation. Proteins containing SH2 domains (shaded in gray) bind to the phosphorylated receptor and trigger activation of RAS and the Raf–MEK–MAPK cascade. In addition, PI3K is activated. AKT leads to inhibition of apoptosis. AKT also stimulates endothelial nitric oxide (NO) synthase and stimulates vascular permeability via NO production. Src is one of several other molecules that are activated by VEGFR-2.

Angiogenic inhibitors

Angiogenic inhibitors normally found in the body (endogenous inhibitors) maintain the angiogenic switch in the "off" position by inhibiting endothelial cell migration and proliferation. Some angiogenic inhibitors are stored as cryptic parts within larger proteins that are not themselves inhibitors (Figure 10.4). Plasminogen can be cleaved by proteinases, including several MMPs, to release the angiogenic inhibitor angiostatin. Angiostatin binds to its endothelial cell surface receptor, annexin II, to exert its inhibitory effects. Endostatin is a fragment of collagen XVIII and

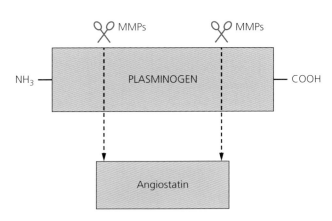

Figure 10.4 Cryptic angiogenic inhibitors.

can be proteolytically released by elastase and cathepsin. It blocks MAPK activation in endothelial cells and also MMPs.

It has been observed that sometimes, when a tumor is removed by surgery or irradiation, dormant metastases are often activated, and growth and angiogenesis are initiated. This phenomenon has been termed "concomitant resistance." Evidence suggests that the production of angiogenic inhibitors, such as angiostatin and endostatin, by certain tumors prevents the growth of remote micrometastases via the blood. When the primary tumor is removed, so are these inhibitors, and the angiogenic switch is activated for the micrometastases. Also, surgery is known to cause induction of angiogenic growth factors and thus may exacerbate malignant disease through this mechanism (Ian Judson, personal communication).

The angiogenic switch is regulated in two ways during tumorigenesis. First, as a tumor grows, it creates conditions of hypoxia (low oxygen concentration), and this induces angiogenesis via the hypoxia-inducible factor-1α (HIF-1α). One target of HIF-1α that is important for angiogenesis is the *VEGF* gene. HIF is actually a heterodimeric transcription factor comprising one HIF-1α and one HIF-1β subunit. The activity of HIF is regulated by oxygen concentration, not at the level of mRNA expression, as both subunit mRNAs are constitutively expressed, but rather at the protein level of HIF-1α (Figure 10.5). Under normoxic conditions (20%

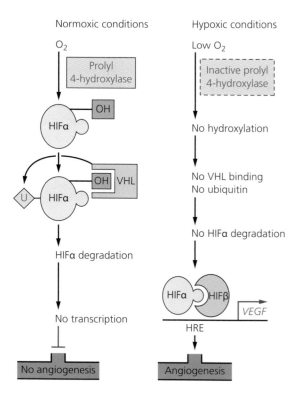

Figure 10.5 The induction of angiogenesis by hypoxia.

oxygen), HIF-1α is rapidly degraded. The von Hippel–Lindau (VHL) tumor suppressor protein is an important regulator of HIF-1α degradation (Kim and Kaelin, 2003). The first step in targeting HIF-1α for degradation under normoxic conditions is modification (hydroxylation) by the enzyme prolyl 4-hydroxylase (shown in red text in Figure 10.5). This enzyme directly binds and links molecular oxygen to specific proline residues on HIF-1α, and thus acts as a direct oxygen sensor in this pathway. VHL binds to hydroxylated HIF-1α and activates a complex of proteins responsible for the addition of ubiquitin (indicated by a "U" in a red diamond, Figure 10.5) that target HIF-1α for proteosomal degradation. In the absence of HIF-1α, HIF target genes cannot be transcriptionally activated, and angiogenesis does not occur.

Under hypoxic conditions, the enzyme prolyl 4-hydroxylase is inactivated; HIF-1α is not hydroxylated, and VHL cannot bind and target HIF-1α for proteosomal degradation. HIF-1α is rapidly stabilized and transported to the nucleus. The heterodimeric HIF transcription factor can then activate its target genes via the hypoxia response element (HRE) containing the binding site 5'-RCGTG-3'. As previously mentioned, the most notable target is the *VEGF* gene that contains a HRE in its promoter region.

Oncogene and tumor suppressor gene products can also lead to increased HIF-1 activity. This is illustrated in Kaposi's sarcoma, a highly vascularized tumor that is caused by a herpesvirus. Three protein products of the viral genome increase HIF-1α half-life, nuclear localization, and transactivation under non-hypoxic conditions, thus mimicking the effects of hypoxia.

PAUSE AND THINK

Allow me to interject at this point to share a personal story of how I have come to appreciate the importance of this pathway. I wear contact lenses and was told, when I purchased them, that I could wear them during sleep. I received a shock during the next visit with my optometrist when he told me that my eyes were not receiving enough oxygen during the night, and blood vessels had begun to grow out into the eye. He explained that my eyelid alone reduces the amount of oxygen to the eye during sleep and that the addition of a contact lens created a hypoxic condition. The hypoxia was sensed and triggered angiogenesis to supply more oxygen to the eye, rather than allow the tissue to become damaged. I no longer wear contact lenses during sleep and have an appreciation of the regulation of prolyl 4-hydroxylase.

Second, the angiogenic switch can also be modified by oncogenic proteins and loss of tumor suppressors. In addition to the well-known direct contribution of oncogenes and tumor suppressors to proliferation, apoptosis, and differentiation, direct roles in angiogenesis are now recognized. Approximately 30 oncoproteins have been shown to tip the balance towards angiogenesis. Aberrant production of a growth factor, in addition to acting in an autocrine manner to stimulate proliferation of tumor cells, can also act in a paracrine manner to stimulate the growth of endothelial cells. "Star" oncogenic proteins, including receptor tyrosine kinases (e.g. EGFR), intracellular tyrosine kinases (e.g. Src), intracellular transducers (e.g. Ras), and transcription factors (e.g. Fos, Jun), have been shown to upregulate the "star" angiogenic inducer VEGF.

Some tumor suppressor proteins normally upregulate/increase angiogenic inhibitors, but, when these are mutated, anti-angiogenic activity decreases. For example, the transcription factor p53 normally binds to, and activates, the promoter of the *thrombospondin-1* gene. Mutations in the *p53* gene, commonly associated with the cancer phenotype, result in a decrease of the angiogenic inhibitor, so that the angiogenic switch favors angiogenesis.

10.2 Cell behavior during angiogenic sprouting

In response to angiogenic inducing signals, endothelial cells extend filopodia and migrate towards the signal. At the location of the highest concentration of VEGF-A, VEGFR-2 is activated. The signal is enhanced by co-receptor neuropilin-1 (Nrp1) and is transduced via the MAPK cascade. This stimulates the formation of a tip cell at the forefront of the sprout (Figure 10.6).

Behind the tip cell are proliferating stalk cells that extend the sprouting blood vessel. The phenotype of the two cell types is not fixed and depends on a competition for VEGFR-2 activation that is regulated by Notch. In brief, upon VEGFR-2 activation, tip cells induce the expression and release of the Notch ligand, Delta-like 4 (DLL4). DLL4 binds to the Notch receptor on neighboring cells. The Notch intracellular domain NICD is

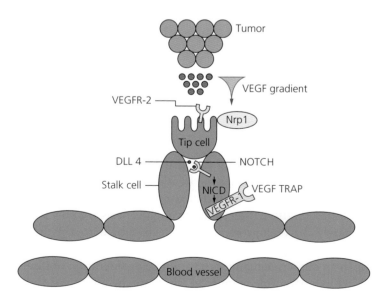

Figure 10.6 Cell and molecular mechanisms of angiogenesis. See text for details.

released and transported to the nucleus where it acts as a transcription factor to repress *VEGFR-2* gene expression and induce *VEGFR-1* gene expression. VEGFR-1 acts as a VEGF trap and reduces the concentration of VEGF that can bind to VEGFR-2. The growing sprout moves along a VEGF gradient. When two tip cells meet, they fuse and allow for a connected lumen, allowing blood to flow through the new vessel.

10.3 Other means of tumor neovascularization

Recent evidence has suggested that, in addition to angiogenesis, vasculogenic mimicry and vasculogenesis contribute to the formation of tumor vessels (Figure 10.7a–c). Vasculogenic mimicry describes the process whereby tumor cells (e.g. melanoma cells; shown as gray circles in Figure 10.7) act as endothelial cells and form vascular-like structures (Figure 10.7b, red arrow). Vasculogenesis involves the differentiation and proliferation of endothelial cells from endothelial progenitor cells. Studies have demonstrated that up to 40% of tumor endothelial cells originated from circulating endothelial progenitor cells (CEPs; shown as a red circle) derived from the bone marrow. Angiogenic factors from the tumor, such as VEGF, are involved in the recruitment of these cells that express VEGFR-2. After reaching the tumor, CEPs differentiate and contribute to the tumor neovasculature (Figure 10.7c, red ovals). It seems likely that different cancers may differ in their requirement for

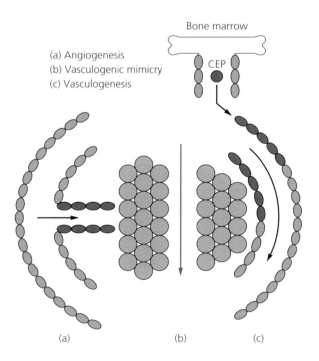

Figure 10.7 Tumor neovascularization.

CEP contributions to the new tumor vasculature. It is known that they are necessary for lymphomas and colon cancer.

◎ Therapeutic strategies

Therapies aimed at the tumor vasculature designed either to halt the angiogenic process (anti-angiogenic drugs) or to destroy the existing new tumor vasculature (vascular targeting) have been developed. Some examples are discussed in Section 10.4 and Section 10.5.

10.4 Anti-angiogenic therapy

Anti-angiogenic therapy is designed to prevent the formation of new blood vessels. Rather than target the tumor cells directly, the aim of anti-angiogenic therapy is to interfere with the responsiveness of normal endothelial cells that is essential to the survival of the tumor. Drugs may be designed to prevent the cells from responding to pro-angiogenic signals or may be targeted to block the activity of the inducers. Overall, these drugs are cytostatic, rather than cytotoxic, and therefore may need long-term continuous administration. Anti-angiogenic therapies, together with vascular targeting (discussed in Section 10.5, "Vascular targeting by vascular disrupting agents" on p. 240), differ from the therapies discussed previously, and the differences have several implications. First, as angiogenesis only occurs on occasion in the adult, drugs that inhibit it are predicted to cause minimal side effects. More importantly, the target endothelial cells recruited during angiogenesis are genetically stable, unlike the tumor cells that have accumulated mutations, and are therefore less likely to develop drug resistance rapidly. There are approximately 95 clinical trials that are currently testing anti-angiogenic cancer drugs (some are listed in Table 10.1). Examples of several different strategies that have been employed are described next.

Drugs that target VEGF or VEGFR

Several strategies can be used to block VEGF signaling in order to prevent angiogenesis (Figure 10.8). Both ligands and receptors may be good targets, and both antibodies and small-molecule inhibitors may be employed (recall therapeutic approaches used to target EGF signaling; see Chapter 4). Since VEGF/VEGFR is a star player in angiogenesis, it was thought that inhibiting this pathway alone would cause a rapid and sustained response. Although some improvement in prognosis is seen, responses are often transient, stop after treatment is stopped, and are not as effective (prolonging

Table 10.1 Anti-angiogenic therapies: examples and status in clinical trials

Drug	Company	Mechanism	Phase of trial
Drugs that block activators of angiogenesis and their receptors			
Aflibercept (VEGF-Trap)	Regeneron Pharm	Soluble decoy VEGFR	Approved
Bevacizumab (Avastin™)	Genetech	Monoclonal Ab to VEGF	Approved
Pazopanib (Votrient™)	GlaxoSmithKline	TKI: VEGFR; PDGFR, FGFR	Approved
Ramucirumab (Cyramza™)	Eli Lilly & Co	Monoclonal Ab to VEGFR-2	Approved
Recentin™ (AZD2171)	AstraZeneca	TKI: VEGFR-1/2	III (failed for CRC, but ongoing for other cancers)
Sorafenib (Nexavar™)	Bayer	TKI: VEGFR, PDGFR, FLT3, Kit, and Raf	Approved
SU5416	Sugen	TKI: VEGFR signaling	Withdrawn
SU6668	Sugen	TKI: VEGFR, FGFR, PDGFR	Withdrawn
Sunitinib (SU-11248)	Pfizer	TKI: VEGFR, PDGFR, FLT3, Kit	Approved
Trebananib	Amgen	Binds angiopoietin 1/2	Phase III
Vandetanib (Zactima™) (ZD6474)	AstraZeneca	TKI: VEGFR-1/2	Approved orphan drug
Drugs that inhibit endothelial-specific integrin signaling			
Cilengitide	Merck KGaA	Antagonist of integrins αvβ3 and αvβ5	Discontinued
Vitaxin II	MedImmune	Inhibitor of integrin αvβ3	Discontinued
Drugs that inhibit endothelial cells			
ABT-510	Abbott Labs	Thrombospondin-1 analog	II
Angiostatin	EntreMed	Inhibition of endothelial cells	I
Endostatin	EntreMed	Inhibition of endothelial cells	II (approved in China, 2005)
Vascular disrupting agents			
Combretastatin	Oxigene	Binds to tubulin; disrupts the cytoskeleton	I/II/III
NPI 2358	Nereus	Binds to tubulin; disrupts the cytoskeleton	I/II

Ab, antibody; CRC, colorectal cancer; TKI, tyrosine kinase inhibitor.

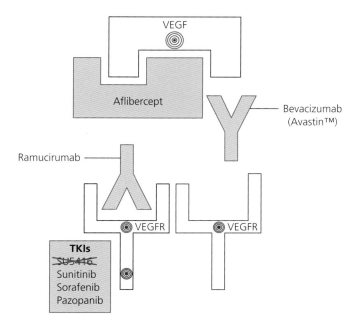

Figure 10.8 Anti-angiogenic therapies that target VEGF or VEGFR. Therapeutic agents are shown in red, and cellular targets are indicated by a (◎) symbol. Clinical development of SU5416, the first VEGFR tyrosine kinase inhibitor (TKI) to enter clinical trials, has been stopped (shown by a red cross). The TKIs shown inhibit multiple targets, including VEGFRs and PDGFRs.

survival only in the order of months) as we had hoped. Broader approaches targeting several angiogenic pathways promise improvements.

One strategy involves targeting angiogenic factors such as VEGF. Bevacizumab (Avastin™), a recombinant humanized monoclonal antibody that recognizes VEGF-A, was first approved for treatment of colorectal cancer. Its use has broadened for several other cancers. Interestingly, the drug failed to show a consistent effect on overall survival in breast cancer trials. Why was a different response observed in the two different cancers? The answer lies in the angiogenic switch: colon tumors are more dependent on VEGF for the induction of angiogenesis, while this is true only for early stages of breast cancer. Advanced breast cancer utilizes a broader arsenal of angiogenic inducers, thus inhibition of just one inducer does not have an effect.

Creating a fusion protein that acts as a decoy receptor to prevent VEGF from interacting with its receptor is another strategy that has been successful. Aflibercept consists of VEGFR-1 and VEGFR-2 ligand-binding domains fused to an IgG constant region (Fc), and can therefore bind more than one VEGF ligand, in contrast to bevacizumab described above. Aflibercept was approved for colorectal cancer.

Antibodies targeted against the extracellular domain of growth factor receptors can be successful drugs, as we have seen earlier in this text. Ramucirumab, a fully humanized antibody that targets the extracellular domain of VEGFR-2, has been approved for advanced gastric cancer.

Small-molecule tyrosine kinase inhibitors have been used to target the VEGFR. Semaxanib (SU5416) was the first VEGFR inhibitor to enter Phase III clinical trials. Its mechanism of action is that it inhibits receptor autophosphorylation. Although semaxanib demonstrated promising results in patients with Kaposi's sarcoma, significant toxicity and poor responses in patients with colorectal cancer led to the withdrawal of the drug. Its further development was also discontinued as a result of unfavorable pharmacology, namely a particularly short half-life of only several hours. As a result of the short half-life of the drug, effective doses were unable to be maintained, even after bi-weekly intravenous administration. Similarly, the drug SU6668, which has a similar mode of action to semaxanib and is orally active, was also withdrawn. However, this strategy has proved successful by the approval of several multi-targeted tyrosine kinase inhibitors. Axitinib, sunitinib, and sorafenib have been approved for the treatment of advanced renal cell carcinoma. All three drugs target both VEGFRs and PDGF-Rs. Axitinib targets VEGFRs, as well as PDGF-R and c-Kit; sunitinib targets VEGFRs, as well as PDGF-R, Kit, and FLT3; sorafenib targets VEGFRs, PDGF-R, Kit, FLT3, and Raf kinase (as mentioned in Chapter 4) (see "Pause and Think"). Other multikinase inhibitors, whose targets include VEGFR and FGFR (pazopanib, nintedanib, regorafenib) or VEGFR and Tie2 (regorafenib, vandetanib, cabozantinib), have also been approved. The success of these recently developed therapeutics is likely due to the fact that they simultaneously target multiple angiogenic regulators, in addition to VEGF. This may prevent alternative pathways from compensating each other as seen when only the VEGF pathway is targeted, and extend the period of the response.

Other strategies for anti-angiogenic drugs

Similar strategies used to target the VEGF/VEGFR pathway can be applied to other angiogenic regulators. One example is the drug trebananib which is a peptide–Fc fusion protein (similar to aflibercept) that binds angiopoietin-1 and -2, ligands of the Tie2 receptor. Trebananib entered Phase III clinical trials.

Administration of recombinant human endogenous inhibitors is another anti-angiogenic treatment strategy that held much promise but which, thus far, has not delivered the expected results. Endostatin, which was discovered in 1996, was the first to enter clinical trials. Although it was demonstrated to be non-toxic, no clinical response was observed. This negative result may again be due to suboptimal design of the clinical trial, rather than inefficacy of the drug, as patients with advanced solid tumors were selected, despite pre-clinical success with early-stage cancer models. Endostatin has been approved in China. The company (EntreMed Inc., Rockville, MD) has announced that it will halt production of endostatin because of financial difficulties.

Antagonists to integrins αvβ3 and αvβ5 would block endothelial integrin–ECM interactions and specifically induce apoptosis of angiogenic vessels with little effect on mature vessels. Two integrin inhibitors have entered clinical trials but are no longer being developed: Vitaxin™, a humanized monoclonal antibody against integrin αvβ3, and cilengitide, a synthetic cyclic peptide antagonist that mimics the Arg–Gly–Asp "ligand" sequence and inhibits integrins αvβ3 and αvβ5.

Although there are several approved drugs that target the HIF pathway indirectly (bortezomib; SAHA), there have been no direct HIF inhibitors approved (see "Activity" at the end of this chapter on p. 242).

PAUSE AND THINK

Do you remember other examples where a small-molecule inhibitor was used to target a tyrosine kinase receptor? Hint: see Chapter 4.

PAUSE AND THINK

Thalidomide, a drug cursed in the past as being teratogenic and causing truncated limbs, and its analogs lenalidomide and pomalidomide are some of the most effective drugs for treating patients with multiple myeloma, a tumor of B-cells. Thalidomide has been shown to inhibit angiogenesis induced by basic fibroblast growth factor (bFGF) or VEGF, and this anti-angiogenic mechanism was long thought to explain its clinical effect. However, recent evidence suggests that its success in treating multiple myeloma is due to its interaction with the E3 ubiquitin ligase complex via a protein called Cereblon, and the subsequent degradation of two important transcription factors, Ikaros and Aiolos, that are important for B-cell development (see references within Stewart, 2014). Note that, since the anti-angiogenic activity of thalidomide is linked to its teratogenicity and limb defects, patient education regarding pregnancy is crucial. Research provides evidence that is constantly changing our knowledge of how things work in biological systems.

Anti-angiogenic effects may be "side effects" of other cancer therapies. Cancer therapies targeted at oncogene products often affect angiogenesis. Herceptin™ (the therapeutic antibody directed against ErbB2; see Chapter 4) has been shown to be anti-angiogenic by inhibiting the production of angiogenic inducers (e.g. TGF-α and angiopoietin-1) by tumor cells and upregulating angiogenic inhibitors (e.g. thrombospondin). Chronic frequent administration of conventional chemotherapy at doses of one-tenth to one-third of the MTD, known as metronomic scheduling, has also resulted in anti-angiogenic effects.

10.5 Vascular targeting by vascular disrupting agents

Vasculature targeting is a therapeutic approach designed to destroy the *existing* neovasculature in the tumor in order to starve it of oxygen and nutrients and lead to tumor regression. This approach is possible owing to the identification of molecular differences between tumor and normal vasculature. Combretastatin, first isolated from the African bushwillow *Combretumcaffrum*, is selectively toxic to neovasculature, and combretastatin and recently developed derivatives are being tested in clinical trials (Marrelli *et al.*, 2011). Combretastatin compounds bind tubulin, cause depolymerization, and disrupt the cytoskeleton. Their effects have been explained by the hypothesis that immature endothelium may have a more intrinsic need for a tubulin cytoskeleton to maintain its shape than stable mature vasculature which is firmly supported by a basement membrane. Loss of shape and rounding up of the endothelial cells in new blood vessels block blood flow and/or lead to vascular collapse, thereby depriving the tumor of oxygen and nutrients (Figure 10.9; compare images before (a) and after (b) treatment with combretastatin A4). As a result, necrosis occurs at the core of the tumor. Unfortunately, a ring of tumor cells at the periphery remains viable. Regrowth of tumor cells from the rim indicates the need for combination therapy. Combinations, such as

Figure 10.9 Effects of combretastatin on tumor neovasculature. Anti-vascular effects were analyzed by magnetic resonance imaging. The image intensity indicates tumor vasculature. A primary tumor before (a) and after (b) treatment with combretastatin A4. Strong anti-vascular effects are seen in the core of the tumor after treatment, but a small, viable rim of tumor tissue can be seen at the periphery. Reprinted by permission from Macmillan Publishers Ltd on behalf of Cancer Research UK: *British Journal of Cancer*, from Beauregard, D.A. *et al.* (1998) Magnetic resonance imaging and spectroscopy of combretastatin A(4) prodrug induced disruption of tumor profusion and energetic status. *Br. J. Cancer* 77: 1761–1767, copyright (1998).

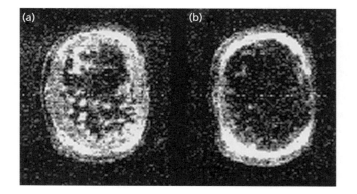

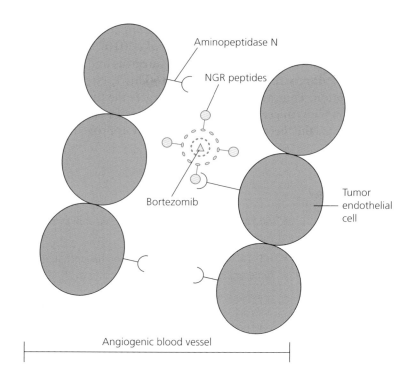

Aminopeptidase N

NGR peptides

Bortezomib

Tumor endothelial cell

Angiogenic blood vessel

Figure 10.10 Vascular targeting by nanoparticle technology. See text for details. NGR, asparagine-glycine-arginine peptide; bortezomib, a drug used to target specific cancers; aminopeptidase N, a tumor endothelial cell marker.

combretastatin with ionizing radiation (Clemenson *et al.*, 2013) or chemotherapies (Zweifel *et al.*, 2011), are being tested. Other mechanisms of action responsible for the cell death effects of combretastatins are also probable. Anti-vascular effects are seen at doses of one-tenth of the MTD. Among several others, NPI 2358 is also a vascular disrupting agent in clinical trials that targets the tubulin cytoskeleton.

Nanoparticle technology, drug delivery, and vascular targeting have come together in an exciting report with hints at future applications. Knowledge of several biochemical areas has been combined specifically to target a drug to the neovasculature of tumors in mice (Zuccari *et al.*, 2015) (Figure 10.10). Liposome-based nanoparticles (Figure 10.10, shaded gray) were coated with NGR (asparagine–glycine–arginine) peptides as ligands to a tumor endothelial cell marker, aminopeptidase N. These tagged liposomes encapsulated a promising drug called bortezomib, a proteasome inhibitor, for neuroblastoma. This formulation was designed to increase the effectiveness and reduce the toxicity seen with "free," non-encapsulated drug. Intravenous delivery to neuroblastoma-bearing mice resulted in treated mice living statistically longer than controls. Also lower toxicity, compared to non-encapsulated drug, was demonstrated.

Furthermore, peptides that target αv integrins that are specifically expressed on tumor endothelial cells have been used to enhance the penetration of drugs into a tumor in mice *without* coupling of the drug to the peptide or other carrier (Sugahara *et al.*, 2010). Similar to RGD peptides, their peptide iRGD binds to these integrins and is cleaved. The fragment gains affinity for the receptor neuropilin-1, Nrp1, and triggers tumor penetration. Thus, this strategy that targets the tumor vasculature in combination with cancer drugs and enhances tumor-specific delivery in mice may increase the therapeutic index of already approved cancer drugs for humans.

CHAPTER HIGHLIGHTS—REFRESH YOUR MEMORY

- Both a primary tumor and metastatic colonization, characterized by progressive growth of a tumor at a distant site, require angiogenesis.

- Tumor vessels differ from healthy vessels in that they are leaky and tortuous and have distinct molecular characteristics.

- The process of angiogenesis is regulated by the angiogenic switch involving a dynamic interplay between pro- and anti-angiogenic factors.

- Members of the VEGF family are specific endothelial cell growth factors that are key players in angiogenesis. Their signals are mediated through transmembrane tyrosine kinase receptors.

- The VEGF signal transduction pathway is similar to the EGF signal transduction pathway.

- Angiogenesis is regulated by hypoxia and oncogene and tumor suppressor gene products.

- HIF is a heterodimeric transcription factor that regulates genes important for angiogenesis such as VEGF.

- Sprouting during angiogenesis is led by the formation of a tip cell via competition for VEGF-A. Neighboring stalk cells proliferate for sprout extension.

- Vasculogenic mimicry and vasculogenesis also contribute to neovascularization of tumors.

- Anti-angiogenic therapy is designed to *prevent the formation of* new blood vessels, while vascular disrupting agents are designed to *destroy* the neovasculature in a tumor.

- Drugs that target either VEGF or its receptor have been approved and are available in the clinic.

ACTIVITY

1. Using the web sites on p. 243, update Table 10.1. Have certain drugs progressed to advanced clinical trials? Have some been terminated? Have new drugs been added? Can you think of any additional strategies not mentioned? Can you think of any strategies that target HIF-1α? (See online references: Burroughs *et al.*, 2013, Zimna and Kurpisz, 2015.)

2. Read about how transcription factors FOXO1 and Myc have been found to regulate the metabolism and proliferation of vascular endothelial cells (Wilhem *et al.*, 2016). Design experiments that may investigate this link with angiogenesis in cancer.

FURTHER READING

Baeriswyl, V. and Christofori, G. (2009) The angiogenic switch in carcinogenesis. *Semin. Cancer Biol.* **19**: 329–337.

Bridges, E.M. and Harris, A.L. (2011) The angiogenic process as a therapeutic target in cancer. *Biochem. Pharmacol.* **81**: 1183–1191.

Cook, K.M. and Figg, W.D. (2010) Angiogenesis inhibitors: current strategies and future prospects. *CA Cancer J. Clin.* **60**: 222–243.

Cross, M.J., Dixelius, J., Matsumoto, T., and Claesson-Welsh, L. (2003) VEGF-receptor signal transduction. *Trends Biochem. Sci.* **28**: 488–494.

Desgrosellier, J.S. and Cheresh, D.A. (2010) Integrins in cancer: biological implications and therapeutic opportunities. *Nat. Rev. Cancer* **10**: 9–22.

Ellis, L.M. and Hicklin, D.J. (2008) VEGF-targeted therapy: mechanisms of anti-tumor activity. *Nat. Rev. Cancer* **8**: 579–591.

Folkman, J. (2006) Antiangiogenesis in cancer therapy—endostatin and its mechanism of action. *Exp. Cell Res.* **312**: 594–607.

Gacche, R.N. and Meshram, R.J. (2014) Angiogenic factors as potential drug target: efficacy and limitations of anti-angiogenic therapy. *Biochem. Biophys. Acta* **1846**: 161–179.

George, D.J. and Moore, C. (2006) Angiogenesis inhibitors in clinical oncology. *Update Cancer Ther.* **1**: 429–434.

Heath, V.L. and Bicknell, R. (2009) Anticancer strategies involving the vasculature. *Nat. Rev. Clin. Oncol.* **6**: 395–404.

Jeltsch, M., Leppanen, V.-M., Saharinen, P., and Alitalo, K. (2013) Receptor tyrosine kinase-mediated angiogenesis. *Cold Spring Harb. Perspect. Biol.* **5**: a009183.

Lippert, J.W. 3rd (2007) Vascular disrupting agents. *Bioorg. Med. Chem.* **15**: 605–615.

McCarty, M.F., Liu, W., Fan, F., Parikh, A., Reimuth, N., Stoeltzing, O., *et al.* (2003) Promises and pitfalls of anti-angiogenic therapy in clinical trials. *Trends Mol. Med.* **9**: 53–58.

Ricci, V., Ronzoni, M., and Fabozzi, T. (2015) Aflibercept a new target therapy in cancer treatment: a review. *Crit. Rev. Oncol. Hematol.* **96**: 569–576.

Ruegg, C., Hasmim, M., Lejeune, F., and Alghisi, G.C. (2006) Antiangiogenic peptides and proteins: from experimental tools to clinical drugs. *Biochim. Biophys. Acta* **1765**: 155–177.

Ruoslahti, E. (2002) Specialization of tumour vasculature. *Nat. Rev. Cancer* **2**: 83–90.

Semenza, G.L. (2010) Defining the role of hypoxia-inducible factor 1 in cancer biology and therapeutics. *Oncogene* **29**: 625–634.

Vasudev, N.S. and Reynolds, A.R. (2014) Anti-angiogenic therapy for cancer: current progress, unresolved questions and future directions. *Angiogenesis* **17**: 471–494.

Welti, J., Loges, S., Dimmeler, S., and Carmeliet, P. (2013) Recent molecular discoveries in angiogenesis and antiangiogenic therapies in cancer. *J. Clin. Invest.* **123**: 3190–3200.

Yilmaz, M., Christofori, G., and Lehembre, F. (2007) Distinct mechanisms of tumor invasion and metastasis. *Trends Mol. Med.* **13**: 535–541.

Zhao, Y. and Adjei, A.A. (2015) Targeting angiogenesis in cancer therapy: moving beyond vascular endothelial growth factor. *Oncologist* **20**: 660–673.

WEB SITES

The Angiogenesis Foundation http://www.angio.org/

National Cancer Institute. Clinical trials http://www.cancer.gov/clinicaltrials

■ SELECTED SPECIAL TOPICS

Beauregard, D.A., Thelwall, P.E., Chaplin, D.J., Hill, S.A., Adams, G.E., Brindle, K.M. (1998) Magnetic resonance imaging and spectroscopy of combretastatin A(4) prodrug induced disruption of tumor profusion and energetic status. *Br. J. Cancer* **77**: 1761–1767.

Bruno, A., Ferlazzo, G., Albini, A., and Noonan, D.M. (2014) A think tank of TINK/TANKs: tumor-infiltrating/tumor-associated natural killer cells in tumor progression and angiogenesis. *J. Natl. Cancer Inst.* **106**: dju200.

Burroughs, S.K., Kaluz, S., Wang, D., Wang, K., Van Meir, E.G., and Wang B. (2013) Hypoxia inducible factor pathway inhibitors as anticancer therapeutics. *Future Med. Chem.* **5**: 553–572.

Clemenson, C., Chargari, C., and Deutsch, E. (2013) Combination of vascular disrupting agents and ionizing radiation. *Crit. Rev. Oncol. Hematol.* **86**: 143–160.

Gupta, G.P., Nguyen, D.X., Chiang, A.C., Bos, P.D., Kim, J.Y., Nadal, C., *et al.* (2007) Mediators of vascular remodeling co-opted for sequential steps in lung metastasis. *Nature* **446**: 765–770.

Kim, W. and Kaelin Jr, W.G. (2003) The von-Hippel-Lindau tumor repressor protein: new insights into oxygen sensing and cancer. *Curr. Opin. Genet. Dev.* **13**: 55–60.

Liu, X., Newton, R.C., and Scherle, P.A. (2009) Developing c-MET pathway inhibitors for cancer therapy: progress and challenges. *Trends Mol. Med.* **16**: 37–45.

Marrelli, M., Conforti, F., Statti, G.A., Cachet, X., Michel, S., Tillequin, F., *et al.* (2011) Biological potential and structure-activity relationships of most recently developed vascular disrupting agents: an overview of new derivatives of natural combretastatin a-4. *Curr. Med. Chem.* **18**: 3035–3081.

Stewart, A.K. (2014). Medicine. How thalidomide works against cancer. *Science* **343**: 256–257.

Sugahara, K.N., Teesalu, T., Karmali, P.P., Kotamraju, V.R., Agemy, L., Greenwald, D.R., *et al.* (2010) Co-administration of a tumor-penetrating peptide enhances the efficacy of cancer drugs. *Science* **328**: 1031–1035.

Wilhelm, K., Happel, K., Eelen, G., Schoors, S., Oellerich, M.F., Lim, R., *et al.* (2016) FOXO1 couples metabolic activity and growth state in the vascular endothelium. *Nature* **529**: 216–220.

Zimna, A. and Kurpisz, M. (2015) Hypoxia-Inducible Factor-1 in Physiological and Pathophysiological Angiogenesis: Applications and Therapies. *Biomed. Res. Int.* **2015**: 549412.

Zuccari, G., Milelli, A., Pastorino, F., Loi, M., Petretto, A., Parise, A., *et al.* (2015) Tumor vascular targeted liposomal-bortezomib minimizes side effects and increases therapeutic activity in human neuroblastoma. *J. Controlled Release* **211**: 44–52.

Zweifel, M., Jayson, G.C., Reed, N.S., Osborne, R., Hassan, B., Ledermann, G., *et al.* (2011) Phase II trial of combretastatin A4 phosphate, carboplatin, and paclitaxel in patients with platinum-resistant ovarian cancer. *Ann. Oncol.* **22**: 2036–2041.

Chapter 11

Nutrient and hormone effects on the genome

Introduction

It is exciting to know that both food components and hormones exert some of their cancer-relevant effects via the genome. This chapter will present the molecular mechanism of nutrients and hormones in the context of cancer. Some components of diets act as cancer-causative factors, and others act as cancer-preventative factors. Diet and exercise also affect cell metabolism, the biochemical pathways involved in obtaining energy from food. The reprogramming of energy metabolism associated with tumor cells, an emerging hallmark of cancer (as stated in Chapter 1), will be described. Upon examination of the mechanisms of action of nutrients, it will become clear that some of these mechanisms parallel the mechanism of action of growth factors and others parallel the mechanism of action of hormones. In fact, some vitamins act in the same manner as hormones—via ligand-dependent transcription factors that control sets of genes. The chapter will conclude with a discussion of the role of hormones in carcinogenesis and the application of our knowledge to therapeutics.

11.1 Introduction to food and cancer

Does our diet influence whether we are the one out of three people who get cancer? Diet plays a significant role in cancer incidence. Many epidemiological studies provide evidence to support the role of diet in both causation and prevention of cancer. Approximately one-third of the variations in cancer incidence between different populations are due to differences in diet. For example, migration studies have demonstrated an increase in colorectal cancer rates in originally low-risk ethnic groups after migration to high-risk locations. An association can be seen in most regions of the world between a consumption of a meat and animal fat dietary pattern and the rate of colorectal cancer (Bishehsari *et al.*, 2014). The knowledge gained from investigations into the role of diet in cancer and cancer prevention should be integrated into lifestyle modifications in order to reduce the occurrence of the disease. Let us examine why we eat food (Figure 11.1). The basic food groups of carbohydrates, fats, and proteins provide us with glucose, fatty acids, and amino acids, respectively, which can be metabolized to produce energy. Food also provides precursors for biosynthetic reactions. For example, proteins provide a source of nitrogen needed for the synthesis of the nitrogenous bases of DNA. Vitamins and minerals provide co-factors that are essential for the function of many enzymes. Additional biologically active microconstituents have been identified in the foods we eat (Table 11.1). Many biologically active microconstituents act as antioxidants, compounds that significantly inhibit or delay the damaging action of ROS (see Chapter 2), often by being oxidized themselves. Plants require many phytochemicals as a defense against excess energy and oxidative damage, as they absorb solar energy for photosynthesis. Many of these phytochemicals provided in the diet are important for the protection of human cells. Although humans can synthesize some required antioxidants, others must be obtained by eating fruit and vegetables. The four major groups of dietary antioxidants–phytochemicals are vitamin C, isoprenoids (e.g. vitamin E), phenolic compounds (flavonoids), and organosulfur compounds. These will be discussed later in this chapter. Lastly, specific nutrients and microconstituents have been shown to affect gene expression (Figure 11.1).

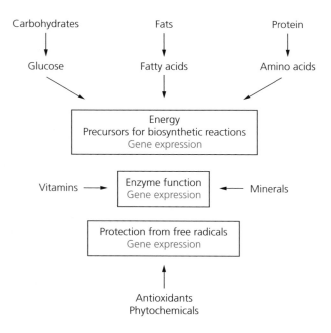

Figure 11.1 Provisions of food.

Information gained about the role of microconstituents in cancer prevention could be applied to the development of chemopreventative supplements, extra sources of dietary components taken in addition to food. However, unraveling the individual contributions of microconstituents as preventative agents against cancer is a challenge for the future. Epidemiological studies strongly suggested that diets rich in β-carotene-containing fruits and vegetables reduced lung cancer risk. Animal studies also generated supportive evidence. This led to the β-Carotene and Retinol Efficacy Trial (CARET) and the Alpha-Tocopherol Beta-Carotene Cancer Prevention Study (ATBC) which tested the effect of β-carotene supplements on smokers and those exposed to asbestos. Surprisingly, β-carotene supplementation increased lung cancer in these high-risk individuals and had no effect on healthy individuals. The results of these trials do not support the initial hypothesis formulated on preclinical findings (see "Pause and think").

Only recently have molecular approaches been used to investigate the molecular mechanisms of dietary constituents involved in the causation

> **PAUSE AND THINK**
>
> Propose a hypothesis to explain these seemingly conflicting results. The most likely explanation is that alternative microconstituents in β-carotene-rich vegetables and fruits may be the active ingredient in reducing lung cancer risk, or perhaps β-carotene works in a synergistic manner with other microconstituents not present in the supplements. Interactions between different dietary constituents must be considered for a complete picture.

Table 11.1 Microconstituents

Food source	Class of compound	Chemical
Cruciferous vegetables	Isothiocyanate	Benzyl isothiocyanate, phenethyl isothiocyanate, sulforaphane
Cruciferous vegetables	Dithiolthione	Ohipraz
Cruciferous vegetables	Glycosinolate	Indole-3-carbinol, 3,3'-diindoylmethane, indole-3-acetonitrile
Onions, garlic, scallions, chives	Allium compound	Diallyl sulfide, allylmethyl trisulfide
Citrus fruit (peel)	Terpenoid	D-limonene, penllyl alcohol, geraniol, menthol, carvone
Citrus fruit	Flavonoid	Tangeretin, nobiletin, ratin
Berries, tomatoes, potatoes, broad beans, broccoli, squash, onions	Flavonoid	Quercetin
Radish, horseradish, kale, endive	Flavonoid	Kaempferol
Tea, chocolate	Polyphenol	Epigallocatechin gallate, epigallocatechin, epicatechin, catechin
Grapes	Polyphenol	Resveratrol
Turmeric	Polyphenol	Curcumin
Strawberries, raspberries, blackberries, walnuts, pecans	Polyphenol	Caffeic acid, ferulic acid, ellagic acid
Cereals, pulses (millet, sorghum, soya beans)	Isoflavone	Genistein
Orange vegetables and fruit	Carotenoid	α- and β-carotene
Tomatoes	Carotenoid	Lycopene
Tea, coffee, cloa, cacao (cocoa and chocolate)	Methylxanthines	Caffeine, theophylline, theobromine

Reprinted from Manson, M.M. (2003) Cancer prevention—the potential for diet to modulate molecular signaling. *Trends Mol. Med.* **9**: 11–18, Copyright (2003), with permission from Elsevier.

or prevention of cancer. One of the most significant insights gained is that *components in foods regulate gene expression*. The power of food has begun to be revealed. This chapter will include a sample of these findings.

11.2 Causative factors

Three main aspects of our diet can be considered as causative factors of cancer. First, any given food is a very complex substance that can carry harmful factors in addition to nutritional value. The consumption of food provides a route for chemical carcinogens to be delivered to the body. Genotoxic agents present as microconstituents in food act as dietary carcinogens. Second, lack of a particular essential nutrient may enhance the risk of cancer. Third, worldwide health issues, such as obesity and chronic alcohol consumption, cause cancer. In this section, we will examine carcinogenic contaminants, nutritional deficiencies, obesity, and chronic alcohol consumption as dietary cancer-causative factors.

Carcinogenic contaminants

The carcinogenic effect of one particular food can be variable. Salmon, rich in omega-3 polyunsaturated fatty acids and known to be an important component of a healthy diet, is one example. Salmon, being fatty carnivorous fish, accumulate pollutants and can pass genotoxic contaminants through the food chain to humans. A study of farmed and wild salmon from around the world found that, in some geographical regions (e.g. Scotland), polychlorinated biphenols (PCBs) and other pesticides are present in quantities that suggested that eating farmed salmon more than once a month could increase cancer risk (Hites *et al.*, 2004). Risk was calculated based on the assumption that the risks of individual carcinogens are additive. This study raises many issues. First, it underscores that differences in the source of food can have varying consequences; farmed salmon has more contaminants than wild salmon, and farmed salmon from Scotland contains significantly more contaminants than farmed salmon available in North American cities. Perhaps the results have a broader implication and point to the suggestion that the source of all food should be properly labeled to allow for consumer choice and to create competition for the production of good products. Data from this study suggested that fish feed (fish meal and fish oils) may be a distinguishing factor for the carcinogenicity of salmon, suggesting that improvements in feed composition are needed. Evaluating cancer risk associated with more than one contaminant at a time, in addition to the benefits of other microconstituents of a particular food, is an area that requires further study. Overall, the study described here underlines the complications that occur when analyzing the relationship between diet and cancer.

Food preparation and storage can contribute to the cancer-causing properties of our diet. Heterocyclic amines produced by cooking meat at high temperatures were discussed as carcinogens in Chapter 2. After metabolic activation, their mechanism of action involves the formation of DNA adducts, resulting in base substitutions, and thus mutations. Similarly, toxins produced by molds that contaminate food form DNA adducts and thus are genotoxic. Aflatoxin B, a fungal product of *Aspergillus flavus*, is a well-known contaminant found on peanuts, and fumonisin B is found on corn. Aflatoxin induces GC→TA transversions and is involved in causing hepatocellular carcinoma. Food preservatives, such as sodium nitrite, are regulated by government agencies, because they too are a risk factor producing carcinogenic *N*-nitroso compounds.

Dietary deficiencies

Evidence is accumulating that supports the concept that micronutrient deficiencies also contribute to cancer risk. The most compelling findings suggest that a deficiency in folate increases the risk of colorectal cancer. Folate, one of the B vitamins, can accept or donate one-carbon units in metabolic reactions. Folate is a critical co-enzyme for nucleotide synthesis and DNA methylation, and these processes can affect carcinogenesis. The enzyme methylenetetrahydrofolate reductase (MTHFR) regulates the balance between nucleotide synthesis and DNA methylation by affecting the relative quantities of 5,10-methylenetetrahydrofolate (5,10-methylene THF) and 5-methyl-tetrahydrofolate (5-methyl THF), the respective precursors of these distinct processes (Figure 11.2a). MTHFR irreversibly converts 5,10-methylene THF to 5-methyl THF. 5,10-methylene THF and deoxyuridylate (dUMP) are reactants for the enzyme thymidylate synthase used for the production of deoxythymidylate (dTMP). 5-methyl THF and homocysteine are reactants used to produce methionine, which regenerates *S*-adenosylmethionine (SAM), the methyl donor for DNA methylation.

The depletion of folate may contribute to tumor development by interfering with both nucleotide synthesis and DNA methylation. A disruption in DNA synthesis leads to DNA instability and fuels mutation, while disruption in DNA methylation may cause genomic hypomethylation (Figure 11.2b). dTMP synthesis is inhibited in conditions of low folate, and the imbalance of the nucleotides results in the incorporation of uracil into DNA. DNA strand breaks occur as a result of attempts to repair this DNA, and these breaks increase cancer risk. Both uracil misincorporation and DNA strand breaks are observed in folate-deficient humans, and both defects are reversed by folate administration. Remember that genomic hypomethylation and specific tumor suppressor gene promoter hypermethylation are characteristic of the epigenetic changes observed in cancer cells (see Chapter 3). As the methyl groups used for DNA methylation

PAUSE AND THINK

Suggest an experiment to examine whether the nutritional benefit of a particular food outweighs its risk as a carcinogen.

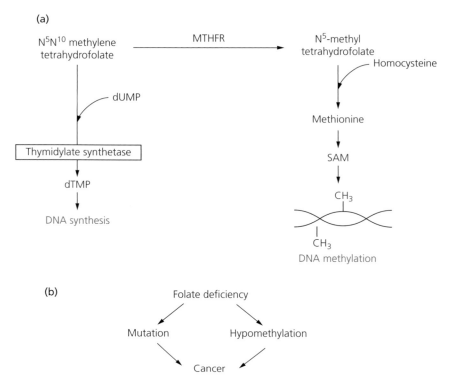

Figure 11.2 (a) Role of folate derivatives in DNA synthesis and DNA methylation. (b) Role of folate deficiency in cancer.

are supplied by folate, a lack of folate causes a decrease in the synthesis of methionine, and subsequently genomic hypomethylation of DNA. Genomic hypomethylation is observed in folate-deficient humans and is reversed upon folate repletion. Hypermethylation at specific 5' gene loci has also been observed during studies of folate depletion. In addition, dietary methyl deficiency has been shown to alter hepatic DNA methylation patterns and induce hepatocarcinogenesis in the absence of a carcinogen in rodents.

Obesity

Obesity, classified as a risk factor for several cancers by the International Agency for Research on Cancer (IARC), is the excessive accumulation of fat that leads to a body weight that is greater than skeletal and physical requirements. Those with a body mass index [weight (in kg)/height (in m) squared] greater than 30 (kg/m^2) are considered obese. It has become a significant problem in the USA, affecting 25% of the population, and it has been suggested that 15–20% of all cancer deaths in the USA can be attributed to being overweight and obese. Based on epidemiological evidence, obesity increases the risk of cancer of the colon, breast, endometrium, kidney, pancreas, liver, and esophagus. Adipose tissue is an endocrine organ that can affect other tissues: it can release free fatty acids, peptide hormones, and steroid hormones. Several mechanisms of action of obesity as a cancer risk factor or tumor promoter, including altered sex hormone metabolism, increased production of fat cell (adipocyte) hormones called adipokines, increased insulin signaling pathways, and dietary alterations in the gut microbiota, have been suggested and are described in the following paragraphs.

Obesity results in high fat deposits in adipose cells. The deposits may be used for the synthesis of estrogen from androgen by aromatase and may contribute to the risk of breast cancer (see Section 11.7, "Hormones and cancer" on p. 265). Elevated levels of cholesterol, hypercholesterolemia, is common in obese people. It has been demonstrated that a metabolite of cholesterol, called 27-hydroxycholesterol (27HC), is a ligand of the estrogen receptor and liver X receptor, and thus may share some mechanisms used by estrogen in cancer such as regulating estrogen-responsive genes (see Figure 11.10 on p. 267) (Nelson *et al.*, 2013). Studies showed that 27HC increased growth and metastasis in mouse models. They also showed that aggressive human breast tumors had higher levels of the enzyme that converts cholesterol to 27HC, the cytochrome P450 oxidase CYP27A1, compared to less aggressive tumors. This suggests new therapeutic strategies that may include the use of statins, drugs that lower cholesterol, and inhibitors of CYP27A1. Another mechanism of obesity may be due to adipokines and inflammation. Evidence from mice demonstrate that obesity causes a chronic inflammatory response with increases in tumor-promoting cytokines, such as interleukin-6 (IL-6) and TNF, and acts as a bona fide liver tumor promoter (Park *et al.*, 2010). Obesity also leads to chronically increased levels of plasma insulin as a result of the release of large amounts of free fatty acids from adipose tissue. Subsequent tumorigenic effects (e.g. promotion of cellular proliferation and inhibition of apoptosis) are mediated through insulin receptors and additional growth factors (e.g. IGF).

More recently, a connection between obesity, our gut bacteria, and liver cancer has emerged. It is known that dietary intake influences the structure and activity of the trillions of microorganisms residing in the human gut. Evidence suggests that obesity promotes liver cancer by causing a change in the gut microbiome (more Gram-positive bacteria), which results in an increase of a bacterial metabolite, deoxycholic acid (Yoshimoto *et al.*, 2013) (see Box "How do we know that?", p. 252). A primary bile acid derived from cholesterol is metabolized by a dehydroxylation reaction in gut bacteria to form deoxycholic acid. This metabolite can circulate to the liver and cause DNA damage. Chronic DNA damage in certain liver cells triggers cell senescence that is characterized by the secretion of inflammatory and tumor-promoting factors, thus inducing the development of obesity-associated liver cancer. These findings from rodent studies may have relevance to humans. The human liver is not efficient at metabolizing deoxycholic acid, and so high levels can accumulate. Higher faecal concentrations of deoxycholic acid have been found in men on high-fat diets. Note that obesity increases the risk of liver cancer by 1.5- to 4-fold.

? HOW DO WE KNOW THAT?

Analysis of the effects of dietary obesity on tumorigenesis by macroscopic examination, microbiome analysis, and analysis of serum metabolites by liquid chromatography mass spectroscopy

(See Yoshimoto et al., 2013.)

In order to examine the effects of dietary obesity on tumorigenesis, mice were treated once with the chemical carcinogen dimethylbenz[a]anthracene (DMBA) at the neonatal stage and then fed either a normal diet or a high-fat diet for 30 weeks. Macroscopic examination revealed that all high-fat diet-fed mice developed liver cancer. In contrast, none of the control mice developed tumors in the liver (Figure 11.3a). Analysis of the microbiome in the faecal community by gene sequence analysis revealed a large increase in Gram-positive bacteria when fed a high-fat diet, and this could be altered when treated with vancomycin (VCM), a Gram-positive-selective antibiotic (Figure 11.3b). Serum metabolites of both sets of mice were analyzed by liquid chromatography mass spectrometry. Deoxycholic acid (DCA) was increased in high-fat diet-fed mice and reduced by: VCM treatment; treatment with an agent (DFAIII) that lowers DCA; and treatment with an agent (UDCA) that stimulates bile secretion (Figure 11.3c).

Figure 11.3 (a) Representative macroscopic photographs of livers. Arrowheads indicate liver cancer. (b) Bacterial diversity analysis of the faecal bacterial community. (c) Serum DCA concentrations (ND, $n = 4$; HFD, $n = 6$; HFD + VCM, $n = 3$; HFD + DFAIII, $n = 3$; HFD + UDCA, $n = 3$). Error bars indicate mean ± s.e.m (standard error of the mean). HFD, high-fat diet; ND, normal diet. Adapted by permission from Macmillan Publishers Ltd: Nature, from Yoshimoto, S. et al. (2013) Obesity-induced gut microbial metabolite promotes liver cancer through senescence secretome. Nature **499**: 97–101, copyright (2013).

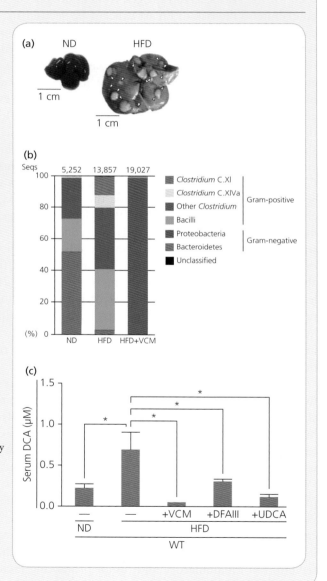

Chronic alcohol consumption

Two billion people consume alcohol. Because of this statistic and the recent classification of alcohol as a carcinogen by the IARC in 2007, there is a great need to publicize the cancer-causing nature of alcohol in order to prevent specific cancers. The mechanism of how alcohol causes cancer is linked to its metabolism in the body (Seitz and Stickel, 2007). Alcohol is metabolized by the enzyme alcohol dehydrogenase to form acetaldehyde that can directly bind to DNA, forming DNA adducts that cause mutations. Acetaldehyde DNA adducts have been found in alcohol consumers. Additional acetaldehyde can be made by bacteria in saliva, such that concentrations can be 10–100 times higher than in blood, accounting for the increased risk of mouth cancer. Acetaldehyde is oxidized in the body by the enzyme acetaldehyde dehydrogenase. A single nucleotide polymorphism of the gene for this enzyme, common in Asians, is responsible for low enzyme activity, and consequently alcohol intolerance. Carriers of the inactive acetaldehyde dehydrogenase allele are at increased risk for alcohol-related esophageal cancer. The association of alcohol consumption with increased risk of breast cancer is thought to be because of its link with estrogen (see Section 11.7 "Hormones and cancer" on p. 266.). Overall, 3.6% of cancers worldwide, including 7–19% of oral cancers, are attributable to drinking alcohol.

11.3 Preventative factors: microconstituents of fruits and vegetables

The Mediterranean diet, which is rich in fruits and vegetables (especially tomatoes, grapes, and garlic), is associated with a reduction in overall cancer incidence and the incidence of specific cancers such as colorectal cancer (Schwingshackl and Hoffman, 2014). The intake of fruits and vegetables as a means of reducing cancer risk is supported by epidemiological studies, but these are not easy studies to perform, as it is difficult to control for other behavioral variables. One report published in 2007 by the World Cancer Research Fund and the American Institute for Cancer Research, called *Food, Nutrition, Physical Activity and the Prevention of Cancer: A Global Perspective*, summarizes an evaluation of a mass of scientific publications with respect to cancer risk. In their report, they conclude that there is probable evidence which supports that intake of fruits and some vegetables decreases the risk of specific cancers. Also, The European Prospective Investigation into Cancer and Nutrition (EPIC) analyzed over 400,000 people and recently concluded that there is a small, but statistically significant, inverse association between intake of total fruits and vegetables and overall cancer risk (Boffetta *et al.*, 2010).

The ability to block DNA damage caused by ROS and/or carcinogens is the most direct strategy for preventing the initiation of cancer and for slowing down the progression of disease. It is here that microconstituents found in fruits and vegetables (Table 11.1) play an important role. This is accomplished either directly by free radical scavengers (see "Free radical scavenging" on p. 255) or indirectly by regulating the expression of genes that code for Phase I (oxidative) and Phase II (conjugative) metabolizing enzymes in the body.

The modulation of Phase I and II metabolizing enzymes is a major defense mechanism against xenobiotics (foreign substances). The cytochrome P450 family of Phase I drug-metabolizing enzymes catalyzes the hydroxylation/oxidation of many drugs, which often has a harmful effect by converting pro-carcinogenic molecules into ultimate carcinogens. Phase I products are often highly electrophilic (e.g. epoxides) and can damage DNA but at the same time induce enzymes required for Phase II. Phase II enzymes, such as UDP-glucuronosyltransferases or glutathione S-transferases, catalyze conjugation reactions of Phase I products to hydrophilic moieties, thus making them more water-soluble and aiding in their innocuous removal from the cell.

 HOW DO WE KNOW THAT?

Analysis of DNA damage by the Comet assay

(See Bub *et al.*, 2003.)

The molecular effects of fruit polyphenols in humans were investigated by analyzing DNA damage in blood samples taken before, during, and after treatment. A precise schedule of fruit juice consumption (330 ml/day; over two 2-week periods) in healthy men was executed. Two juices were tested; both contained apple, mango, and orange juice, but one was enhanced with berries rich in anthocyanin and the other with green tea, apricot, and lime, rich in flavanols. The single-cell microgel electrophoresis, or Comet, assay was used to detect oxidized DNA bases. Blood cells were embedded onto agarose-coated slides, lysed, treated with alkali (for unwinding) and a specific endonuclease (III) to detect oxidized pyrimidine bases, and subjected to electrophoresis. Comet-like images, indicating single-strand breaks and resulting from the extension of DNA into the agarose, were computer-analyzed after neutralization and ethidium bromide staining. The data showed significantly lower levels of DNA base oxidation for both juices over the last treatment period (see Figure 1 in Bub *et al.*, 2003). That is, the effect was observed after the second 2-week period of consumption, but not the first. The effect was not permanent, as levels returned to baseline when tested 11 weeks after the experiment was terminated. The time delay indicated by this experiment suggests that ROS scavenging is not the prime mechanism and that protective detoxifying enzymes (described in "Regulation of genes that code for drug-metabolizing and antioxidant enzymes" on p. 255) are induced.

PAUSE AND THINK

In general, how would you modify Phase I and II enzyme activity to reduce tumor formation? Inhibit Phase I, and induce Phase II.

Let us look at the modifications of the carcinogen aflatoxin B1 (AFB1) made by Phase I and II metabolizing enzymes. First, AFB1 undergoes oxidation by Phase I cytochrome P450 to form the potent genotoxic metabolite AFB1–8,9-epoxide. AFB1–8,9-epoxide is conjugated to glutathione by the Phase II enzyme glutathione S-transferase to produce AFB1–glutathione. This leads to detoxification and facilitates easy excretion.

Free radical scavenging

Several microconstituents in fruits and vegetables act as antioxidants that scavenge ROS. Water-soluble vitamin C (Figure 11.4) can donate an electron to a free radical directly, thus inhibiting its reactivity and blocking free radical chain reactions. Oxidized vitamin C forms an ascorbyl radical that is fairly stable and unreactive because of electron delocalization or resonance. An enzyme called vitamin C reductase can regenerate vitamin C from the ascorbyl radical for reuse, or the ascorbyl radical may lose another electron and become degraded. Consequently, vitamin C reserves need to be replenished daily. Lipid-soluble vitamin E (Figure 11.4) acts as a free radical scavenger in a similar manner. A resonance-stabilized structure called the α-tocopheryl radical is produced after vitamin E donates an electron to a free radical (e.g. singlet oxygen) and helps to terminate chain reactions of free radicals in membranes.

Regulation of genes that code for drug-metabolizing and antioxidant enzymes

"You are what you eat" is a common expression. It has been given greater significance recently by its translation into molecular terms and the study of nutrigenomics: some dietary constituents can affect the expression of our genes. The molecular mechanisms employed are common to those discussed in previous chapters. The crucial link between some dietary constituents and their role in preventing DNA damage via gene expression was made by the identification of an antioxidant response element (ARE) (5'-A/G TGA C/T NNNGC A/G-3') in the promoter region of several genes encoding detoxification and antioxidant enzymes (e.g. glutathione S-transferase and NADPH:quinine oxidoreductase 1). The ARE confers antioxidant-dependent regulation of target genes. In fact, the ARE is transcriptionally activated not only by food antioxidants such as isothiocyanates (e.g. sulforaphane) in cruciferous vegetables and epigallocatechin-3-gallate (EGCG) in green tea, but also in response to reactive electrophilic intermediates and H_2O_2 (i.e. carcinogens).

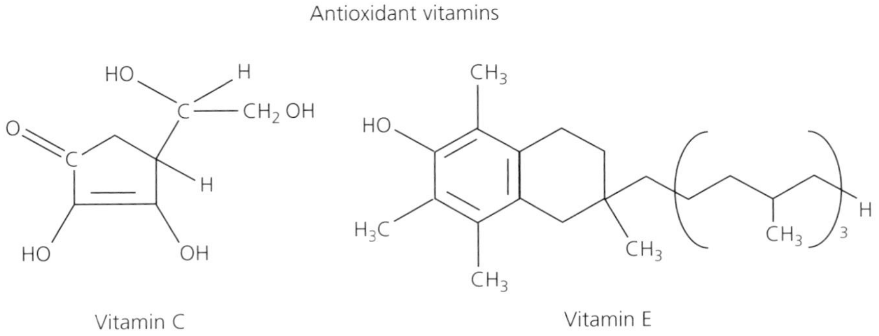

> **PAUSE AND THINK**
>
> Is the hydroxyl radical described in Chapter 2 likely to be scavenged by these microconstituents? No. In fact, the menacing reactive hydroxyl radical is unlikely to be scavenged by these microconstituents because of its extremely rapid reaction time. It would take exceptionally high concentrations to prevent such interactions with molecules immediately surrounding it.

Antioxidant vitamins

Vitamin C

Vitamin E

Figure 11.4 The structure of the antioxidant vitamins, vitamin C and vitamin E.

Activation of gene expression through the ARE is mediated by the transcription factor Nrf2 and co-activator Maf, a member of the basic leucine zipper family. Nrf2 is a main defense mechanism in cells because of its crucial role of inducing genes coding for detoxification and antioxidant enzymes. KEAP1 is an important repressor of Nrf2 via protein–protein interactions. Thus, Nrf2 has both KEAP1 and DNA-binding domains. In non-stressed cells, KEAP1 forms a complex with Nrf2 and Cul3-E3 ubiquitin ligase. This leads to Nrf2 ubiquitination and degradation by proteasomes (Figure 11.5a). KEAP1 contains cysteines that are important for its inactivation. It is these cysteines that normally act as sensors of the redox status in the cell. Electrophiles or ROS can react with these cysteines, causing a conformational change that prevents Nrf2 ubiquitination. KEAP1 molecules become saturated with Nrf2 protein that is not modified by the addition of ubiquitin, and newly synthesized Nrf2 translocates into the nucleus. Here, Nrf2 induces gene expression via the ARE (Figure 11.5b). This is an important molecular mechanism that the cell possesses to limit damage from oxidative and xenobiotic stresses: these stresses stimulate induction of enzymes that will make them less toxic and modify them for excretion.

Some food components can also inactivate KEAP1 (Figure 11.5c). As mentioned earlier, sulforophane found in broccoli also activates Nrf2-regulated transcription. Sulforophane (a hydrolysis product of sulfur-containing glucosinolates) contain sulfhydryl groups that are able to react with cysteine residues within the Nrf2 repressor KEAP1. The direct interaction of sulforophane with the cysteine residues of KEAP1 causes a conformational change in KEAP1 that blocks its action and leads to the translocation of Nrf2 into the nucleus. Nrf2 is then able to bind to the ARE, dimerize with Maf, and induce the expression of genes encoding Phase II detoxification enzymes.

EGCG, the major polyphenol found in green tea, is another dietary constituent that activates the Nrf2 pathway. Several possibilities for its mode of action have been proposed (Na and Surh, 2008): KEAP1 may be inhibited by metabolites of EGCG or by EGCG-generated ROS (Figure 11.5c). Alternatively, evidence suggests that EGCG may activate a distinct MAPK signal transduction pathway (not shown). Intracellular kinases may lead to phosphorylation of Nrf2 and its subsequent stabilization, resulting in gene expression.

Therefore, some dietary microconstituents mimic carcinogenic ROS or electrophiles that inactivate KEAP1 and stabilize Nrf2. This molecular pathway is important for protecting the cell from DNA damage caused by oxidants by regulating the expression of genes that code for detoxification enzymes.

In sharp contrast to the role of Nrf2 in protecting normal cells, Nrf2 has a "dark side" in that it may also protect cancer cells from chemotherapeutic agents during treatment and contribute to drug resistance

> ### LIFESTYLE TIP
>
> We should use the knowledge that we have gained about the preventative role of particular foods and beverages to make better choices about what we ingest. Green tea is the second most popular beverage in the world, after water. A high intake of green tea is associated with a low incidence of several cancers (e.g. gastric and colorectal cancer). Green tea is a better choice than soda.

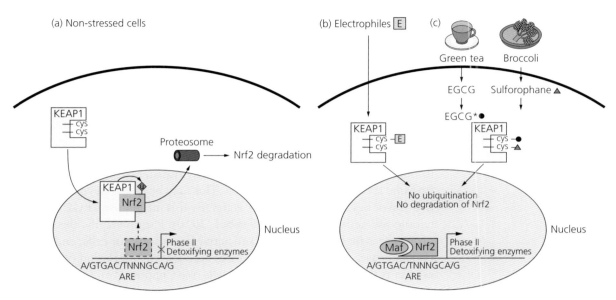

Figure 11.5 The Nrf2–ARE signaling pathway. (a) In non-stressed cells, KEAP1 (shaded in yellow) forms a complex with Nrf2 and Cul3-E3 ubiquitin ligase. This promotes Nrf2 ubiquitination and subsequent degradation by the proteasomes. (b) Dangerous electrophiles interact with cysteine (cys) residues present in KEAP1, causing a conformational change that prevents Cul3-E3 ligase from modifying Nrf2. Upon saturation of KEAP1 with unmodified Nrf2, newly formed Nrf2 translocates to the nucleus. Here, Nrf2 can bind to the ARE, heterodimerize with its co-activator Maf, and stimulate transcription of detoxification enzymes. (c) Several dietary constituents regulate gene expression via the transcription factor Nrf2. Broccoli is a rich source of sulforophane. Sulforophane reacts with the cysteine residues in KEAP1, causing a conformational change and inhibition of ubiquitination. As in (b) above, Nrf2 translocates into the nucleus where it can bind to the ARE, heterodimerize with its co-activator Maf, and stimulate transcription of detoxification enzymes. Also, green tea contains EGCG. It has been suggested that an electrophilic metabolite of EGCG, EGCG*, may bind and inactivate KEAP1, as above, allowing Nrf2 to translocate to the nucleus and induce transcription of detoxification enzymes via the ARE.

(Jaramillo and Zhang, 2013). This is accomplished by inducing genes that are involved in drug metabolism and xenobiotic transporters such as multi-drug resistance protein 1. Some cancers have been found to have constitutively elevated levels of Nrf2. Somatic mutations, including loss-of-function mutations of KEAP1 and Cul3-E3 ligase and gain-of-function mutations of Nrf2 genes, have been reported. Epigenetic silencing of KEAP1 has also been shown to give rise to an accumulation of Nrf2.

Additional mechanisms of dietary microconstituents

Current evidence suggests several mechanisms for the cancer-preventative role of fruits and vegetables. As we have seen earlier, one mechanism is the ability to decrease oxidative DNA damage by free radical scavenging or inducing protective enzymes. Two other mechanisms for the role of particular vegetables in cancer prevention are modulation of apoptosis and/or cell proliferation. The microconstituents of garlic utilize all three mechanisms (see references within Yun *et al.*, 2014). The antioxidant properties of organosulfur compounds in garlic include the induction of

Phase II enzymes and scavenging. Ajoene, a major compound in garlic, has been shown to induce apoptosis of leukemic cells in patients with leukemia. Particular caspases (3 and 8) and transcriptional regulators (IκB) are activated, and peroxide is produced. Allicin, another major compound in garlic, has been shown to inhibit the proliferation of human mammary, endometrial, and colon cancer cells. It is suggested that some of these effects are mediated via inhibition of the NF-κB signaling pathway (see Chapter 13).

Another protective mechanism of EGCG found in green tea is its ability to bind to the catalytic site of DNA methyltransferase, blocking entry of cytosine and preventing methylation. This may be able to reverse the inactivation of key tumor suppressor genes. By this mechanism, EGCG blocks telomerase activity (Li and Tollefsbol, 2010). Paradoxically, the hTERT promoter is highly methylated and active in most tumor cells, and, in this unusual case, demethylation represses activity. The inhibition of telomerase limits the replicative capacity of cells (see Chapter 3) and, in this study, correlates with a decrease in tumor size in mouse models.

Although fiber is usually included in discussions of preventative agents of cancer, I have chosen to omit this topic here owing to the inconsistencies of recent large studies (Romaneiro and Parekh, 2012). However, one EPIC study shows a strong protective effect of dietary fiber against colorectal cancer. The data suggest that preventative effects were not seen in some previous studies, because the range of fiber intake was much lower than those in the EPIC study (see Activity 2 at the end of this chapter on p. 273).

In conclusion, a brief examination of several different foods demonstrates that the molecular mechanisms by which nutrients affect carcinogenesis are beginning to be revealed.

11.4 Reprogramming energy metabolism in tumor cells—an emerging hallmark of cancer

The digestion of food provides many of the compounds required for metabolism, the sum of the biochemical reactions in the body. Thus, diet (and exercise) affects cellular metabolism. Reprogramming energy metabolism is an emerging hallmark of cancer (see Chapter 1). Some tumor cells seem to be addicted to increased glucose uptake and glycolysis. The observation that cancer cells carry out aerobic glycolysis, converting glucose to lactate in the presence of oxygen, was made in the 1920s and is called the Warburg effect. This metabolic alteration differs from both anaerobic (without oxygen) glycolysis and aerobic metabolism used by differentiated cells that proceeds through the Krebs cycle and electron transport chain (see Box "A quick review about glucose metabolism" on p. 262). This area of study has been revisited, and differing viewpoints have not yet settled. Some suggest that 60–90% of tumors shift to glycolysis; others propose that the shift is time-dependent, and others again suggest it may be cell type-dependent (stem cells versus differentiated cells). It has been proposed that the metabolism of cancer cells resembles the metabolism of proliferating cells. Both are adapted to meet the anabolic needs of proliferation. Making new cells requires proteins, nucleotides, and lipids. Aerobic glycolysis provides anabolic substrates and ATP from enhanced glucose uptake. Note that metabolism is linked to important signaling pathways involving oncogenes and tumor suppressor genes. Inactivation of p53 results in many biological effects that stimulate the Warburg effect. Many metabolites play a role in signaling pathways, especially ATP and acetyl CoA. And as we saw in Chapter 7, cytochrome *c* which plays a role in electron transport in the mitochondria is also important for apoptosis. Recently, a new role for the reprogramming of a tumor to Warburg metabolism has been uncovered—tumor immunosuppression. Glycolysis causes a depletion of extracellular glucose, and this reduces glycolysis and effector function of tumor-infiltrating T-cells (Chang *et al.*, 2015; Ho *et al.*, 2015). Thus, tumor cells can alter the metabolic characteristics of the microenvironment to affect immune responses against the tumor. The Warburg effect is only one of many metabolic alterations associated with carcinogenesis.

Epigenetic regulation by metabolites and exercise

A link between intermediary metabolites and epigenetic regulation of gene expression has been demonstrated (Gut and Verdin, 2013). Almost all chromatin-modifying enzymes use intermediary metabolites as co-factors or substrates. Nutrients are converted to intermediary metabolites during metabolism. Glucose is converted into acetyl CoA during glycolysis. Acetyl CoA is the universal donor for acetylation reactions, including histone acetylation reactions. The shift in metabolism of cancer cells can alter the abundance of metabolites like acetyl CoA and may induce the transcription of genes that regulate cellular proliferation.

The family of NAD^+-dependent HDACs, called sirtuins, provides additional links between metabolites, gene expression, and cancer (Chalkiadaki and Guarente, 2015; Morris, 2013). This family consists of seven members, each with different complex functions. They mediate responses to changes in the energy status of cells by using NAD^+ as a co-substrate for enzyme function. These enzymes deacetylate histones and other proteins such as transcription factors (e.g. p53). The excitement around these proteins came from the demonstration that overexpression of some members can extend life span in rodents and was linked to anti-aging and anti-cancer effects. Resveratrol, a polyphenol present in grapes and wine, may exert some of its cancer preventative effects by allosterically activating SIRT1. SIRT6 acts as a tumor suppressor via its inhibition of the Warburg effect and role in DNA damage response. Sirtuins are also thought to be the molecular mediators of the beneficial effects of exercise (Pucci *et al.*, 2013)—an exciting hypothesis to support the cancer preventative effects of exercise. There seems to be little doubt that sirtuins have roles in carcinogenesis, but the field is new, and contrasting experimental results need to be resolved.

It is of interest that the Warburg effect is the basis for an important imaging technique used to detect tumors in the clinic. Positron emission tomography (PET) scans work on the basis that tumor cells exhibit a greater uptake of glucose than most normal cells. A glucose analog, [18F] fluoro-2-deoxyglucose (FDG), is injected into the bloodstream and is converted by the glycolytic enzyme hexokinase to FDG phosphate, which can be visualized (Figure 11.6).

Data suggest that both genetic alterations (oncogenes and tumor suppressor genes) and a response to hypoxia via HIF-1α contribute to the altered metabolism observed in cancer cells. Let us examine the evidence for both mechanisms.

 Understanding of the pathway of a key metabolic enzyme, AMP-activated protein kinase (AMPK), is uncovering links between energy metabolism and cancer. AMPK senses the energy state of a cell and is activated under conditions of increased AMP and decreased ATP in response to starvation, hypoxia, and exercise. It is composed of an α catalytic subunit and β and γ regulatory subunits. In the current model of AMPK activation (Figure 11.7), AMP binds to the γ subunit of AMPK and causes a conformational change, exposing a potential phosphorylation site

PAUSE AND THINK

You may remember from Chapter 10 that HIF is composed of two subunits; the α subunit is regulated at the protein level via regulated degradation, and the β subunit is constitutive.

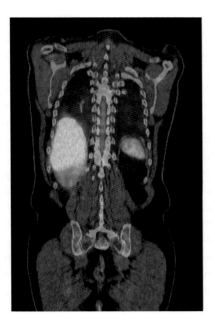

Figure 11.6 A positron emission tomography (PET) image. Picture courtesy of Siemens Medical Solutions.

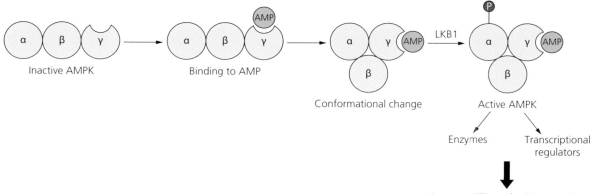

Figure 11.7 Activation of AMP-activated protein kinase AMPK. When the energy of a cell decreases, AMP increases. AMP binds to inactive AMPK and causes a conformational change that allows phosphorylation of the α catalytic subunit by LKB1. The activating phosphate is shown in red.

within the α subunit of AMPK. The serine/threonine kinase LKB1 phosphorylates (Thr 172) and activates AMPK. AMPK, also a serine/threonine kinase, exerts its effects through the regulation of target enzymes and transcription factors. In general, ATP-generating pathways are stimulated (e.g. fatty acid oxidation) and ATP-consuming pathways are inhibited (e.g. fatty acid synthesis). A link between AMPK and cancer comes from the finding that LKB1 and a downstream effector called TSC2 are tumor suppressor genes (Figure 11.8). Germline mutations in both have been identified in syndromes that predispose patients to cancer (Peutz–Jeghers syndrome and tuberous sclerosis, respectively). Although more studies are required, inhibition of growth has been reported for tumor cells in culture upon activation of AMPK. In addition, AMPK activates the "star" tumor suppressor p53. In this capacity, p53 acts as a metabolic checkpoint and induces cell cycle arrest in response to low cellular energy.

Target genes of tumor suppressor p53 include inhibitors of glycolysis (e.g. TIGAR) and stimulators of oxidative phosphorylation (e.g. cytochrome *c* oxidase 2) (Matoba *et al.*, 2006). Thus, mutations in p53 that drive cancer also cause altered metabolism that leads to the Warburg effect. (Note: in contrast to the effect of p53 in inhibiting of glycolysis, one report suggests that AMPK stimulates glycolysis in some cell types, and thus the role of AMPK in the regulation of glycolysis requires further study.)

Both hypoxia, and oncogenic mutations in the absence of hypoxia, can activate HIF-1α (Figure 11.8). The inactivation of the tumor suppressor VHL stabilizes HIF-1α in the presence of oxygen. Also, HIF-1α is increased in cells transformed with oncogenes such as *Src* and *Ras*. In addition to regulating VEGF transcription, HIF-1α upregulates the expression of nearly all of the core enzymes of glycolysis and inhibits the expression of

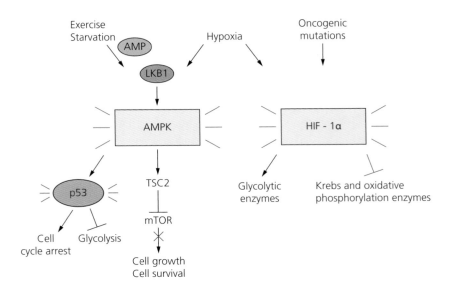

Figure 11.8 Pathways of metabolic regulators AMPK and HIF-1α (shown in yellow).

some enzymes involved in the Krebs cycle and oxidative phosphorylation. Thus, the increased glycolysis observed in some tumors is regulated by a specific transcriptional program, rather than only a lack of oxygen.

Warburg's original hypothesis implied a dysfunction of mitochondria in cancer cells, but we now know that most cancer cells do not have defects in mitochondrial metabolism (with some rare exceptions: succinate dehydrogenase and fumarate hydratase; germline mutations in both can lead to cancer). Rather, complex metabolic pathways, such as those discussed earlier, underlie the Warburg effect. Further investigations regarding the mechanism of this long observed phenomenon are needed and may have additional therapeutic implications.

A quick review of glucose metabolism

In normal cells, the glycolysis pathway involves the breakdown of glucose in the absence of oxygen (anaerobic metabolism) and the production of pyruvate. Pyruvate is converted into lactate by the enzyme lactate dehydrogenase. In the presence of oxygen, pyruvate from glycolysis forms acetyl CoA, which is shuttled through the Krebs cycle, and NADH and $FADH_2$ are generated. These electron carriers shuttle their electrons through the electron transport chain located in the inner mitochondrial membrane. Oxygen is the terminal electron acceptor. The resulting proton motor force generates ATP.

11.5 Genetic polymorphisms and diet

It seems that some people can do all the "wrong things," such as drink excessive amounts of alcohol and smoke heavily, and still live long, healthy lives. We often hear that this is due to an individual's metabolism. Cancer risk associated with diet is influenced by an individual's metabolism. Metabolic reactions are catalyzed by enzymes. Enzyme activities may vary among individuals because of small variations, often single nucleotide changes, in the genes that code for them. The study of genetic polymorphisms that alter the response to a particular dietary constituent is called nutrigenetics. Here are two examples. A polymorphism (C→T transition at nucleotide 677) in the *MTHFR* gene reduces its enzyme activity, and homozygotes for this polymorphism have a 50% decreased risk of colorectal cancer, compared with those with wild-type alleles. These individuals have an increased availability of 5,10-methylene THF and a lower chance of disrupting nucleotide and subsequently DNA synthesis; these conditions deter mutation and carcinogenesis (Figure 11.2). However, the polymorphism increases the risk of cancer if these individuals become deficient in folate. Under these conditions, methyl-THF becomes depleted, and DNA methylation is altered in a manner that is characteristic of carcinogenesis. Note: chronic alcohol consumption is the most common cause for impairment of folate uptake. Recommendation of higher intake of folate and lower intake of alcohol is especially important for those who are homozygous for the T allele at nucleotide 677 of the *MTHFR* gene (Kim, 2007). Polymorphisms in the gene that codes for *N*-acetyltransferase modify the risk of specific cancers in response to the consumption of red meat. This enzyme is involved in the metabolic activation of carcinogenic heterocyclic amines produced by cooking meat at high temperatures. Individuals with the "rapid variant" of the enzyme (a fast acetylator) who consume large amounts of red meat have an increased risk of colon cancer, compared with those who have this variant and do not consume much red meat, or those who possess the "slow variant" polymorphism who do. Therefore, the response to red meat intake with respect to an increased risk of cancer depends on a person's genotype in combination with exposure to carcinogens that result from cooking.

Inherited metabolic diseases can illustrate a more obvious role of metabolism in carcinogenesis. Here are two examples resulting from blocks in tyrosine metabolism pathways. Albinos have an inherited deficiency of the enzyme tyrosinase and are unable to produce melanin, causing the characteristic lack of pigment in their skin. The lack of pigment causes albinos to be more sensitive to the sun and results in an increased risk of skin carcinoma. Tyrosinemia type I, another disorder of tyrosine metabolism, results from a deficiency of fumarylacetoacetate hydrolase. As a result of this metabolic block, the metabolites fumarylacetoacetate and maleylacetate accumulate. Both are alkylating agents and cause DNA mutations and tumorigenesis. In brief, tyrosinemia type I is characterized by the synthesis and accumulation of carcinogens.

LIFESTYLE TIP

Vitamin D deficiency is a worldwide problem. However, there is controversy on the daily requirement: The Endocrine Society set the daily requirement of vitamin D to 1000–2000 IU, while the Institute of Medicine recommends 600–800 IU. Exposure to a "sensible" amount of sunlight (depending on skin type) may also be encouraged.

11.6 Vitamin D: a link between nutrients and hormone action

Vitamin D_3, a precursor or pre-vitamin to biologically active vitamin D, can be obtained through the diet (fortified dairy products and seafood) or produced in the skin from 7-dehydrocholesterol upon exposure to sunlight. The pre-vitamin, regardless of its source (skin or diet), must be metabolized first in the liver to form 25-hydroxyvitamin D (biologically inert) and then in the kidney to form the biologically active form 1,25-dihydroxyvitamin D_3, also known as the steroid hormone calcitriol. The UVB–vitamin D–cancer hypothesis suggests that sunlight is linked to a reduction in cancer risk due to the effects of vitamin D production. Although some strong evidence exists in the literature, this topic is not conclusive, and we await strong data from ongoing and future human randomized clinical trials (such as the VITAL trial). Below is some evidence that supports the UVB–vitamin D–cancer hypothesis.

Epidemiological evidence has demonstrated that there is an increased risk of several cancers (particularly prostate, colon, and breast; note that prostate, colon, and breast cells contain the enzyme needed to produce 1,25-dihydroxyvitamin D) in people living at higher latitudes (Grant, 2012). It has been proposed that vitamin D deficiency underlies this effect, although additional effects of sun exposure may also play a role. The association between circulating vitamin D levels and colorectal cancer was examined in a meta-analysis, and results showed a 30–40% reduction in colorectal cancer risk in those patients with high circulating levels of vitamin D, compared to those with low levels (Lee *et al.*, 2011), and these findings are supported by other studies (Ying *et al.*, 2015). Pre-clinical studies include examining the growth of colon cancer cells in **xenografts** (human cells implanted in the back of immunodeficient mice) in vitamin D-deficient versus vitamin D-sufficient mice (Tangpricha *et al.*, 2005). Tumors were 80% larger on average in mice that were deficient in vitamin D, compared with vitamin D-sufficient mice. Collectively, some epidemiological and *in vivo* evidence support a link between vitamin D deficiency and increased cancer risk.

A little lesson about the history of vitamin D fortification of milk . . .

Synthesis of the pre-vitamin in the skin upon UV exposure accounts for 90–95% of an individual's requirement for vitamin D; very few foods (except oily fish) naturally contain vitamin D. This is one reason why some countries fortify milk and other foods.

Rickets, a debilitating bone-deforming disease, affected more than 80% of children living in industrialized cities in northeastern USA and northern Europe at the beginning of the 20th century. Vitamin D fortification was introduced to try to reduce the incidence of rickets. In the USA, milk, orange juice, and some cereals are fortified with vitamin D. In Europe, fortification of foods was carried out until the late 1940s. Over-fortification of milk caused an outbreak of vitamin D intoxication in Great Britain and led to regulations ➡

→ that ablated vitamin D fortification throughout Europe. These remain active today. Over-fortification most likely resulted from human error or inaccurate food analysis. As food analysis has become more reliable over the last 50 years, should vitamin D fortification be reconsidered in northern Europe and elsewhere, especially in light of the possible effect of reducing cancer risk? This is currently a controversial issue.

The link between nutrients and gene expression became apparent upon the discovery that the receptors for vitamins A and D are members of the steroid hormone receptor superfamily.

Let us examine the molecular mechanisms of vitamin D action. The active form of vitamin D, calcitriol (1,25-hydroxyvitamin D_3), acts as a ligand for the vitamin D receptor, a member of the steroid hormone receptor superfamily. This receptor which is present in most cells of the body recognizes the vitamin D response element in gene promoter regions and regulates transcription of its target genes.

Current data suggest that vitamin D is a chemopreventative agent that inhibits growth and induces differentiation and apoptosis through several molecular mechanisms. Here are a few examples. Vitamin D can act as a dominant negative ligand for EGFR (see Chapter 4). That is, vitamin D can bind to the ligand-binding domain of EGFR, instead of EGF, and prevent the binding of EGF to EGFR. As a result, vitamin D can inhibit growth. Second, upon binding to the vitamin D receptor, the active form of vitamin D can directly activate specific tumor suppressor genes, such as *BRCA1* and *p21*, through a vitamin D response element in their promoter regions. You may recall that the p21 protein is an inhibitor of cdk and that it can induce cell cycle arrest. Vitamin D promotes apoptosis through mitochondrial signaling, independent of caspase activation. It induces the redistribution of two pro-apoptotic proteins BAK and BAX from the cytosol to the mitochondria. These two proteins form channels in the mitochondrial membrane and facilitate the release of cytochrome *c* and apoptosome assembly (see Chapter 7). Simultaneously, Bcl-2 and IAPs, inhibitors of apoptosis, are downregulated. The vitamin D receptor also plays a role in signaling pathways in the colon. Recall that the Wnt pathway is often a target of oncogenic activation in colon cancer. The vitamin D receptor regulate this pathway (Larriba *et al.*, 2013): it can directly bind and inhibit nuclear translocation of β-catenin and also induces the expression of Dickkopf (DKK)-1, an endogenous inhibitor of Wnt signaling.

PAUSE AND THINK

How do steroid hormone receptors function? Remember from Chapter 3 that they are ligand-dependent transcription factors.

11.7 Hormones and cancer

There is group of cancers linked by a common mechanism of carcinogenesis that involves endogenous hormones as initiators, rather than chemicals, viruses, or radiation. The hormone-related cancers include breast, endometrium, ovary, prostate, testis, and thyroid cancer. We will examine breast cancer as a paradigm for hormonal carcinogenesis. (Note: links between hormones and some other hormone-related cancers are not as straightforward.)

Breast cancer is the most common type of cancer in women. Estrogens (estradiol and estrone) appear to play a central role in the initiation and progression of breast cancer. Life events that prolong exposure to estrogens are considered risk factors for breast cancer (Figure 11.9). Early menarche (the start of the menstrual cycle) and late menopause indicate an extended length of time when the ovaries are producing estrogen. Pregnancy affects the exposure time to estrogens. The principal site of estrogen synthesis in the body changes with increased age: the ovaries are the main source in premenopausal women, and adipose tissue (fat) is the main source in post-menopausal women. Obesity in post-menopausal women is a risk factor, because adipose cells are the primary source of estrogen at this stage in life. Obese women have an increased number of fat cells and therefore produce increased amounts of estrogen. Adipose cells use the enzyme aromatase to produce estrogen from androgens. Thus, obesity increases the risk of breast cancer through increased estrogen production. It is thought that alcohol consumption increases the risk

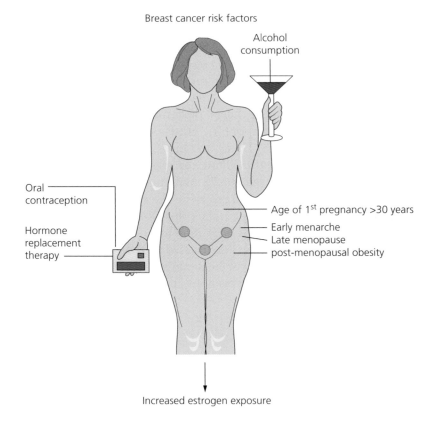

Figure 11.9 Risk factors for breast cancer.

of breast cancer because of its association with increased concentrations of estrogen. The mechanism for the increase in estrogen concentration is not fully known, but it is known that alcohol dehydrogenase can also metabolize estrogens. It has been suggested that alcohol may compete with estrogens in this metabolic pathway and lead to an increase in estrogen concentration. Some data suggest that one alcoholic drink per day causes a 7% increase in the risk of breast cancer. The use of exogenous hormones in oral contraceptives and in hormone replacement therapy has also been linked to increased risk of breast cancer. Conversely, factors that interrupt the menstrual cycle, such as pregnancy, lactation, and physical activity, are considered protective factors. Although a rare event, breast cancer can occur in men. Men also produce some amounts of estrogen, and those with high levels appear to be more at risk.

Two predominant models for the mechanisms by which estrogens exert their effects have been proposed (Figure 11.10). Evidence suggests that both mechanisms contribute to breast cancer.

One model is that estrogens promote cell proliferation of the breast, and the high division rate allows less time for DNA repair, thus creating an opportunity for errors to occur during DNA replication. The increase in error rate translates into an increase in somatic mutations that lead

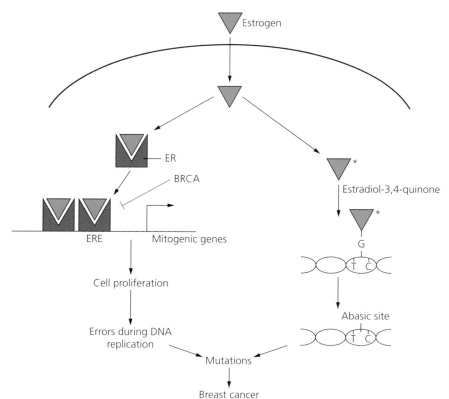

Figure 11.10 Carcinogenic mechanisms of estrogens: mitogenic and/or genotoxic.

to carcinogenesis. This model contrasts chemical- and radiation-induced carcinogenesis, as no specific initiator, other than errors in replication, is required. Estrogen does indeed act as a mitogen for cells in the breast that contain estrogen receptors (ERs). During pregnancy, estrogen levels increase and cause the mammary ducts to grow and the breasts to nearly double in size. Effects of estrogen are mediated through ERs, ER-α and ER-β, members of the steroid hormone receptor superfamily (see Chapter 3). The receptors bind as dimers to the estrogen response element (ERE) and regulate estrogen-responsive genes. In addition to this genomic pathway, non-genomic signaling pathways outside the nucleus have also been documented. The expression of the ER isoforms changes during carcinogenesis of the breast. ER-α is significantly upregulated and ER-β is downregulated in the majority of breast cancers. The reasons for these alterations are unknown at present. However, as we shall see in Section 11.10, "Drugs that target estrogen" on p. 270, blocking ER function has proved to be a successful strategy for the treatment of breast cancer.

The involvement of estrogen signaling in increasing susceptibility to breast cancer is supported by studies of the breast cancer susceptibility genes (*BRCAs*). About 5–10% of all cases of breast cancer are a result of an inherited predisposition, 85% of which result from germline mutations in either the *BRCA1* or *BRCA2* gene. The *BRCA* gene products are nuclear tumor suppressor proteins that play a role in transcriptional regulation, DNA repair, and regulation of the cell cycle. BRCA1 inhibits the transcriptional activation activity of ERs, and this suggests that it suppresses the proliferating effects of estrogen signaling. It has been suggested that the loss of modulation of estrogen signaling contributes to the increased risk to breast cancer in patients with an inherited mutation of *BRCA1*.

Another model suggests that estrogen and its metabolites are genotoxic. This model is consistent with the mechanism by which chemicals, viruses, and radiation initiate carcinogenesis. Estradiol is metabolized to form estradiol-3,4-quinone in cells. This metabolite covalently binds to adenine or guanine bases. The resulting adducts destabilize the bonds linking the base to the DNA backbone and result in abasic sites, and ultimately mutations. (Note: as adenine and guanine are purines, abasic sites that involve the loss of adenine or guanine are referred to as apurinic sites.) Estradiol quinones are present in human breast tissue. ER knockout mice were used as an animal model to test the effect of estrogen in the absence of ERs. Results demonstrated that genotoxic effects occurred in the absence of ERs (Santen *et al.*, 2015; Yue *et al.*, 2003). Genotypic changes were not blocked by inhibitors of the ER, showing that this effect was not receptor-mediated. Removal of the ovaries caused a reduced incidence of tumors and delayed their onset, whereas this effect was reversed by adding exogenous estradiol. The aromatase inhibitor letrozole also delayed the onset of tumor formation. Together, these results support the concept that estrogen metabolites contribute to breast carcinogenesis.

For this second model, predispositions to breast cancer may involve germline mutations in genes involved in estrogen biosynthesis and metabolism. Loss of the BRCA tumor suppressor proteins may leave breast cells more susceptible to the genotoxic effects of estrogen metabolites because of the normal role of BRCA tumor suppressor proteins in DNA repair.

A third model combines the two models just described, and suggests that the ER acts as a "Trojan Horse" and carries carcinogenic estrogen metabolites directly to estrogen-sensitive genes (Bolton and Thatcher, 2008).

 Therapeutic strategies

11.8 "Enhanced" foods and dietary supplements for chemoprevention

Chemoprevention is the use of naturally occurring or synthetic agents to prevent, inhibit, or reverse the process of carcinogenesis in pre-malignant cells. Foods are not yet widely accepted as preventative agents against cancer. However, as we learn more about the important role of nutrients in cancer and as our skills for manipulating food composition increase, this concept is set to change. The development of enhanced food products (foods that have altered levels of particular microconstituents), derived in some cases from genetically modified crops, will begin to flood the market. Foods have already been produced to have increased levels of antioxidants and may be used in future chemopreventative diets. Tomatoes have been classically bred to be bright red for consumer appeal and, as a result, contain more lycopene. They have also been engineered to contain increased levels of zeaxanthin by overexpressing enzymes utilized in its synthesis. Recently, broccoli containing high levels of glucosinolates (which are hydrolyzed to isothiocyanates, including sulforophane; discussed in Section 11.3, "Preventative factors: microconstituents of fruits and vegetables" on p. 255) has been developed by a traditional plant-breeding program and licensed to Seminis Inc., the world's largest developer and grower of vegetable and fruit seeds. A balance of potential advantage from enhanced foods must be tempered by the likelihood that we may have adverse reactions to food constituents at abnormally high concentrations.

Although we discussed the disappointing results of β-carotene supplementation with respect to lung cancer, an interesting observation resulting from the ATBC trial prompted further study into the effects of vitamin E supplementation on the risk of prostate cancer. A significantly decreased rate of prostate cancer was observed in patients taking vitamin E (11.7% versus 17.8%) in the ATBC trial. A similar observation was observed for selenium in a separate study. The Selenium and Vitamin E Cancer Prevention Trial (SELECT), a Phase III randomized placebo-controlled trial, has recruited over 35,000 male participants to test the effects of selenium and vitamin E supplements after a minimum of 7 years. Again, the results were disappointing in that they found that dietary supplementation with vitamin E increased the incidence of prostate cancer by 17% (see web site reference at the end of the chapter on p. 274). This supports the World Cancer Research Fund recommendation not to use supplements to protect against cancer.

The synthetic agent oltipraz was examined as a chemopreventative agent. Similar to sulforophane, it induces Phase II enzymes via transcription factor Nrf2. It was tested in a clinical trial for lung cancer prevention in smokers, but the trial was terminated early due to toxicity. Food-based approaches (e.g. using broccoli sprout teas) may be better alternatives, and these are also being investigated.

As we learn more about the role of nutrients in cancer prevention, it is certain that the list of trials for testing the effects of diet and supplements as chemopreventative agents will expand.

11.9 Drugs that target energy pathways

Although a fairly new strategy, the potential of targeting energy pathways may be worth exploring (Gatenby and Gillies, 2007). As some tumor cells exhibit increased aerobic glycolysis (the Warburg effect), one obvious approach is to target hexokinase, the enzyme that catalyzes the first and rate-limiting step in glycolysis. Pre-clinical experiments testing the effects of a hexokinase inhibitor, 3-bromopyruvate (3-BrPA), demonstrated anti-tumor effects, but clinical development has been discontinued, partly because of questions of enzyme selectivity. Numerous approaches that inhibit HIF are being developed. The small molecule inhibitor PX-478 is one example that showed promising results in a Phase I study but has been discontinued for undisclosed reasons. Metformin is widely used for the treatment of type 2 diabetes. It interferes with mitochondrial respiration and activates AMPK via LKB1. Results from a pilot study have suggested that cancer risk is reduced in type 2 diabetics taking metformin (Evans *et al.*, 2005). Additional studies are ongoing. See Galluzzi *et al.* (2013) and Vander Heiden (2011) for tables of strategies to target metabolic enzymes for cancer therapy.

11.10 Drugs that target estrogen

There are two strategies for the design of drugs that target estrogen action (Figure 11.11). The first is to design drugs that antagonize the actions of estrogens by interacting with ERs in order to block growth in

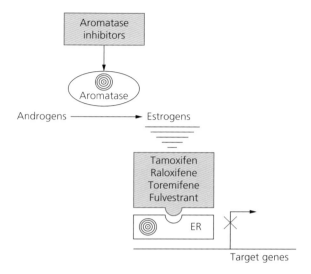

Figure 11.11 Drugs that target estrogen action (shown in red). Molecular targets are indicated by a (◎) symbol.

"ER-positive" tumors. They are called selective ER modulators (SERMs), because they show cell type specificity. Tamoxifen, used in the clinic for over 30 years, is the most widely used SERM to treat ER-positive pre-menopausal breast cancer. Tamoxifen is a competitive inhibitor that alters the folding of the ligand-binding domain of the ER and blocks its ability to transactivate and initiate transcription of its target genes. Raloxifene is another SERM used to treat breast cancer. Interestingly, both SERMs have been shown to prevent breast cancer in high-risk women (Sestak and Cuzick, 2015). Fulvestrant, a new addition to the drug arsenal, can bind and accelerate the degradation of the ER and is called a selective ER downregulator (SERD).

The second strategy for blocking the function of estrogen in breast cancer is to design drugs that interfere with estrogen synthesis. Drugs that target aromatase, the enzyme that converts androgens into estrogens, in post-menopausal women have been developed. Targeting this enzyme does not interfere with the synthesis of other steroids but is rate-limiting for estrogen production. Aromatase is the main source of estrogen production in post-menopausal women, because the ovaries no longer produce it. Aromatase inhibitors can be prescribed to reduce the level of estrogen in the body of post-menopausal breast cancer patients, making it appropriate for over 50% (ER-positive and post-menopausal) of all breast cancers. These drugs are not an option for premenopausal patients, as aromatase is not the major estrogen producer (see "Pause and think").

Three aromatase inhibitors have gone through clinical trials and have been approved in the USA: exemestane, anastrozole, and letrozole. Exemestane (Aromasin™) is a steroidal compound that competes with endogenous androgens for enzyme binding. Intermediates are formed upon binding to aromatase. These intermediates bind irreversibly and block enzyme

PAUSE AND THINK

How would you design an aromatase inhibitor? One way is to model your compound on androstenedione, the endogenous substrate for the enzyme. These drugs should interact and inactivate the steroid-binding domain of aromatase.

function. Anastrozole (Arimidex™) and letrozole (Femara™) are non-steroidal compounds that bind reversibly to the active site of aromatase (Hiscox *et al.*, 2009). A 10-year large-scale trial called the ATAC trial (Arimidex™, Tamoxifen Alone or in Combination) compared anastrozole and tamoxifen as adjuvant treatment (treatment given after surgery to prevent recurrence) for early-stage breast cancer. Results supported long-term superior efficacy and safety of anastrozole over tamoxifen as initial adjuvant therapy for post-menopausal women with early-stage breast cancer (Cuzick *et al.*, 2010). The result from the ATAC trial is to be expected when considering the two proposed mechanisms of action of estrogen. Tamoxifen blocks ER-mediated effects only. Aromatase inhibitors reduce total estrogen concentrations, and therefore block both ER mechanisms and non-receptor-mediated genotoxic events. Molecular knowledge advances drug design.

Tamoxifen has potent chemopreventative activity that may be greater than its therapeutic activity. Several studies have demonstrated that tamoxifen reduces the incidence of invasive breast cancer by 49%. As a result, the US FDA has approved tamoxifen for breast cancer risk reduction. However, the perceived risk of taking tamoxifen for the recommended 5 years (increased risk for endometrial cancer and stroke) and its associated side effects have not made it a popular option. The ER modulator raloxifene has shown equivalent efficacy to tamoxifen and has fewer side effects, and thus may become a better alternative. In fact, both SERMs and aromatase inhibitors have been shown to prevent breast cancer in high-risk women (Sestak and Cuzick, 2015). Future chemopreventative strategies need to consider the risk–benefit ratio and also improved means for the identification of high-risk individuals.

▓ CHAPTER HIGHLIGHTS—REFRESH YOUR MEMORY

- Diet plays a role in both the causation and prevention of cancer.

- In contrast to the association of a β-carotene-rich diet with reduced lung cancer incidence, β-carotene supplementation increases lung cancer in smokers.

- Diet contributes to carcinogenesis by the delivery of carcinogenic contaminants.

- Folate deficiency may affect nucleotide synthesis and DNA methylation.

- Obesity and chronic alcohol drinking are causative factors in cancer, and the mechanisms that underlie their effects may include chemical carcinogenesis

- Nutrients may work as chemopreventative agents by blocking DNA damage via scavenging or the induction of

metabolizing enzymes, inducing apoptosis, or inhibiting cell proliferation.

- The ARE is found in the gene promoters of detoxifying and antioxidant enzymes.

- Some dietary constituents regulate gene expression of detoxifying enzymes via the transcription factor Nrf2 and AREs.

- The major polyphenol of green tea, EGCG, inhibits telomerase.

- The Warburg effect describes the increase in aerobic glycolysis that occurs in some tumor cells.

- Metabolites (and exercise) can affect signaling pathways and epigenetic regulation.

- Genetic polymorphisms can interact with nutrient status and affect cancer risk.

- Vitamin D is a nutrient that acts through a member of the steroid hormone receptor family.

- Breast cancer is a paradigm for hormonal carcinogenesis.

- Estrogen acts as a mitogen for cells in the breast.

- Estrogen and its metabolites may damage DNA directly to initiate carcinogenesis.

- Dietary factors, such as obesity and alcohol intake, increase breast cancer risk by increasing estrogen.

- Germline mutations in the *BRCA1* and *BRCA2* genes predispose patients to breast cancer.

- Enhanced food and dietary supplements are being investigated as chemopreventative agents.

- Tamoxifen is a breast cancer drug that acts as a SERM and blocks estrogen binding to its receptor.

- Aromatase inhibitors, such as anastrozole, act by inhibiting the enzyme aromatase that converts androgens to estrogen.

ACTIVITY

1. Record what you ate and drank for dinner last night. Give a critical account of how the meal contributed either to reducing or enhancing your cancer risk. Bon appétit!

2. By reviewing the literature and using experimental evidence critically, discuss the role of fiber as a cancer-preventative factor in our diet. You may wish to start with Romaneiro and Parekh (2012).

FURTHER READING

Aggarwal, B.B. and Shishodia, S. (2006) Molecular targets of dietary agents for prevention and therapy of cancer. *Biochem. Pharmacol.* **71**: 1397–1421.

Cairns, R.A., Harris, I.S., and Mak, T.W. (2011) Regulation of cancer cell metabolism. *Nat. Rev. Cancer* **11**: 85–95.

Feldman, D., Krishnan, A.V., Swami, S., Giovannucci, E., and Feldman, B.J. (2014) The role of vitamin D in reducing risk and progression. *Nat. Rev. Cancer* **14**: 342–357.

Gut, P. and Verdin, E. (2013) The nexus of chromatin regulation and intermediary metabolism. *Nature* **502**: 489–498.

Key, T.J., Allen, N.E., Spencer, E.A., and Travis, R.C. (2002) The effect of diet on risk of cancer. *Lancet* **360**: 861–868.

Lamprecht, S.A. and Lipkin, M. (2003) Chemoprevention of colon cancer by calcium, vitamin D and folate: molecular mechanisms. *Nat. Rev. Cancer* **3**: 601–614.

Levine, A.J. and Puzio-Kuter, A.M. (2010) The control of the metabolic switch in cancers by oncogenes and tumor suppressor genes. *Science* **330**: 1340–1344.

Lumachi F., Brunello, A., Maruzzo, M., Basso, U., and Basso, S.M. (2013) Treatment of estrogen receptor-positive breast cancer. *Curr. Med. Chem.* **20**: 596–604.

Nguyen, T., Nioli, P., and Pickett, C.B. (2009) The Nrf2-antioxidant response element signaling pathway and its activation by oxidative stress. *J. Biol. Chem.* **284**: 13291–13295.

Pool-Zobel, B., Veeriah, S., and Bohmer, F.-D. (2005) Modulation of xenobiotic metabolizing enzymes by anticarcinogens—focus on glutathione S-transferases and their role as targets of dietary chemoprevention in colorectal carcinogenesis. *Mut. Res.* **591**: 74–92.

Renehan, A.G., Zwahlen, M., and Egger, M. (2015) Adiposity and cancer risk: new mechanistic insights from epidemiology. *Nat. Rev. Cancer* **15**: 484–498.

Ross, S.A. (2010) Evidence for the relationship between diet and cancer. *Exp. Oncol.* **32**: 137–142.

Ruiz, R.B. and Hernandez, P.S. (2014) Diet and cancer: risk factors and epidemiological evidence. *Maturitas* **77**: 202–208.

Shaw, R.J. (2006) Glucose metabolism and cancer. *Curr. Opin. Cell Biol.* **18**: 1–11.

Surh, Y.-J. (2003) Cancer chemoprevention with dietary phytochemicals. *Nat. Rev. Cancer* **3**: 768–780.

Vander Heiden, M.G. (2009) Understanding the Warburg effect: the metabolic requirements of cell proliferation. *Science* **324**: 1029–1033.

World Cancer Research Fund International/American Institute for Cancer Research. (2007) Food, Nutrition, Physical Activity and the Prevention of Cancer: a Global Perspective. AICR, Washington DC. http://wcrf.org/int/research-we-fund/continuous-update-project-cup/second-expert-report

Yager, J.D. and Davidson, N.E. (2006) Estrogen carcinogenesis in breast cancer. *N. Engl. J. Med.* **354**: 270–282.

▨ WEB SITES

Key findings of the EPIC study http://epic.iarc.fr/highlights/highlights.php

World Cancer Research Fund International http://www.wcrf.org/

Selenium and Vitamin E Cancer Prevention Trial (SELECT) http://www.cancer.gov/types/prostate/research/select-trial-results-qa#4

▨ SELECTED SPECIAL TOPICS

Bishehsari, F., Mahdavinia, M., Vacca, M., Malekzadeh, R., and Mariani-Costantini, R. (2014) Epidemiological transition of colorectal cancer in developing countries: environmental factors, molecular pathways, and opportunities for prevention. *World J. Gastroenterol.* **20**: 6055–6072.

Boffetta, P., Couto, E., Wichmann, J., Ferrari, P., Trichopoulos, D., Bueno-de-Mesquita, H.B., *et al.* (2010) Fruit and vegetable intake and overall cancer risk in the European Prospective Investigation into Cancer and Nutrition (EPIC). *J. Natl. Cancer Inst.* **102**: 529–537.

Bolton, J.L. and Thatcher, G.R.J. (2008) Potential mechanisms of estrogen quinone carcinogenesis. *Chem. Res. Toxicol.* **21**: 93–101.

Bub, A., Watzl, B., Blockhaus, M., Briviba, K.L., Liegibel, U., Muller, H., *et al.* (2003) Fruit juice consumption modulates antioxidative status, immune status, and DNA damage. *J. Nutr. Biochem.* **14**: 90–98.

Chalkiadaki, A. and Guarente, L. (2015) The multifaceted functions of sirtuins in cancer. *Nat. Rev. Cancer* **15**: 608–624.

Chang, C.H., Qiu, J., O'Sullivan, D., Buck, M.D., Noguchi, T., Curtis, J.D., *et al.* (2015) Metabolic competition in the tumor microenvironment is a driver of cancer progression. *Cell* **162**: 1229–1241.

Cuzick, J., Sestak, I., Baum, M., Buzdar, A., Howell, A., Dowsett, M., *et al.* (2010) Effect of anastrozole and tamoxifen as adjuvant treatment for early-stage breast cancer: 10 year analysis of the ATAC trial. *Lancet Oncol.* **11**: 1135–1141.

Evans, J.M., Donnelly, L.A., Emslie-Smith, A.M., Alessi, D.R., and Morris, A.D. (2005) Metformin and reduced risk of cancer in diabetic patients. *BMJ.* **330**: 1304–1305.

Galluzzi, L., Kepp, O., Vander Heiden, M.G., and Kroemer, G. (2013) Metabolic targets for cancer therapy. *Nat. Rev. Drug Discov.* **12**: 829–846.

Gatenby, R.A. and Gillies, R.J. (2007) Glycolysis in cancer: a potential target for therapy. *Int. J. Biochem. Cell Biol.* **39**: 1358–1366.

Grant, W.B. (2012) Ecological studies of the UVB-vitamin D-cancer hypothesis. *Anticancer Res.* **32**: 223–236.

Hiscox, S., Davies, E.L., and Barrett-Lee, P. (2009) Aromatase inhibitors in breast cancer. *Maturitas* **63**: 275–279.

Hites, R.A., Foran, J.A., Carpenter, D.O., Hamilton, M.C., Knuth, B.A., and Schwager, S.J. (2004) Global assessment of organic contaminants in farmed salmon. *Science* **303**: 226–229.

Ho, P.C., Bihuniak, J.D., Macintyre, A.N., Staron, M., Liu, X., Amezquita, R., *et al.* (2015) Phosphoenolpyruvate is a metabolic checkpoint of anti-tumor T cell responses. *Cell* **162**: 1217–1228.

Holick, M.F. (2006) Vitamin D: its role in cancer prevention and treatment. *Prog. Biophys. Mol. Biol.* **92**: 49–59.

Jaramillo, M.C. and Zhang, D.D. (2013) The emerging role of the Nrf2-Keap1 signaling pathway in cancer. *Genes Dev.* **27**: 2179–2191.

Kim, D.-H. (2007) The interactive effect of methyl-group diet and polymorphism of methylenetetrahydrofolate reductase on the risk of colorectal cancer. *Mut. Res.* **622**: 14–18.

Larriba, M.J., González-Sancho, J.M., Barbáchano, A., Niell, N., Ferrer-Mayorga, G., Muñoz, A. (2013) Vitamin D Is a Multilevel Repressor of Wnt/b-Catenin Signaling in Cancer Cells. *Cancers (Basel)* **5**: 1242–1260.

Lee, J.E., Li, H., Chan, A.T., Hollis, B.W., Lee, I.M., Stampfer, M.J.,*et al.* (2011) Circulating levels of vitamin D and colon and rectal cancer: the Physicians' Health Study and a meta-analysis of prospective studies. *Cancer Prev. Res. (Phila)* **4**: 735–743.

Li, Y. and Tollefsbol, T.O. (2010) Impact on DNA methylation in cancer prevention and therapy by bioactive dietary components. *Curr. Med. Chem.* **17**: 2141–2151.

Manson, M.M. (2003) Cancer prevention—the potential for diet to modulate molecular signaling. *Trends Mol. Med.* **9**: 11–18.

Matoba, S., Kang, J.-G., Patino, W.D., Wragg, A., Boehm, M., Gavrilova, O., *et al.* (2006) P53 regulates mitochondrial respiration. *Science* **312**: 1650–1653.

Morris, B.J. (2013). Seven sirtuins for seven deadly diseases of aging. *Free Radic. Biol. Med.* **56**: 133–171.

Na, H.K. and Surh, Y.-J. (2008) Modulation of Nrf2-mediated antioxidant and detoxifying enzyme induction by the green tea polyphenol EGCG. *Food Chem. Toxicol.* **46**: 1271–1278.

Nelson, E.R., Wardell, S.E., Jasper, J.S., Park, S., Suchindran, S., Howe, M.K., *et al.* (2013) 27 Hydroxycholesterol links hypercholesterolemia and breast cancer pathophysiology. *Science* **342**: 1094–1098.

Park, E.J., Lee, J.H.Yu, G.-Y., He, G., Ali, S.R., Holzer, R.G., *et al*. (2010) Dietary and genetic obesity promote liver inflammation and tumorigenesis by enhancing Il-6 and TNF expression. *Cell* **140**: 197–208.

Pucci, B., Villanova, L., Sansone, L., Pellegrini, L., Tafani, M., Carpi, A., *et al*. (2013) Sirtuins: the molecular basis of beneficial effects of physical activity. *Intern. Emerg. Med*. **8**: 23–25.

Revankar, C.M., Cimino, D.F., Sklar, L.A., Arterburn, J.B., and Prossnitz, E.R. (2005) A transmembrane intracellular estrogen receptor mediates rapid cell signaling. *Science* **307**: 1625–1630.

Romaneiro, S. and Parekh, N. (2012) Dietary Fiber Intake and Colorectal Cancer Risk: Weighing the Evidence from Epidemiologic Studies. *Top. Clin. Nutr*. **27**: 41–47.

Santen, R.J., Yue, W., and Wang J.P. (2015) Estrogen metabolites and breast cancer. *Steroids* **99**: 61–66.

Schwingshackl, L. and Hoffmann, G. (2014) Adherence to Mediterranean diet and risk of cancer: a systematic review and meta-analysis of observational studies. *Int. J. Cancer* **135**: 1884–1897

Seitz, H.K. and Stickel, F. (2007) Molecular mechanisms of alcohol-mediated carcinogenesis. *Nat. Rev. Cancer* **7**: 599–612.

Sestak, I. and Cuzick, J. (2015) Update on breast cancer risk prediction and prevention. *Curr. Opin. Obstet. Gynecol*. **27**: 92–97.

Tangpricha, V., Spina, C., Yao, M., Chen, T.C., Wolfe, M.M., and Holick, M.F. (2005) Vitamin D deficiency enhances the growth of MC-26 colon cancer xenografts in Balb/c mice. *J. Nutr*. **135**: 2350–2354.

Vander Heiden, M.G. (2011) Targeting cancer metabolism: a therapeutic window opens. *Nat. Rev. Drug Discov*. **10**: 671–684.

Ying, H.-Q., Sun, H.-L., He, B.-S., Pan, Y.-Q., Wang, F., Deng, Q.-W., *et al*. (2015) Circulating vitamin D binding protein, total, free and bioavailable 25-hydroxyvitamin D and risk of colorectal cancer. *Sci. Rep*. **5**: 7956.

Yoshimoto, S., Loo, T.M., Atarashi, K., Kanda, H., Sato, S., Oyadomari, S., *et al*. (2013) Obesity-induced gut microbial metabolite promotes liver cancer through senescence secretome. *Nature* **499**: 97–101.

Yue, W., Yager, J.D., Wang, J.P., Jupe, E.R., and Santen, R.J. (2013) Estrogen receptor-dependent and independent mechanisms of breast cancer carcinogenesis. *Steroids* **78**: 161–170.

Yun, H.-M., Ban, J.O., Park, K.-R., Lee, C.K., Jeong, H.-S., Han, S.B., *et al*. (2014) Potential therapeutic effects of functionally active compounds isolated from garlic. *Pharmacol. Ther*. **142**: 183–195.

Chapter 12

Tumor immunology and immunotherapy

Introduction

The immune system is a network of cells, signals, and organs that helps protect against dangerous pathogens and cancer. It is also important in tissue repair. All types of white blood cells have roles in immune responses (Table 12.1). Cells of the immune system are derived from hematopoietic stem cells (HSCs) in the bone marrow. As illustrated in Figure 8.7, the immune cells arise from two types of progenitor cells: myeloid progenitor cells and lymphoid progenitor cells. Immune cells that respond early and non-specifically to infection are part of the innate immune response and include neutrophils, eosinophils, basophils, dendritic cells, natural killer cells, and macrophages derived from monocytes (important in inflammation, as we will see later in Chapter 13). Immune cells that respond later during infection and specifically to antigens presented by antigen-presenting cells (APCs) (e.g. dendritic cells and macrophages) are involved in the adaptive immune response and include B cells and T cells (helper T cells and cytotoxic T cells).

Perhaps surprisingly, the immune system plays a dual role in cancer: it has both anti-tumor and pro-tumor effects. On one hand, immune cells can recognize and eliminate tumor cells, including the killing of virus-infected cells. On the other hand, the immune system may exert selective pressure that shapes the antigens displayed by the tumor cells, leading to evasion of the immune system and cancer promotion. Recall that one hallmark of cancer (see Chapter 1) is the escape from the immune response. Overall, the role of the immune system is influenced by the tumor, and the tumor is affected by the immune system.

This chapter will discuss tumor suppression and tumor promotion mechanisms of the immune system. It will describe regulatory mechanisms that are used to control the immune response and factors that play a role in immune evasion. Finally, the chapter will conclude with a report on exciting new applications of this knowledge towards developing cancer immunotherapies, therapies that are demonstrating unprecedented results. Immunotherapy is becoming an additional mode of standard treatment joining the ranks of surgery, radiation, chemotherapy, and targeted therapy.

12.1 Lymphocytes: B cells and T cells

In this chapter, we will focus on lymphocytes (B and T cells), cells of specific immunity involved in the adaptive response. The main function of B cells is to synthesize and secrete antibodies. The aspect of immunity mediated by antibodies is called humoral immunity. Antibodies can recognize almost any antigen encountered by the immune system. Once they have recognized an antigen, antibodies can coordinate cell-mediated cell lysis.

An antibody is a "Y"-shaped molecule that consists of two identical heavy chains and two identical light chains (Figure 12.1). The two heavy chains are linked to each other by disulfide bonds, and each heavy chain is linked to a light chain by a disulfide bond. Antibodies contain an antigen-binding domain (an **antigen** can be defined as any molecule that is able to generate an immune response) located in the Fab (fragment, antigen binding) region. The Fab contains the variable regions, regions of amino acids that vary greatly among different antibodies and are responsible for antigen specificity. The Fc (fragment crystallizable) region makes up the stem of the antibody and functions in regulating the immune response. The constant regions found in both Fab and Fc are common among each class of antibody (IgG, IgM, IgA, IgD, and IgE).

Table 12.1 Immune cell function in cancer: a short guide

Cell type	Function
B cells	Production of tumor-specific antibodies; can trigger cell lysis
T cells:	
CD4+	Helper cells for cytotoxic T-cell response → tumor rejection. Also supports the B-cell response
CD8+	Principal effector cells for cytotoxic T-cell response → cell lysis
Treg	Immunosuppression → tumor-promoting
Dendritic cells	Antigen-presenting cells (APCs)
Macrophages	Important phagocytic cells of the inflammatory response; also APCs
Natural killer cells	Innate immune cells → rejection of viral-infected cells and tumor cells; release cytotoxic enzymes or secrete cytokines (e.g. interferon-γ)
Basophils	
Neutrophils	Innate immune cells → respond early and non-specifically to infection
Eosinophils	

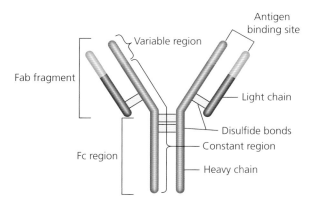

Figure 12.1 Structure of an antibody.

Antibody production may be triggered by protein, lipid, or polysaccharide antigens and requires B cell–T cell interaction in lymphoid follicles, triggering differentiation of B cells into long-lived memory B cells and plasma cells that produce immunoglobulin G (IgG) antibodies. Since many B cells respond to an antigen, a mixture of antibodies is produced by many clones (polyclonal). Experimentally, we can grow a single clone of a specific B lymphocyte by creating a hybridoma, a hybrid cell formed by fusing a B cell with a B cell cancer cell (myeloma), in order to produce quantities of a specific monoclonal antibody.

T cells, mainly CD4[+] helper T cells and CD8[+] effector T cells, are the principal cells responsible for **cell-mediated immunity**. They have membrane-bound T-cell receptors that can "see" antigen that is presented to them via specialized antigen-presenting cells. Maturation of T cells occurs in the thymus—thus the "T" in T cells. T cells coordinate an immune response and eliminate virally infected cells and tumor cells. Long-lived memory T cells are also produced. Memory provides a faster and stronger response to re-exposure to an antigen. **Cytokines**, small secreted proteins, are the major cell-signaling proteins of both B and T cells (e.g. interferon-γ, interleukins, chemokines). Note that T regulatory (Treg) cells suppress the function of immune cells and are important regulators of the immune response. To summarize, B cells can secrete antibodies that are capable of recognizing soluble antigens while T cells must receive antigen from an antigen-presenting cell via their T cell receptors. Immunological memory is a unique feature of the immune system.

12.2 The tumor suppressive roles of the immune system

The immune system protects against cancer in three ways. Most obviously, it protects us from viral and bacterial infection that may be causative agents of cancer. It also helps resolve inflammation, a promoter of cancer (see Chapter 13). And most directly, it can recognize and kill tumor cells. The concept that the immune system recognizes cancer cells as foreign agents and eliminates them is called **immunosurveillance** and was first hypothesized in the mid-20th century. In support of this hypothesis, the use of mouse models has shown that mice lacking **interferon-γ** responsiveness or adaptive immunity (that is mice lacking B cells, T cells, and natural killer T cells) were more susceptible to both spontaneous and carcinogen-induced cancer. In humans, it was noticed that immunodeficient patients and patients who received organ transplants and were prescribed immunosuppressive drugs had a higher incidence of cancer than the general population. Thus, the immune system can act as a tumor-suppressive defense (reviewed in Vesely *et al.*, 2011).

The cancer immunity cycle

The steps involved in an immune response against a tumor have been defined in the cancer immunity cycle (Chen and Mellman, 2013) (Figure 12.2). Seven steps are listed below:

1. Release of cancer cell antigens during tumor cell death
2. Presentation of cancer antigens (AG) on APCs
3. Priming and activation of T cells in lymph nodes
4. Transport to tumors via the bloodstream

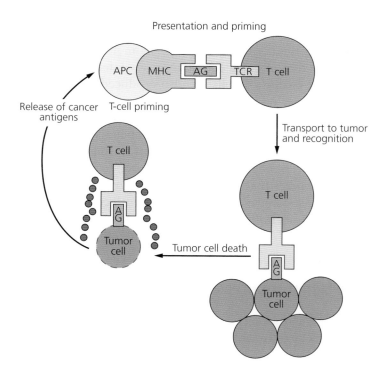

Figure 12.2 The cancer immunity cycle.

5. Infiltration of T cells into tumors

6. Recognition of cancer cells

7. Tumor cell death.

The immune system can recognize tumor-specific antigens (molecules that are unique to cancer cells) or tumor-associated antigens (molecules that are differentially expressed in cancer and normal cells), and it is because of these antigens that cancer cells can be distinguished from "self." These types of antigens arise because of mutations that either produce altered protein products or result in mis-regulated gene expression, respectively. Examples include mutant p53 peptides and HER2 (overexpressed cellular protein). There is an ongoing effort to identify tumor-specific antigens for therapeutic purposes. Also, tumors caused by viruses exhibit viral antigens.

Tumor antigens released by dying cells can be presented to T cells via the major histocompatibility complex (MHC) on APCs (dendritic cells and macrophages). The antigen is recognized by a specific T-cell receptor (TCR). TCRs are generated by random shuffling of gene segments, and this produces a diversity of receptors capable of recognizing every possible foreign substance. The TCR–antigen–MHC interaction is the initiating trigger of T-cell activation, though there are many co-modulators that can modify the downstream response, as discussed in Section 12.3, "Immune checkpoints". The events initiating the transition of conventional naive T cells from a quiescent to an activated state are referred to as T-cell priming. The net activity of co-modulators regulates full activation. T cells then traffic to, and infiltrate into, tumors where they recognize and bind to cancer cells through interaction of their TCR and cognate antigen. Upon full activation, T cells release molecules, such as perforin and granzymes, and express ligands for death receptors on the target cancer cell. These events have been described as the kiss of death, as these events kill the targeted tumor cells. The cancer immunity cycle does not operate optimally in cancer patients and can be modulated at several steps. The tumor microenvironment also presents a source of factors of immunosuppression.

Note, the cytotoxic T-cell response is one of the most important anti-tumor defenses of the body, and it has a memory. During a primary immune response, T cells strongly proliferate to increase the number of effector cells. Some of these cells will die, and about 10% will differentiate into memory T cells to give a faster and stronger response in the future.

12.3 Immune checkpoints

It is important that the duration and strength of the immune response are regulated to maintain self-tolerance (that is, not raising an immune response to self-antigens) and to prevent tissue damage. The immune response is regulated by inhibitory pathways known as immune checkpoints that are triggered by receptor–ligand interactions. Let us examine the molecular interactions that occur during the initiation of a T-cell response upon the presentation of an antigen to a TCR. Note that T cells cannot recognize free antigens, in contrast to the ability of antibodies to recognize free antigen. T cells recognize antigens associated with MHC molecules on the surface of cells (see Box "A little lesson about major histocompatibility complex (MHC) molecules", p. 283).

First, an antigen that is bound to MHC molecules is presented by an APC to the TCR on a T cell. This is the primary signal for initiating the downstream signaling pathways of an immune response. But full activation only occurs after exposure to a second co-stimulatory signal. CD28 is a co-stimulatory molecule that is constitutively expressed on T cells. A second co-stimulatory interaction occurs between CD28 on the T cell and B7 ligands on the APC. This initiates a full T-cell response (Figure 12.3a). Note that most cancer cells (excluding some lymphomas) do not express B7 and hence are largely invisible to the immune system. (It is the presentation of tumor antigens by APCs to T cells that make tumor cells "visible" to the immune system.)

 Following activation, cytotoxic T-lymphocyte antigen 4 (CTLA-4), a co-inhibitory molecule, is rapidly mobilized from intracellular vesicles within the T cell to the cell membrane. Here it competes with co-stimulator

(a)

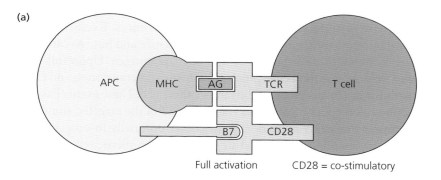

(b)

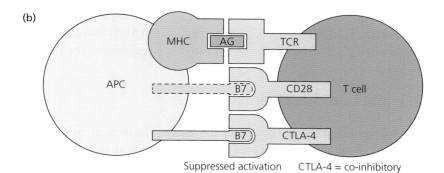

Figure 12.3 The CTLA-4 immune checkpoint. (a) Activation by B7. (b) Suppression by CTLA-4.

CD28 receptor for binding to its ligand B7 (Figure 12.3b). CTLA-4 binds more strongly than CD28 to ligand B7. Thus, CTLA-4 binding serves as a negative feedback signal by sequestering B7, preventing it from binding to CD28 and suppressing T-cell activation (see "Pause and think").

PAUSE AND THINK

As stated in Section 12.3, "Immune checkpoints" on p. 281, CTLA-4 is a key negative regulator of T-cell activation. What experiments could you propose to illustrate this? Think about the different types of evidence: correlative, "gain of function," "loss of function." How can one examine "loss of function" *in vivo*? Yes, the creation of a CTLA-4 knock-out mouse would be a good suggestion. What type of phenotype would you predict? See the classical *Science* paper by Waterhouse *et al.* (1995).

A little lesson about major histocompatibility complex (MHC) molecules . . .

The MHC is also called the human leukocyte antigen (HLA) in humans. A description of Class I MHC molecules and Class II MHC molecules are noted below.

Class I MHC: expressed on most cells

- Antigens bound by Class I are recognized by CD8+ T cells.
- Involved in the presentation of intracellular antigen: cytoplasmic proteins are degraded into peptide fragments by proteasomes and shuttled to the rough endoplasmic reticulum (RER) where MHC I molecules are produced.
- The MHC I–antigen complex is then transported to the cell surface to present antigen to CD8+ T cells.

Class II MHC: restricted mainly to immune cells such as macrophages

- Antigens bound by Class II are recognized by CD4+ T cells.
- Involved in the presentation of extracellular antigen: extracellular antigens are taken inside cells by endocytosis and are degraded by lysosomal enzymes. MHC II molecules are transported to the lysosomes.
- MHC II–antigen complexes are formed in lysosomes and then transferred to the cell surface to present to CD4+ T cells.

Interaction between receptors and ligands lead to intracellular signal transduction pathways. Kinases (PI3K, protein kinase C) and phosphatases (SHP phosphatases) are commonly involved in activating and deactivating signal transduction pathways, respectively, which indicates the importance of phosphorylation/dephosphorylation in the regulation of the T-cell response. Many other co-stimulatory receptors (e.g. inducible co-stimulator (ICOS); OX40; GITR) and co-inhibitory receptors (e.g. lymphocyte activation gene-3 (LAG-3); TIM3; VISTA) have been identified. Regulation of ligands present on APCs provides another level of regulation of T-cell function.

Programmed death 1 protein (PD-1) and its ligands PD-L1 and PD-L2 are additional co-inhibitory molecules. PD-L2 is expressed on APCs. PD-L1 is expressed on many cell types, including tumor cells, after exposure to interferon-γ produced by activated T cells and thus can modulate the T-cell response at the tumor interface (Figure 12.4). The PD-1 axis interferes with signaling mediated by the T-cell antigen receptor. The interaction of PD-L1 on tumor cells and its receptor on activated effector T cells causes the recruitment of SHP-2 phosphatases and inactivation

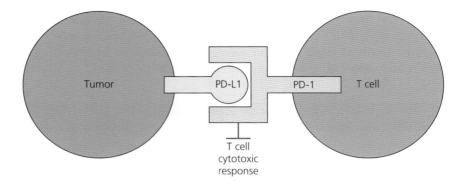

Figure 12.4 Tumor cell evasion of a T-cell response by PD-L1.

of the PI3K cascade, blocking the production and secretion of molecules required for a cytotoxic response. This block is reversible. Note that PD-1 engagement by PD-L1 occurs after T-cell activation by APC cells (and later than CTLA-4) in the process of the cancer immunity cycle.

12.4 Cancer immunoediting and tumor promotion

The immune system is not always successful in inhibiting tumor formation, and, in fact, the immune system can also promote tumor formation by selecting for tumor cells that are best adapted to survive in an immunocompetent host or by creating conditions within the microenvironment of the tumor that facilitate tumor growth. The concept of cancer immunoediting integrates the dual role of the immune system in cancer. Immunoediting is the term used to describe how the host immune system shapes tumor immunogenicity and clonal selection, and the way that cancer cells constantly modulate and "edit" the host anti-tumor immune system. Immunoediting may go through three distinct phases (the three "Es" of immunoediting):

(1) Elimination of the cancer

(2) Equilibrium, the selection of less immunogenic tumor cells during an anti-tumor response

(3) Escape whereby the tumor evades the immune system.

The process begins with the elimination phase. During the elimination phase, the immune system is alerted to a developing tumor, most likely by cytokines that are induced during early tumor development. As a consequence, an anti-tumor immune response, carried out by both innate and adaptive immunity, is initiated. If the immune system is successful at eliminating the tumor cells, then the host is cancer-free, and the next phase is not initiated. Immunoediting is complete. But if only a portion of tumor cells are eliminated, then the process enters a phase of equi-librium between the immune system and the developing tumor. During equilibrium, rare tumor cell variants that escape the elimination are kept at bay by the adaptive immune response. This may account for decades of immune dormancy. The immune system may also help shape tumor immunogenicity. This phase may provide pressure to select for tumor cells that acquire immunoevasive mutations, leading to lower quantities of tumor antigens or less potent antigens. These tumors cells may then enter the final phase, escape.

Evidence for cancer immunoediting

Supporting evidence that the immune system can affect the immunogenic-ity of a tumor is that tumors formed in a deficient immune system are more immunogenic and are hence described as "unedited," whereas tumors formed in immunocompetent hosts are termed "edited." External factors, such as stress, aging, and medical intervention, can also affect the process.

The strongest evidence for cancer immunoediting in humans comes from the studies that show that the profile (quantity, type, and location) of tumor-infiltrating lymphocytes correlate with patient survival and prog-nosis for several different cancers. For example, melanoma patients with high levels of $CD8^+$ cell infiltration survive longer than those with tumors containing low numbers of these cells. Conversely, tumor infiltration by immune inhibitory cells, such as Tregs, have a negative prognosis. In fact, some studies report that the type and density of tumor-infiltrating lym-phocytes were a more accurate prognostic indicator than tumor staging by pathology and analysis of oncogene expression. The characterization of T-cell infiltration into the tumor as a biomarker or prognostic indica-tor has been termed the "immunoscore" and is being validated in clinical studies. In addition, as mentioned earlier, patients with immunodeficien-cies or who have had organ transplants and are taking immunosuppres-sive drugs are more likely to develop some types of cancer (see Roithmaier *et al.*, 2007 as one example).

12.5 Mechanisms of avoiding immune destruction

As discussed in Chapter 1, avoiding immune destruction is a new emerg-ing hallmark of cancer. Mechanisms for tumor evasion include loss of tumor antigens, downregulation of antigen-presenting molecules, and tumor resistance to cytotoxic pathways, including overexpression of immune checkpoint proteins (e.g. PD-L1 described above in Section 12.3 "Immune checkpoints" on p. 283) and anti-apoptotic molecules.

Several factors in the tumor microenvironment have been shown to play a role in tumor escape. Tumors secrete immunosuppressive molecules such as TGF-β, IL-10, VEGF, and indolamine-2,3-dioxygenase. TGF-β binds to the TGF-β receptors on lymphocytes and can alter their phenotype and cytokine secretion profiles. They have been implicated in the generation of Treg cells that actively inhibit other T cells. IL-10, a cytokine, exerts its effects by binding to its cell surface receptor expressed on immune cells. The signal transduction pathway involves tyrosine kinase phosphorylation and activation of the STAT transcription factors, leading to the upregulation of hundreds of genes. IL-10 is able to prevent the maturation of dendritic cells, the most important APC involved in primary immune responses. In addition, IL-10 has a direct effect on CD4[+] T cells; it inhibits their proliferation and their production of cytokines. Indoleamine-2,3-dioxygenase (IDO) was first known for its role in maternal tolerance to antigens from the fetus and is thus involved in evading the immune response. The expression of this enzyme by cells in the tumor stroma causes a reduction in the availability of the amino acid tryptophan to T cells and inhibits their activation, proliferation, and survival. Preclinical studies suggest that inhibitors of IDO may restore immunity and permit immune destruction of the tumor, and small-molecule inhibitors are being tested in clinical trials.

◎ Therapeutic strategies

Boosting the immune system to treat cancer has been a long-standing goal of medicine. The therapeutic strategy of immunotherapy is to target the host immune system, not the tumor. One aim is to use recombinant antibodies to target oncoproteins on the cancer cell surface and relevant soluble factors, such as growth factors, or to use them to deliver chemotherapeutic agents. Another aim of some immunotherapies is to generate a population of T cells with long-term memory, so that these cells can attack the tumor and metastases before they become successfully colonized. Treating metastasis has been the Achilles heel for cancer treatments. Further still, another aim of some immunotherapies is to reverse the immunosuppression mechanisms exerted by tumor cells. Some immunotherapies are yielding unprecedented outcomes with durable responses rarely seen before and can be applied to many different cancer types.

12.6 Therapeutic antibodies

The use of therapeutic antibodies has been discussed throughout this text, and, by now, you are familiar with drugs like Herceptin™ and Avastin™ that target key molecules. In fact, there are over 13 antibodies that have been approved by the FDA for cancer (see Table 3 in Scott *et al.*, 2012 and Table 1 in Sliwkowski and Mellman, 2013). Many antibodies may directly interfere with signal transduction and/or engage with receptors on immune cells, and lead to antibody-dependent cellular cytotoxicity (ADCC) or phagocytosis. The Fc portion of the antibody is important for ADCC. Modification of the Fc fragment can either eliminate or potentiate ADCC.

Development of these drugs required antibody engineering, so that antibodies created in mice can be "humanized" so that they would not be immunogenic in humans. Initially, this was done by grafting complementarity-determining regions (CDRs) from the mouse antibody raised against a specific antigen onto a recombinant human immunoglobulin backbone. More recently, improved procedures are being used such as making antibodies in rodents with reconstituted human genes and cells, or using phage or yeast display libraries to create antibodies without the need for immunization of mice. Panitumumab is a monoclonal antibody against EGFR (discussed in Chapter 4) which was developed in a transgenic mouse that was capable of synthesizing human antibodies.

Antibodies have been linked to chemotherapeutic agents to create antibody–drug conjugates (ADCs) that can deliver drugs specifically to the tumor. Two ADCs, brentuximabvedotin (a monoclonal antibody against CD30 modified with monomethyl auristatin E, a microtubule drug) and T-DM1 (trastuzumab (T) conjugated to a maytansine derivative (DM1)), have been approved.

12.7 Cancer vaccines

Our ability to harness the immune system to prevent and/or kill tumor cells is becoming evident. Vaccination is called active immunization, because it tries to stimulate the individual's own immune effector cells. A vaccine is composed of antigen(s) and adjuvant(s). Adjuvants are vaccine additives that enhance the immune response to an antigen. This contrasts passive immunization, which involves the transfer of effectors of the immune system into the patient, such as T cells (see Section 12.9, "Adoptive T-cell transfer, modified T-cell receptors, and chimeric antigen receptors" on p. 296) or secreted products of lymphoid cells (discussed in Box "Passive immunization").

Passive immunization

Antibody-based drugs, discussed in Section 12.6, "Therapeutic antibodies" on p. 286 and throughout this text, are a successful form of passive immunity. Note that early attempts at cancer immunotherapy utilized passive immunization strategies that involved the transfer of cytokines. Interferon-γ, a cytokine with promising ability to modulate the immune response, was studied extensively in Phase I, II, and III trials. In general, however, poor clinical responses were observed. TNF also gave disappointing results in clinical trials. However, cytokines may be used in the future in combination immunotherapy approaches.

PAUSE AND THINK

The term vaccination comes from the Latin word "vacca" or cow, because the first vaccine, reported in 1798 by Edward Jenner, used cowpox virus for immunization against smallpox. As a doctor in the English countryside during a smallpox outbreak, Jenner noticed that milk maids were less likely to contract smallpox, although these women were often exposed to cowpox infection. He hypothesized that cowpox infection was the cause of the resistance to smallpox and carried out experiments that supported his hypothesis.

Cancer vaccines can either be designed to stimulate the immune system in order to cause tumor regression in a patient with cancer, a *therapeutic vaccine*, or they can prepare the immune system prior to getting cancer for cancer prevention, so-called *prophylactic vaccines*. Most cancer vaccines are designed to be therapeutic vaccines, though we will consider both types here.

Therapeutic vaccines

The production of a vaccine involves the selection of appropriate antigens that will stimulate an effective anti-tumor response. Tumor-associated antigens may be derived from either degradation and processing of unfolded intracellular proteins that are shuttled to the surface of the tumor cell or from damaged or dying tumor cells. These may include oncoproteins arising from oncogenic mutations or chromosomal translocations. As T cells are the main effectors of an anti-tumor response, antigens from the vaccine must be displayed eventually on the surface of APCs. A series of cellular events characterize an immune response upon administration of a cancer vaccination (Figure 12.5). APCs, such as dendritic cells that reside in the tissue, are at the heart of signaling for the mission of eliciting T-cell-mediated immunity. It is the dendritic cells that (1) *acquire* and (2) *process* the antigens, and, upon maturation, migrate to the lymphoid organs to (3) *present* the antigens to the main effector T cells. The uptake of antigens by the dendritic cells is primarily by endocytosis. Antigen processing involves cleavage of the antigen into small peptides by proteases (depicted by scissors in Figure 12.5). The adjuvant in a cancer

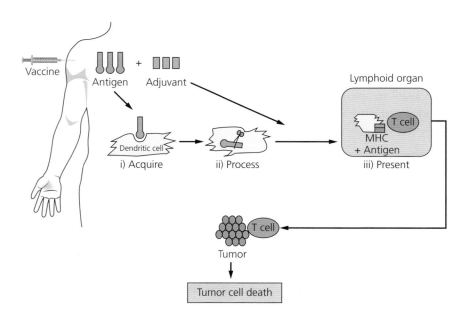

Figure 12.5 Cellular events in an immune response following vaccination.

vaccine induces the maturation of the APCs and their migration to the lymphoid organs. Processed antigen is translocated to the cell surface for presentation in association with proteins from the MHC (see Box "A little lesson about major histocompatibility complex (MHC) molecules . . . ", p. 283, for a review). CD8$^+$ T cells are first activated by the presentation of antigen by the APCs and then traffic into the tumor. It is the CD8$^+$ cyto-toxic T cells that recognize the antigen on the tumor cell membrane and proceed to kill the tumor cells by releasing cytotoxic granules or inducing apoptosis.

Cancer vaccines are required to overcome tumor-protective and immune-suppressive mechanisms. As discussed in Section 12.4, "Cancer immunoediting and tumor promotion" on p. 284, tumor cells may undergo immunoediting which allows them to evade and suppress the immune system. For example, TGF-β obstructs dendritic cell maturation and interferes with antigen presentation to T cells. Let us examine several strategies used for producing cancer vaccines.

Whole-cell vaccines

Vaccines against infectious diseases are composed of bacteria or viruses whose ability to produce disease has been reduced or attenuated by different processes such as passage through an unnatural host, chemical treatment, or irradiation. The first cancer vaccines were composed of irradiated tumor cells, being modeled after successful attenuated pathogen vaccines.

All of the antigens expressed by a specific tumor are included in the whole-cell vaccine design. These first cancer vaccines demonstrated immune responses in mouse models but were disappointing in clinical trials, causing either a weak response from the immune system (a weak immunogenic response) or a response against normal cells (autoimmunity). This may be because of the under-representation of immunogenic antigens relative to the total number of antigens and stimulation against normal gene products, respectively. For example, vitiligo, an autoimmune disease that targets melanocytes, was observed in studies of a melanoma vaccine, suggesting that the induced immune response also targeted normal antigens, and thus normal cells. Some modifications of whole-cell vaccines are being pursued. For example, gene-modified tumor cells that express stimulatory molecules for T cells double as antigens and adjuvants. However, regardless of their degree of success, whole-cell vaccines have been important stepping stones towards antigen-specific vaccines.

Peptide-based vaccines

Another strategy for the development of cancer vaccines is to use tumor-associated antigens to generate an immune response. This involves the identification and characterization of specific molecules on the tumor cells

PAUSE AND THINK

What is the difference in methods between making synthetic and recombinant proteins? Synthetic peptides are made in the laboratory by linking amino acids together in a specific order, while recombinant proteins are synthesized *in vivo* from genetically engineered molecules that include a DNA sequence encoding the specified amino acids.

that are recognized by T cells, rather than using whole cells from tumors, as was described earlier. Tumor-specific antigen molecules have qualitative or quantitative differential expression patterns in tumor cells, compared with normal cells. Many of these antigens elicit an immunogenic response without autoimmunity. This has led to the production of antigen-specific peptide vaccinations. The peptides used are short sequences of amino acids that code for a part of the tumor-associated antigen and can be produced as synthetic or recombinant proteins.

A growing list of breast tumor antigens, including HER2, mucin1, and carcinoembryonic antigen (CEA), provide the basis for the production of breast cancer vaccines. Several melanoma tumor antigens have also been characterized. A peptide-based vaccine targeting the melanoma-associated antigen glycoprotein 100 (gp100) has been developed to treat melanoma patients. The gp100 antigen is an antigen that is expressed in normal melanocytes, melanomas, and pigmented retinal cells. The gp100 peptide vaccine, along with IL-2, has been examined in a Phase III clinical trial, and results demonstrated a higher response rate than IL-2 alone (Schwartzentruber *et al.*, 2011). This was the first Phase III trial to demonstrate a clinical benefit for a peptide vaccine in melanoma.

Dendritic cell vaccines

Vaccines may also be composed of dendritic cells, cells that are the most potent antigen-presenting and stimulatory cells for the induction of a T-cell-dependent immune response. Dendritic cells originate in the bone marrow and reside in an immature state in peripheral tissues. As described earlier, after antigen processing and receiving inflammatory signals, they differentiate or mature and migrate to lymph nodes where antigens are presented and the T-cell response is initiated. *In vivo*, tumors secrete several factors that suppress dendritic cell differentiation and migration, and may contribute to the immunosuppression observed in cancer patients.

For the purpose of vaccination, dendritic cells must be isolated from an individual patient and cultured *in vitro*, during which time they can be loaded or pulsed with specific antigens, DNA, or RNA via their high capacity for endocytosis (Figure 12.6) (or other means of transfection such as electroporation). Subsequently, they are reintroduced into the patient. Thus, dendritic cell vaccines are labor-intensive and expensive. Initial clinical trials using antigen-loaded dendritic cells have shown positive clinical responses and no significant toxicity. An antigen-loaded dendritic cell vaccine called Provenge™ (sipuleucel-T; Dendreon Corporation, Seattle, WA) was produced for the treatment of prostate cancer by the following steps:

(1) A dendritic cell precursor-enriched fraction was isolated.

(2) The cells were matured *in vitro* by incubation with a recombinant fusion protein (consisting of prostatic acid phosphatase linked to

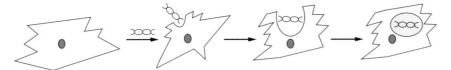

Figure 12.6 Dendritic cell loading.

granulocyte–macrophage colony-stimulating factor (GM-CSF)) that targets the GM-CSF receptor present on dendritic cells.

(3) The mature dendritic cells, now carrying the prostate cancer antigen, are administered.

After positive Phase I, II, and III trials, Provenge™ (sipuleucel-T) was approved by the US FDA in 2010 (Thara *et al.*, 2011).

Dendritic cell vaccines continue to be actively investigated. Expanding and loading dendritic cells *in vivo* would eliminate the dangers (e.g. contamination) associated with cell transfers and reduce labor and costs. Chimeric proteins composed of an antibody specific for a dendritic cell receptor fused to a specific antigen could be used to deliver tumor antigens to dendritic cells *in vivo* in the presence of dendritic cell activating agents. After uptake by endocytosis and antigen presentation, dendritic cells could then initiate an immune response.

Vaccine CDX-1401, composed of an antibody specific for a dendritic cell receptor (DEC-205) fused to a specific tumor antigen (NY-ESO-1), demonstrated proof of concept, safety, and induction of immunity in a first-in-human study (Dhodapkar *et al.*, 2014).

Overall, the optimization of therapeutic vaccines needs to be pursued, as there is a lack of evidence for reproducible clinical responses. The correct patient population (i.e. with respect to age, cancer stage, molecular signature of tumors) and timing of vaccine administration need to be a focus. Side effects of vaccines are usually minimal, in contrast to conventional chemotherapies.

These vaccines may become an important partner in combination therapies that include checkpoint blockades (see Section 12.8, "Immune checkpoint blockades" on p. 293).

Vaccines for cancer prevention

Vaccines generated from shared tumor antigens have been successful as prophylactic vaccines in animal models but have not been tested in humans. As we will see in Chapter 13, there are a few select types of cancer that are caused by pathogenic carcinogens (i.e. bacteria or viruses), and, in the case of viruses, conventional prophylactic vaccines that target the pathogen can be produced.

Large strides are being made in the development of prophylactic vaccines for breast cancer. As mentioned earlier, several promising breast

PAUSE AND THINK

Note that the Nobel Prize for Medicine or Physiology in 2011 was awarded to Ralph Steinman for his discovery of dendritic cells. He, along with colleagues, demonstrated *in vivo* dendritic cell loading.

cancer antigens (e.g. α-lactalbumin expressed in the majority of human breast carcinomas and only in healthy mammary cells during lactation; Jaini *et al.*, 2010) have been characterized. Prophylactic breast cancer vaccines are likely to be an important alternative to prophylactic mastectomies and/or oophorectomies or chemoprevention in women who carry germline mutations in the *BRCA1* and *BRCA2* genes (see Section 11.10, "Drugs that target estrogen" on p. 270). Safety and immune responses have been demonstrated for therapeutic vaccines in several Phase I and II trials in patients with breast cancer, but prophylactic vaccine trials are needed. Reluctance to carry out large-scale trials comes from the fear of autoimmunity against normal breast tissue, though autoimmune attack of normal breast tissue may be tolerable and may not have more severe consequences than mastectomies.

Coley's toxin of the 1880s and mixed bacterial vaccines

It is shocking to know that a treatment used in the 1880s was more successful for treating a type of sarcoma than modern therapy. The treatment was called Coley's toxin developed by a New York surgeon, William Coley, who developed the bacterial vaccine after observing that sarcoma patients who had concurrent infections showed cancer regression. **Coley's toxin** was a mixture of heat-inactivated *Streptococcus pyogenes* and *Serratia marcescens*. Half of Coley's patients lived for 10 years, compared to 38% of those treated with today's treatment. Bacteria produce pattern recognition receptor ligands that activate dendritic cells, and this is required for full activation of T cells. Further, these ligands, when bound to their receptors, induce an inflammatory response, including cytokine secretion. Therefore, this vaccine is considered the origin of modern cancer immunotherapy. Maletzki *et al.* (2012) report on studies that re-evaluate Coley's method, and others in the biotechnology field are developing vaccines based on this method, with one in early clinical trials.

Hurdles to jump

There are several problems that need to be overcome for the full potential of vaccine development to be reached. Vaccines target the immune system, and not the tumor; they need time for an immune response to develop and may have long-term clinical significance such as improved overall survival with minimal reduction in tumor size. These facts are leading to changes in the conventions by which clinical trials are evaluated. It must be remembered that the immune system becomes less effective with aging and is suppressed by conventional chemotherapy. It is rare that pre-clinical studies are performed in old mice or mice that have been pre-treated with chemotherapy, and this may help to explain the discrepancies between outcomes in mice and humans; positive immunological responses in mice are often not reproducible in humans. It may be that therapeutic cancer vaccines may be more successful in pediatric cancer patients than older patients. Such comparisons need to be carried out. Second, many vaccines may be most effective in early-stage cancer patients, although trials using such patients are unlikely to receive approval. In addition, resistance against therapeutic vaccines may arise. Antigen-negative tumor cell clones evolve as a result of selective pressure exerted by the vaccine. Mutations that alter antigen expression will allow tumor cells to evade the immune response and survive. This suggests that combinations of immune therapies should be investigated further, and several combinations are in clinical trials.

Also, vaccines need to be tested in all appropriate contexts. Vaccines against tumor-specific antigens are being tested in humans exclusively as therapeutic agents, and not as prophylactics, even though the success of these agents in pre-clinical trials has been demonstrated almost exclusively as prophylactics. Note that, as human tumors can only be grown in immune-deprived mice (e.g. nude mice), immunotherapy studies on human tumors cannot be performed in existing pre-clinical models, and results from animal models may be species-specific. We cannot assume that what is successful in mice will be successful in humans, because some aspects of physiology between the two are different. Prophylactic vaccines aimed at tumor-specific antigens (not including those directed against pathogens, e.g. HPV) have not been tested in clinical trials, because the test population will be healthy individuals and the consequences and/or side effects are unknown. However, at some point, vaccines as prophylactics need to be tested in humans.

12.8 Immune checkpoint blockades

Immune checkpoint therapy are drugs designed to inhibit immune checkpoints, the mechanisms used to limit and control an immune response (Sharma and Allison, 2015). Using a common analogy, these drugs release the brakes on the immune system. The specific molecules of these checkpoints are good candidate targets for cancer therapy.

The first therapeutic checkpoint blockade approved was ipilimumab (Yervoy™; Bristol-Myers Squibb), a fully human monoclonal antibody against CTLA-4 (Figure 12.7a). Although the drug was approved by the US FDA in 2011 for patients with advanced melanoma, the drug is in clinical trials for many other types of cancer, including lung, prostate, and ovarian cancers (Graziani *et al.*, 2012; Wolchok *et al.*, 2013).

The data from clinical trials of checkpoint therapies are changing the shape of previously known survival curves for cancer patients undergoing treatment. In one randomized, double-blind Phase III study (MDX010-20), overall survival in patients with metastatic melanoma treated with 3 mg per kilogram of body weight of ipilimumab, with or without a glycoprotein100 (gp100) peptide vaccine, was compared to that in those treated with vaccine alone (Hodi *et al.*, 2010). gp100 is a melanoma antigen that was used to develop a peptide vaccine (gp100 peptide vaccine),

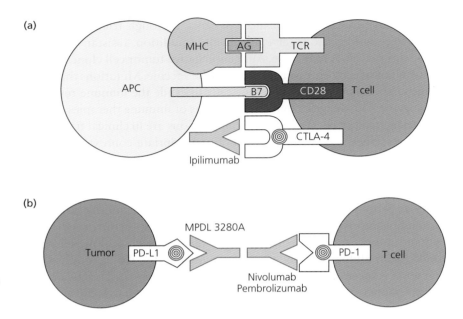

Figure 12.7 Immune checkpoint blockades. Therapeutic agents are shaded in red. (a) Ipilimumab. (b) Drugs that target PD-1 or PDL-1.

PAUSE AND THINK

Study Figure 12.8. First, orientate yourself with each axis and the meaning of the different curves. What is the proportion of alive patients treated with control gp100 peptide vaccine at 3.5 years? What is the proportion of alive patients treated with ipilimumab at 3.5 years? Does this change at 4.5 years? What is the proportion of alive patients at 1 year for gp100 versus ipilimumab? This is a near doubling. These are significant results that were the basis for the regulatory approval of ipilimumab.

PAUSE AND THINK

Are the mechanisms for the breakpoint molecules CTLA-4 and PD-1 the same? No, because CTLA-4 works via competition with co-stimulatory factor CD28 and not by interfering with signaling, as PD-1 does.

but it showed limited anti-tumor activity, and so it was used as an active control in this study. Data from this trial are illustrated in the graph in Figure 12.8, which shows overall survival curves for patients receiving a control (gp100 peptide vaccine alone; blue line) versus those receiving ipilimumab with or without the gp100 peptide vaccine (red and purple lines, respectively).

Anti-tumor responses with ipilimumab have been rapid and significant: results demonstrate greater than 20% of treated patients living for more than 4 years, and long-term survival greater than 10 years has been demonstrated in a subset of patients (Schadendorf *et al.*, 2015). It is important to note that there are associated side effects that include severe autoimmune and inflammatory responses, though the management of such responses with agents, such as steroids, is improving. Since the CTLA-4 molecule is on T cells, and not the tumor, it is likely that this approach could be applied to other cancers.

Checkpoint PD-1 and its ligand PDL-1 are additional checkpoint targets of drugs recently approved or currently in clinical trials (Figure 12.7b; Garon *et al.*, 2015; Topalian *et al.*, 2012).

Anti-PD-1 agents pembrolizumab (Keytruda™; Merck) and nivolumab (Opdivo™; Bristol-Myers Squibb) are approved for melanoma and melanoma and lung cancer, respectively. MPDL3280A (Genentech) is one of several drugs in clinical trials that target PD-L1. One large clinical Phase I trial of pembrolizumab showed a response rate of approximately 37% (Hamid *et al.*, 2013). The safety profiles of these drugs are generally better than drugs that target earlier acting checkpoints such as CTLA-4.

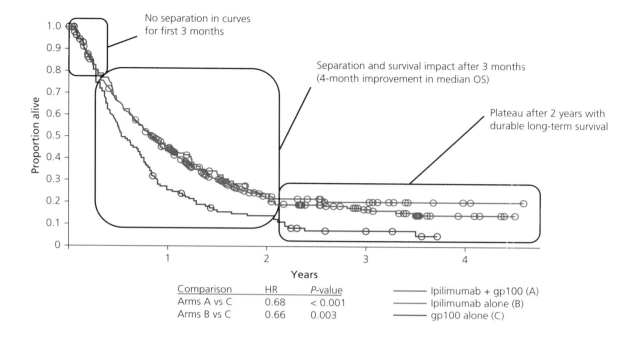

Figure 12.8 Phase III clinical trial data for ipilimumab. Kaplan–Meier analysis of overall survival in the phase III study MDX010-20. Separation of the Kaplan–Meier survival curves was not observed until three months, at which time an OS benefit of approximately four months was observed with ipilimumab treatment compared with the gp100 control. Survival in the ipilimumab monotherapy group reached a plateau after two years, indicating durable response and long-term survival benefits. The table shows data for the primary endpoint of OS, and secondary endpoints of one- and two-year survival rates. Adapted from Hodi, F.S. *et al.* (2010) Improved Survival with Ipilimumab in Patients with Metastatic Melanoma. *N. Engl. J. Med.* **363**: 711–723. Copyright © 2010 Massachusetts Medical Society. Reprinted with permission from Massachusetts Medical Society.

Since only a fraction of patients benefit from checkpoint therapy, there is a desire to identify biomarkers that can be used to select patients who are likely to respond to a specific type of checkpoint therapy. Expression of PD-L1 was an obvious candidate as a biomarker for selecting patients who would respond to antibodies against PD-1 or PD-L1. Surprisingly, data showed that patients who had PD-L1-negative tumors can also respond to anti-PD-1 or PD-L1 therapy. The concept of "hot" immunogenic tumor microenvironments that are rich in infiltrating T cells and specific cytokines versus "cold" non-immunogenic tumor environments is proposed to correlate with clinical benefit for patients treated with PD-1 and PD-L1 therapies. Studies to identify robust biomarkers are ongoing.

There are many roads to follow for the future of immunotherapy. It can be predicted that many other immune checkpoint molecules will be targets for immunotherapy. Candidates include LAG-3 and VISTA and co-stimulatory molecules ICOS, OX40, and 4-1BB. Combinations of different checkpoint blockades or checkpoint blockades with other types of immune/targeted therapy are also being tested and producing promising results. For example, a trial investigating nivolumab plus ipilimumab in melanoma demonstrated 53% of patients had an objective response, all with tumor reduction of 80% or more (Wolchok *et al.*, 2013; Figure 12.9). Targeted therapies that kill tumor cells cause a release of tumor antigens that may be presented by APCs to initiate an immune response. This immune response may be enhanced by checkpoint therapy. Additive or synergistic effects may result from combinations.

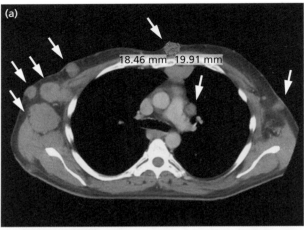

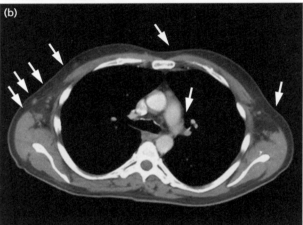

Figure 12.9 Computed tomographic (CT) scans of the chest showing tumor regression in a patient who received the concurrent regimen of nivolumab and ipilimumab. The CT scan obtained at week 12 shows a marked reduction in all areas of disease (Panel B). Arrows indicate locations of metastatic disease. Adapted from Wolchok, J.D. *et al.* (2013) Nivolumab plus ipilimumab in advanced melanoma. *N. Engl. J. Med.* **369**: 122–133. Copyright © 2013 Massachusetts Medical Society. Reprinted with permission from Massachusetts Medical Society.

12.9 Adoptive T-cell transfer, modified T-cell receptors, and chimeric antigen receptors

The idea to harvest a patient's T cells, expand them outside of the body (*ex vivo*), and return an expanded "army of T cells" back to the patient has been a therapeutic aim for many years. Treatment involving the transfer of immune cells that have been expanded *ex vivo* and can attack tumors is called adoptive cell transfer. The transferred cells can proliferate in the body, producing long-lasting effects, and have been referred to as a "living" treatment (see Rosenberg and Restifo, 2015 for review). Early attempts of this strategy were generally unsuccessful until a specific subset of T cells was used.

Immunotherapy based on adoptive cell transfer of naturally occurring T cells found in a tumor, called tumor-infiltrating lymphocytes (TILs), has demonstrated promising results in the clinic. These cells have been activated in response to tumor-associated antigens, but, in the body, they are restrained by immunosuppressive molecules such as CTLA-4 and

PD-1. TILs are removed from this immunosuppressive environment when removed from the tumor. The approach involves excising the tumor and growing small fragments in culture to produce pure cultures of lymphocytes. After selecting for anti-tumor activity, they can be expanded *ex vivo* and transferred back to patients. Results are best if patients receive lymphodepletion treatment before transfer.

Rosenberg and his group have demonstrated promising results after the transfer of autologous TILs into melanoma cancer patients who underwent lymphodepleting chemotherapy (destruction of endogenous lymphocytes by cytotoxic drugs with or without radiation) (Rosenberg *et al.*, 2011). TILs were expanded *ex vivo* and transferred to patients, along with the cytokine adjuvant IL-2 (a T-cell growth factor), for immunization. Complete tumor regression was observed in 22% of patients with metastatic melanoma.

Adoptive cell transfer strategies can also involve genetically modifying T cells from blood, so that they express T-cell receptors that target specific tumor antigens. Positive clinical responses have been demonstrated when transferring high-affinity T-cell receptors.

Genetic engineering is being used to increase the affinity of T cells for tumors by creating chimeric antigen receptors (CARs). Many CARs are created by combining a single chain of sequences encoding the variable regions involved in antigen binding of monoclonal antibodies with sequences that code for the TCR complex (TCR plus accessory co-receptors) intracellular domains that induce T-cell activation (Figure 12.10). T cells genetically modified to express CARs can recognize surface structures

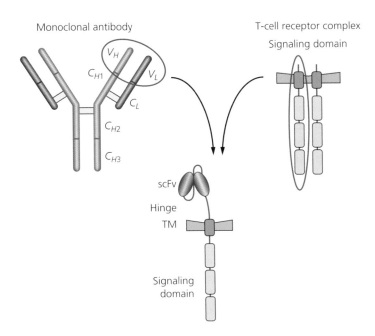

Figure 12.10 Structure of a chimeric antigen receptor. scFv, single chain variable fragment; TM, transmembrane.

on tumors without the need for presentation by MHC molecules. CARs have been successful in the treatment of hematologic cancers because of the identification of suitable targets that are expressed in these cancers (and non-essential normal tissue) such as CD19. CD19 is expressed on more than 90% of B-cell cancers and on B cells. Adoptive cell transfer of cells expressing anti-CD19 CAR has demonstrated positive results for lymphoma and lymphocytic leukemia patients.

Further still, genetic engineering is being used to create bispecific antibodies whose aim is to link T cells to tumors without TCR–MHC interactions. In concluding this section, it is important to note that adoptive cell transfer is highly personalized for each patient and requires specialized facilities for production and delivery.

12.10 Oncolytic viruses and virotherapy

Oncolytic viruses, either natural or engineered, are viruses that can selectively infect and kill tumor cells. Selection of tumor cells by these viruses may be due to dysregulated cell growth and metabolism and/or altered antiviral response pathways of tumor cells. Upon viral replication within the tumor cell and subsequent lysis of the cell, the virus can spread within and between tumors, thus amplifying its effect. Normal tissues are less susceptible to infection due to intact antiviral pathways.

The application of oncolytic viruses as a means of selectively lysing tumor cells has evolved into the application of genetically altered viruses for generating an immune response. Viral lysis of tumor cells release tumor antigens that can be presented to T cells by APCs and can serve as "*in situ* vaccination."

Oncolytic virotherapy has been administered to thousands of patients in clinical trials, and there have been no oncolytic virotherapy-related deaths. Although no oncolytic therapies have been approved in the USA, several are in clinical trials, including T-Vec (talimogene laherparepvec; Amgen) for melanoma and Pexa-Vec (pexastimogene devacirepvec; Silla-Jen Biotherapeutics) for liver cancer. Both viruses have been engineered to express the transgene for an immunostimulatory cytokine (e.g. GM-CSF) in order to help stimulate an immune response. H101, a recombinant adenovirus that replicates in, and kills, cells with an inactive p53 pathway, is approved in China for the treatment of head and neck cancer.

▦ CHAPTER HIGHLIGHTS—REFRESH YOUR MEMORY

- The immune system plays a dual role in cancer: it can both suppress and promote tumor growth.

- B cells synthesize antibodies, and T cells are responsible for cell-mediated immunity.

- Immunosurveillance is the concept that the immune system can recognize tumor-specific and tumor-associated antigens as "foreign" and can eliminate cancer cells.

- The cancer immunity cycle describes the steps involved in generating an immune response against a tumor.

- The immune response is regulated by immune checkpoints involving specific receptor–ligand interactions.

- CTLA-4 is a co-inhibitory molecule that suppresses T-cell initiation, and PD-1 and its ligand PD-L1 are co-inhibitory molecules that regulate T-cell responses at the tumor interface.

- Immunoediting describes the interaction and effects of the immune system on the tumor, and vice versa.

- Immunoediting has three stages: elimination, equilibrium, and escape.

- Tumors may evade the immune system by loss of tumor antigens, downregulation of antigen-presenting molecules, overexpression of immune checkpoint proteins and anti-apoptotic molecules, and secreting tumor-suppressive proteins.

- Immunotherapy targets the host immune system, not the tumor.

- Immunotherapy strategies include: therapeutic antibodies, cancer vaccines, immune checkpoint blockades, adoptive T-cell transfer and modified TCRs, and oncolytic viruses.

- Some immunotherapies are yielding unprecedented outcomes with durable responses rarely seen before, and can be applied to many different cancer types.

▨ ACTIVITY

1. Compose a list of co-inhibitory molecules of immune checkpoints other than CTLA-4 and PD-1, and report on any progress of using them as targets for new cancer immunotherapeutic agents. Begin with Chen and Flies (2013).

▨ FURTHER READING

Bell, J. and McFadden, G. (2014) Viruses for tumor therapy. *Cell Host Microbe* **15**: 260–265.

Gajewski, T.F., Schreiber, H., and Fu, Y.-X. (2013) Innate and adaptive immune cells in the tumor microenvironment. *Nat. Immunol.* **14**: 1014–1022.

Gao, J., Bernatchez, C., Sharma, P., Radvanyi, L.G., and Hwu, P. (2013) Advances in the development of cancer immunotherapies. *Trends Immunol.* **34**: 90–98.

Grivennikov, S.I., Greten, F.R., and Karin, M. (2010) Immunity, inflammation, and cancer. *Cell* **140**: 883–899.

Lichty, B.D., Breitbach, C.J., Stojdl, D.F., and Bell, J.C. (2014) Going viral with cancer immunotherapy. *Nat. Rev. Cancer* **14**: 559–567.

Palucka, K. and Banchereau, J. (2012) Cancer immunotherapy via dendritic cells. *Nat. Rev. Cancer* **12**: 265–277.

Pardoll, D.M. (2012) The blockade of immune checkpoints in cancer immunotherapy. *Nat. Rev. Cancer* **12**: 252–264.

Restifo, N.P., Dudley, M.E., and Rosenberg, S.A. (2012) Adoptive immunotherapy for cancer: harnessing the T cell response. *Nat. Rev. Immunol.* **12**: 269–281.

Schlom, J. (2012) Therapeutic cancer vaccines: current status and moving forward. *J. Natl. Cancer Inst.* **104**: 599–613.

Schreiber, R.D., Old, L.J., and Smyth, M.J. (2011) Cancer immunoediting: integrating immunity's roles in cancer suppression and promotion. *Science* **331**: 1565–1570.

Sharma, P. and Allison, J.P. (2015) The future of immune checkpoint therapy. *Science* **348**: 56–61.

Swann, J.B. and Smyth, M.J. (2007) Immune surveillance of tumors. *J. Clin. Invest.* **117**: 1137–1146.

Tan, T.-T. and Coussens, L.M. (2007) Humoral immunity, inflammation and cancer. *Curr. Opin. Immunol.* **19**: 209–216.

Topalian, S.L., Drake, C.G., and Pardoll, D.M. (2012) Targeting the PD-1/B7-H1 (PD-L1) pathway to activate anti-tumor immunity. *Curr. Opin. Immunol.* **24**: 207–212.

Wolchok, J.D., Hodi, F.S., Weber, J.S., Allison, J.P., Urba, W.J., Robert, C., *et al.* (2013) Development of ipilimumab: a novel immunotherapeutic approach for the treatment of advanced melanoma. *Ann. N.Y. Acad. Sci.* **1291**: 1–13.

Zhu, Y., Yao, S., and Chen, L. (2011) Cell surface signalling molecules in the control of immune responses: a tide model. *Immunity* **34**: 466–478.

▨ WEB SITE

Understanding the immune system http://www.nsta.org/publications/press/extras/files/debatable/theimmunesystem.pdf

▨ SELECTED SPECIAL TOPICS

Chen, L. and Flies, D.B. (2013) Molecular mechanisms of T cell co-stimulation and co-inhibition. *Nat. Rev. Immunol.* **13**: 227–242.

Chen, D.S. and Mellman, I. (2013) Oncology meets immunology: the cancer-immunity cycle. *Immunity* **39**: 1–10.

Dhodapkar, M.V., Sznol, M., Zhao, B., Wang, D., Carvajal, R.D., Keohan, M.L., *et al.* (2014) Induction of antigen-specific immunity with a vaccine targeting NY-ESO-1 to the dendritic cell receptor DEC-205. *Sci. Transl. Med.* **6**: 232ra 51.

Garon, E.B., Rizvi, N.A., Hui, R., Leighl, N., Balmanoukian, A.S., Eder, J.P., *et al.* (2015) Pembrolizumab for the treatment of non-small cell lung cancer. *N. Engl. J. Med.* **372**: 2018–2028.

Graziani, G., Tentori, L., and Navarra, P. (2012) Ipilimumab: a novel immunostimulatory monoclonal antibody for the treatment of cancer. *Pharmacol. Res.* **65**: 9–22.

Hamid, O., Robert, C., Daud, A., Hodi, F.S., Hwu, W.J., Kefford, R., *et al.* (2013) Safety and tumor responses with lambrolizumab (anti-PD-1) in melanoma. *N. Engl. J. Med.* **369**: 134–144.

Hodi, F.S., O'Day, S.J., McDermott, D.F., Weber, R.W., Sosman, J.A., Haanen, J.B., *et al.* (2010) Improved survival with ipilimumab in patients with metastatic melanoma. *N. Engl. J. Med.* **363**: 711–723.

Jaini, R., Kesaraju, P., Johnson, J.M., Altuntas, C.J., Jane-wit, D., and Tuohy, V.K. (2010) An autoimmune-mediated strategy for prophylactic breast cancer vaccination. *Nat. Med.* **16**: 799–803.

Maletzki, C., Klier, U., Obst, W., Kreikemeyer, B., and Linnebacher, M. (2012) Reevaluating the concept of treating experimental tumors with a mixed bacterial vaccine: Coley's Toxin. *Clin. Develop. Immunol.* **2012**: 230625.

Roithmaier, S., Haydon, A.M., Loi, S., Esmore, D., Griffiths, A., Bergin, P., *et al.* (2007) Incidence of malignancies in heart and/or lung transplant recipients: a single institution experience. *J. Heart Lung Transplant.* **26**: 845–849.

Rosenberg, S.A. and Restifo, N.P. (2015) Adoptive cell transfer as personalized immunotherapy for human cancer. *Science* **348**: 62–68.

Rosenberg, S.A., Yang, J.C., Sherry, R.M., Kammula, U.S., Hughes, M.S., Phan, G.Q., *et al.* (2011) Durable complete responses in heavily pretreated patients with metastatic melanoma using T-cell transfer immunotherapy. *Clin. Cancer Res.* **17**: 4550–4557.

Schadendorf, D., Hodi, F.S., Robert, S., Weber, J.S., Margolin, K., Hamid, O., *et al.* (2015) Pooled analysis of long-term survival data from phase II and phase III trials of ipilimumab in unresectable or metastatic melanoma. *J. Clin. Oncol.* **33**: 1889–1894.

Schwartzentruber, D.J., Lawson, D.H., Richards, J.M., Conry, R.M., Miller, D.M., Treisman, J., *et al.* (2011) gp100 peptide vaccine and interleukin-2 in patients with advanced melanoma. *N. Engl. J. Med.* **364**: 2119–2127.

Scott, A.M., Wolchok, J.D., and Old, L.J. (2012) Antibody therapy of cancer. *Nat. Rev. Cancer* **12**: 278–287.

Sharma, P. and Allison, J.P. (2015) Immune checkpoint targeting in cancer therapy: toward combination strategies with curative potential. *Cell* **161**: 205–214.

Sliwkowski, M.X. and Mellman, I. (2013) Antibody therapeutics in cancer. *Science* **341**: 1192–1198.

Thara, E., Dorff, T.B., Pinski, J.K., and Quinn, D.I. (2011) Vaccine therapy with sipuleucel-T (Provenge) for prostate cancer. *Maturitas* **69**: 296–303.

Topalian, S.L., Hodi, S., Brahmer, J.R., Gettinger, S.N., Smith, D.C., McDermott, D.F., *et al.* (2012) Safety, activity, and immune correlates of anti-PD-1 antibody in cancer. *N. Engl. J. Med.* **366**: 2443–2454.

Vesely, M.D., Kershaw, M.H., Schreiber, R.D., and Smyth, M.J. (2011) Natural innate and adaptive immunity to cancer. *Ann. Rev. Immunol.* **29**: 235–271.

Waterhouse, P., Penninger, J.M., Timms, E., Wakeham, A., Shahinian, A., Lee, K.P., *et al.* (1995) Lymphoproliferative disorders with early lethality in mice deficient in CTLA-4. *Science* **270**: 985–988.

Wolchok, J.D., Kluger, H., Callahan, M.K., Postow, M.A., Rizvi, N.A., Lesokhin, A.M., *et al.* (2013) Nivolumab plus ipilimumab in advanced melanoma. *N. Engl. J. Med.* **369**: 122–133.

Chapter 13

Infectious agents and inflammation

Introduction

One-sixth of all cancers are caused by infectious agents and inflammation. This may be a surprising fact and suggests the logical question: "Can we 'catch' cancer?" The answer to this question is not straightforward. One does not "catch" cancer in the same way as one "catches a cold." Exposure to an infectious agent does not immediately trigger cancer. However, we now know that long-term exposure to specific infectious agents, some causing chronic inflammation, can lead to cancer. In addition, there is evidence that **chronic** inflammation, in the absence of an infectious agent, leads to an increased risk of cancer. An inflammatory microenvironment is characteristic of all tumors, and tumor-promoting inflammation has been named as an emerging hallmark of cancer (see Chapter 1). As we saw in Chapter 12, the immune system plays a dual role in cancer: it can both promote and inhibit cancer progression. Immune cells can recognize and eliminate viral-infected tumor cells. On the other hand, immune responses, such as long-term inflammation, can promote carcinogenesis.

This is good news for cancer prevention and treatment, as we have learned a lot about the prevention and treatment of some types of infection related to other diseases and also about prevention and treatment of chronic inflammation. Infection and inflammation may be major preventable causes of human cancer.

Infectious agents involved in carcinogenesis include DNA and RNA viruses and bacteria. DNA viruses contain viral genes that do not have cellular homologs. As we discussed in Chapter 6 (Figure 6.9), they act by producing proteins that interact and inhibit tumor suppressor genes (e.g. *p53* and *Rb*) to promote cell proliferation. RNA viruses, or retroviruses, carry altered forms of cellular genes, called oncogenes. Some DNA and RNA viruses disrupt normal gene expression via insertional mutagenesis (discussed in Chapter 4). The mouse mammary tumor virus (MMTV), an oncogenic retrovirus, is an important model system that induces breast cancer in mice through insertional mutagenesis. Conflicting evidence for a role of MMTV-like sequences in human breast cancer has been reported. Bacteria products can be transferred into cells, and interfere with important signaling pathways and cause abnormal gene expression.

Inflammation is a physiological response against infectious agents and injury, and is a consequence of wound healing. Under normal conditions, it is highly regulated and short-lived: such acute inflammation typically resolves itself with the help of anti-inflammatory factors. By contrast, recent evidence suggests that it is lingering, chronic inflammation that plays an important role in causing cancer.

Some viral and bacterial infections induce a chronic inflammatory response that contributes to the process of carcinogenesis. Inflammation caused by an external agent is often referred to as extrinsic inflammation. Intrinsic inflammation refers to the body's response to a tumor, resulting in the creation of an inflammatory micro-environment around it. Cancer has been referred to as a "wound that never heals." Inflammatory cells, growth factors, and reactive oxygen/nitrogen species characterize the site of the inflammatory response. These factors of inflammation set the stage for cell proliferation, mutagenesis, angiogenesis, and metastasis.

In this chapter, infectious agents that are considered to be carcinogens will be identified, and several modes of action of these infectious agents will be discussed. The molecular mechanisms of chronic inflammation (in the presence or absence of an infectious agent) that contribute to carcinogenesis will also be described. Finally, the chapter will conclude with a report on the major therapeutic applications of this knowledge.

13.1 Identifying infectious agents as carcinogens

In order to identify which infectious agents actually *cause* cancer, several general criteria are applied. First, there must be a consistent association between infection and the cancer that is supported by either epidemiological or molecular evidence. In addition, cell transformation or induction of tumors in animal models by the infectious agent must be demonstrated. In some instances, infectious agents may not reside in the transformed cells but may cause cancer indirectly, through inflammation and chronic tissue injury or paracrine growth stimulation (see "Kaposi's sarcoma-associated herpesvirus (KSHV)" on p. 307). A list of some of the infectious agents that are known to be carcinogenic is shown in Table 13.1 and discussed in the following sections.

Epstein–Barr virus (EBV)

EBV, a DNA virus, is a good starting point for examining the role of infectious agents as carcinogens. It causes several types of lymphoma, including Burkitt's lymphoma and nasopharyngeal cancer. There is sufficient

Table 13.1 Infectious agents and their role as cancer-causing agents

Infectious agent	Type	Type of cancer	Major effector
Human papillomavirus	DNA virus	Cervical, throat, anogenital	E6, E7
Epstein–Barr virus	DNA virus	Nasopharyngeal, some lymphomas (including Burkitt's)	LMP1
Kaposi's sarcoma-associated herpesvirus	DNA virus	Kaposi's sarcoma	LANA
Human T-cell leukemia virus	RNA virus	T-cell leukemia	TAX protein
Hepatitis B (and C) virus	DNA virus	Liver	HBV X
Helicobacter pylori	Bacterium	Gastric	Cag A
Schistosoma haematobium	Liver fluke	Bladder	
Opisthorchisviverrini	Liver fluke	Liver	
Clonorchissinensis	Liver fluke	Liver	

(Data from IARC. *Monographs on the evaluation of carcinogenic risks to humans, volume 100. A review of carcinogen—Part B: biological agents.* International Agency for Research on Cancer, Lyon; 2011.)

evidence in humans for the carcinogenicity of EBV. It has been demonstrated that EBV can transform lymphoid cells in culture, and it has been documented that all cases of nasopharyngeal carcinoma and endemic Burkitt's lymphoma are associated with EBV infection. Malaria may be a co-factor for Burkitt's lymphoma.

EBV encodes several viral proteins that affect host gene expression. One of these, oncoprotein LMP1, is able to transform cells in culture. Its multi-functional effects include the activation of genes that are important for promoting cell proliferation (e.g. *EGFR*) and inhibiting apoptosis (e.g. *Bcl-2*). It also activates a nuclear transcription factor called NF-κB. As we will see later, activation of cell growth pathways and anti-apoptotic pathways, as well as activation of inflammation by key player NF-κB, are important molecular themes that run throughout this chapter.

Human papillomavirus (HPV)

HPV infections are the most common sexually transmitted viral infections, and the risk of infection increases with the number of sexual partners. HPV infection is usually cleared by the immune system and does not affect a person's health, but chronic infection over years causes cancer. HPV has been found to be associated with several types of cancer, though cervical cancer is the most common. Virtually 100% of all cervical cancer cells contain HPV. There is a lag time between HPV infection and cervical cancer of at least 10 years. This DNA virus requires access to the proliferating cells of the cervical epithelium. It uses entry points created by micro-erosion of the overlying cell layers to reach the basal layer of the epithelium where stem cells and progenitor cells reside. The development of cancer involves expression of viral genes, integration of viral DNA into the host chromosomes, and changes to host cell genes and gene products. Genomic hotspots have been identified for HPV chromosomal integration. Viral integration may use a host DNA repair mechanism and occurs in the initiation stage of cervical cancer (Hu *et al.*, 2015).

The HPV gene products E6 and E7 are major players in carcinogenesis that target tumor suppressor proteins in the host cell. E7 binds to, and triggers, the degradation of RB, thus preventing the sequestration of E2F. This leads to the constitutive expression of E2F-responsive genes, including cyclin A and cyclin E that are important for cell cycle progression. E6 forms a complex with ubiquitin ligase. The complex binds p53, and p53 degradation is triggered (see Chapter 6). E6 and E7 can also induce genomic instability by targeting telomerase, γ-tubulin (a centrosome regulator), and proteins involved in DNA damage pathways. Both E6 and E7 have been shown to induce DNA damage. Viral products E6 and E7 are able to transform cultured human cells and induce tumors in mice.

Note that only a subset of the total of 130 types of HPV that have been identified is considered "high risk" for cervical cancer (types 16, 18, 31, 33, 35, 39, 45, 51, 52, 56, 58, 59; Figure 13.1). HPV16 and HPV18 account for about 70% of the global distribution of HPV types in cervical cancer. Viral genotype analysis, distinguishing between the different types, was crucial in establishing the link with cervical cancer; if all HPVs were considered as one agent, the link would not have been established. High-risk HPVs also contribute to over 50% of oral/throat cancers (likely to be transmitted by oral sex), and penile, vulvar, and anal cancers. The incidence of HPV-associated throat (oropharyngeal) cancers has increased in the USA over recent years.

Human T-cell lymphotropic virus type 1 (HTLV-1)

Almost all cases of adult T-cell leukemia exhibit molecular evidence for the presence of HTLV-1. This is the only virus known to have a causal link to human leukemia. Also, it is the only retrovirus that is linked to human cancer. HTLV-1 infection is prevalent in Japan, the Caribbean, South

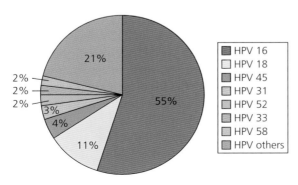

Figure 13.1 A pie chart showing the distribution of HPV types in cervical cancer. (Data from Munoz *et al.*, 2003.)

America, and Central Africa. About 2–5% of infected individuals develop T-cell leukemia/lymphoma.

Transmission of the virus is through intimate contact and includes transmission through breast milk, semen, unscreened blood, and contaminated needles among drug users. In Japan, breast-feeding for more than 6 months has been discouraged; as a result, transmission of HTLV-1 to infants has been greatly reduced.

As a retrovirus, HTLV-1 genomic RNA is copied into DNA by reverse transcriptase before viral proteins are synthesized by the host cell's machinery. The TAX protein of HTLV-1 is a key player in the mechanism of HTLV-1-induced carcinogenesis. TAX exerts its functions by protein–protein interactions with over a hundred cellular proteins generating a plethora of effects, including stimulation of cell growth, inhibition of DNA repair, and alteration of the cell cycle by activating cdks (Boxus and Willems, 2009). It is a molecule that activates host cell genes through the transcription factors NF-κB (see "NF-κB is a key player in the inflammatory response" on p. 312) and AP1, and interferes with several tumor suppressor proteins (e.g. p53) to promote cell proliferation and, ultimately, leukemia.

◉ Hepatitis B virus (HBV) and liver cancer

HBV, a DNA virus, is the most common carcinogen after tobacco. People who are chronically infected with HBV have a high risk of developing liver cancer (hepatocarcinoma), as demonstrated by the fact that HBV carriers have a 10- to 25-fold greater risk of developing liver cancer than uninfected individuals (Guerrieri *et al.*, 2013).

Both indirect and direct mechanisms contribute to the ability of HBV to cause liver cancer. Host–viral interactions evoke an immune response that results in chronic inflammation, liver necrosis, and regeneration. It has been proposed that increased proliferation and/or oxidative stress from inflammation may lead to oncogenic mutations. Inflammation caused by HBV is important for the development of hepatocellular carcinoma, in a manner that parallels mechanisms observed in gastric cancer (see "*Helicobacter pylori* infection and gastric cancer" on p. 308).

In addition to these indirect mechanisms, recent evidence suggests several direct mechanisms for the role of HBV in liver cancer. One direct mechanism is the integration of viral DNA into the host chromosome and insertional mutagenesis at specific sites. Integration into the human telomerase reverse transcriptase (hTERT) gene occurs frequently and results in upregulation due to close proximity with the viral enhancer.

Another direct mechanism of HBV is thought to be carried out by the multi-functional viral protein HBV X. HBV X is important for HBV-induced carcinogenesis, and it was shown to induce liver cancer in transgenic mice. It functions by activating proto-oncogenes via various signaling cascades, including many kinase cascades (e.g. the RAS–Raf–MAPK pathway; see Chapter 4), interacting with NF-κB (see "NF-κB is a key player in the inflammatory response" on p. 312), and binding to and inactivating p53. HVB X also physically binds to several nuclear proteins involved in epigenetic and transcriptional regulation such as chromatin-modifying enzymes and transcription factors (e.g. DNA methyltransferases, TFIIB, CREB). Markers of stem cells are increased by inhibiting Suz12—a component of the polycomb repressive complex (see Chapter 8). In summary, HBV can lead to transcriptional activation and repression of important oncogenes and tumor suppressor genes, respectively.

> Note that hepatitis C virus is also a risk factor for hepatocellular carcinoma, but details will not be included here.

Kaposi's sarcoma-associated herpesvirus (KSHV)

KSHV, also known as herpesvirus 8 (HHV8), is a DNA virus. It is causally linked with a vascular tumor of heterogeneous cell composition usually associated with the skin, called Kaposi's sarcoma (>95%), and primary effusion lymphomas (100%). The virus is thought to infect circulating endothelial precursor cells or endothelial cells. Although Kaposi's sarcoma is associated frequently with AIDS patients, there are types found in different populations where AIDS is not prevalent. Classic Kaposi's sarcoma is seen in elderly men of Mediterranean or eastern European Jewish ancestry, while an endemic type is prevalent in parts of Central and Eastern Africa, described before the HIV pandemic. Kaposi's sarcoma is also seen in immunosuppressed patients who have received an organ transplant. It is thought that immunodeficiency is the main mechanism by which HIV infection acts as a co-factor for KSHV. KSHV infection in the general population rarely leads to Kaposi's sarcoma.

KSHV is thought to contribute to tumorigenesis through several mechanisms. First, several viral gene products have been shown to drive cell proliferation and prevent apoptosis. KSHV produces viral cyclin, viral anti-apoptotic proteins (e.g. vBcl-2), viral-encoded miRNAs, and a viral protein called LANA that interferes with the function of RB and p53. The mechanism of LANA of targeting two crucial tumor suppressor proteins is shared with HPV viral proteins.

However, note that Kaposi's sarcoma cells are not fully transformed in that they are not able to induce tumors in nude mice like other tumor cells, and they require external growth factors and cytokines to grow *in vitro*. It has been suggested that KSHV tumorigenesis involves a paracrine process (coined paracrine neoplasia) whereby KSHV-infected cells induce neighboring uninfected cells to produce cytokines and growth factors that are necessary as autocrine and paracrine factors for driving tumorigenesis—tumorigenesis by "remote control."

◎   *Helicobacter pylori* infection and gastric cancer

H. pylori is a bacterium capable of inducing chronic inflammation in the stomach and initiating carcinogenesis. It has been identified as a carcinogen for humans by the IARC/World Health Organization. It is estimated to be responsible for over 5% of all cancers worldwide and the major agent of infection-related cancers. Gastric carcinogenesis depends on the bacterial strain, host response, and environmental factors. It is a multi-step process where inflammation is considered the initial and required step in the process: *H. pylori* infection → chronic superficial gastritis → atrophic gastritis → intestinal dysplasia → gastric carcinoma.

Gastric atrophy is characterized by the loss of normal glandular cells and results in a decrease of acid production. These conditions allow additional bacteria to colonize the stomach and trigger an inflammatory response.

Epidemiological studies show that the prevalence of gastric cancer corresponds with the prevalence of particular strains of *H. pylori* in specific geographical locations. Overall, it is estimated that about 75% of all gastric cancers are caused by *H. pylori* infection. Experimental evidence includes the induction of gastric cancer in animals by *H. pylori* infection and reduction in gastric cancer risk and the prevention of precancerous lesions by eradication of *H. pylori* infection.

Let us examine the proposed mechanism by which this bacterium can cause cancer. Many high-risk strains of *H. pylori* code for a protein called cytotoxin-associated antigen A (Cag A). Epidemiological studies show that Cag A-positive strains are predominant in regions with a high prevalence of gastric cancer. The bacterial Cag A protein is an effector protein that is injected into cells by the bacterial secretion system to elicit cellular effects such as stimulation of cell growth. Integrin receptors on host cells represent a portal of entry for Cag A injection. Cag A is phosphorylated by the cellular Src and Abl family of tyrosine kinases. Phosphorylated Cag A interacts with SH2 domain-containing proteins such as SHP-2 (an oncogenic tyrosine phosphatase; see Chapter 4) and Grb2. SHP-2 normally maintains an inactive conformation by a mechanism similarly described for Src (Figure 4.8): an amino-SH2 domain of SHP-2 blocks substrate access by an intramolecular interaction. Binding of Cag A may cause a conformational change that relieves the intramolecular inhibition and leads to the stimulation of SHP-2 phosphatase activity (Figure 13.2a; note the change in shape of SHP-2 shown in red). Thus, Cag A binding of SHP-2 mimics a gain-of-function oncogenic mutant form of SHP-2. Signal transduction pathways, including MAPK, are initiated and affect cell mitosis, migration, and adhesion. As mentioned in Chapter 9, E-cadherin is important for cell–cell contacts, and loss of E-cadherin is associated with gastric cancer. Both the E-cadherin gene and protein are targets of *H. pylori*. *H. pylori*-induced inflammation stimulates

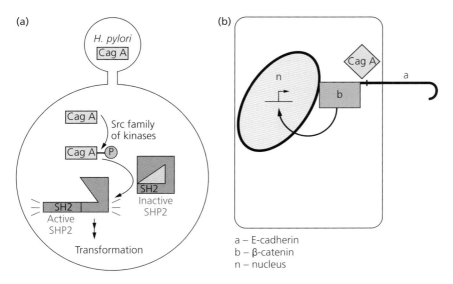

(a)

(b)

a – E-cadherin
b – β-catenin
n – nucleus

Figure 13.2 Molecular mechanisms involved in *H. pylori* infection that contribute to carcinogenesis. (a) Phosphorylated cytotoxin-associated antigen A (Cag A) causes a conformational change and activates an oncogenic protein tyrosine phosphatase called SHP-2. (b) Cag A interrupts the E-cadherin/β-catenin interaction, releasing β-catenin for transport into the nucleus so that it can regulate gene expression.

DNA methyltransferases and results in epigenetic changes by hyper-methylation, including the methylation of the promoter of E-cadherin, and this reduces E-cadherin gene expression. In addition, unphosphorylated Cag A interacts with E-cadherin protein and causes it to release β-catenin for transport into the nucleus (Figure 13.2b). β-catenin is an important host transcriptional regulator of genes that are required for the development of metaplasia. Another *H. pylori* protein linked with gastric malignancy is Vac A. Vac A is a secreted protein that induces vacuolation and apoptosis by inserting into gastric epithelial cells and mitochondrial membranes. Dietary factors may also interact with *H. pylori* carcinogenesis. Salt has been defined by epidemiology as a risk factor for gastric cancer. Studies show that Cag A expression is induced when *H. pylori* is cultured in high salt concentrations. Further studies of the effects of dietary factors and *H. pylori* carcinogenesis are needed.

In addition to the effect of bacterial proteins on cell signaling and cell biology, other mechanisms seem to be involved in carcinogenesis caused by *H. pylori*. These include effects of chronic inflammation and stem cell recruitment (see "Inflammation and tissue injury in gastric cancer recruits bone marrow stem cells" on p. 315). In addition to Cag A, *H. pylori* also delivers proteins to cells that induce pro-inflammatory mediators through the NF-κB pathway. Inflammation may induce oxidative stress and ultimately increase the rate of mutations. *H. pylori* may also induce oxidative stress by production of superoxide and reduction of the antioxidant vitamin C.

Note that, although causality has not yet been proven, associations between other bacterial infections and cancer have been noted. These include *Salmonella typhi* infection and gallbladder cancer, *Streptococcus bovis* and colon cancer, *Bartonella* and vascular tumors, and *Chlamydia pneumoniae* and lung cancer.

LIFESTYLE TIP

Indications of a stomach ulcer should be brought to the attention of a doctor. Early eradication of *H. pylori* infection, especially in high-risk locations such as East Asian countries, may help protect against gastric cancer.

PAUSE AND THINK

Try to make a list of the infectious agents discussed in Section 13.1, "Identifying infectious agents as carcinogens" on p. 303, and describe how their major effector molecule(s) exert its (their) effects. Check your answers with Table 13.1.

13.2 Inflammation and cancer

Infection is a major inducer of inflammation-driven carcinogenesis. The link between inflammation and cancer is clearly illustrated by the association of chronic inflammation in both HBV- and *H. pylori*-induced cancers, as seen in Section 13.1, "Identifying infectious agents as carcinogens" on p. 303. However, inflammation may also play a causative role, independent of infection. There are a number of chronic inflammatory conditions caused by non-infectious agents that are associated with cancer.

- In addition to being a tumor initiator because of its 81 carcinogens, tobacco smoke also acts as a tumor promoter by virtue of its ability to induce chronic inflammation (Takahashi *et al.*, 2010). (Tumor promotion is the process of tumor growth from a single initiated cell into a primary tumor.)
- It has been proposed that asbestos acts as an inflammatory stimulus in the lungs, predisposing individuals to bronchial carcinoma.
- Similarly, esophageal reflux causes injury of the esophagus and may induce an inflammatory response that increases the risk of esophageal carcinoma.
- Inflammatory bowel disease greatly increases the risk of colorectal cancer.
- In an animal model prototype of inflammation-associated cancer, genetically altered mice that develop liver inflammation also subsequently develop cancer (this model is used in the study by Pikarsky *et al.* (2004), discussed in Box "How do we know that?", on p. 314).

Cancer cells can stimulate the immune system, and the immune system can enable most, if not all, of the hallmarks of cancer. Oncogenes, activated upon cancer initiation, may regulate the expression of tumor-promoting inflammatory cytokines, generating an inflammatory microenvironment for tumors that do not develop from an inflammatory environment. *RAS* and *myc* are two examples of important oncogenes that induce inflammatory cytokines.

 Let us examine the cellular and molecular events of the inflammatory response and how they contribute to the hallmarks of cancer. The key cells of a chronic inflammatory response are macrophages. These cells produce cytokines, including TNF-α, that help orchestrate the inflammatory response by inducing a range of effector molecules, some of which help perpetuate the inflammatory response. Many inflammatory pathways involved in carcinogenesis lead to the activation (via inflammatory cytokines) of two important transcription factors: STAT and NF-κB (discussed "NF-κB is a key player in the inflammatory response" on p. 312). STAT promotes cell growth by inducing the *cyclin D*, *cyclin B*, and *myc* genes, and increases survival by inducing anti-apoptotic genes such as

Bcl-2. In this way, proliferation, survival, and apoptotic pathways are affected.

On the other side of the coin, the expression of the genes that code for pro-inflammatory factors is regulated by the transcription factor NF-κB. As NF-κB is induced by carcinogens, including oncogenic viral products, it is an important link between inflammation and cancer. In summary, inflammation can activate transcription factors such as NF-κB, and NF-κB can activate pro-inflammatory products.

In addition, leukocytes produce reactive oxygen and nitrogen species (ROS and NOS, respectively) to help fight infection, but these products (via the formation of peroxynitrite) also cause DNA damage. Chronic inflammation is associated with an increased production of ROS and NOS, increased risk of DNA damage, and increased mutation rates, leading to genomic instability. Tissue regeneration and cell renewal induced by HBV and *H. pylori* infections, respectively, involve cell division, during which DNA is most susceptible to damage. Thus, the interplay between infection and an immune response can be seen.

Once a tumor has developed, the immune response continues to play a role in tumor progression. Both the tumor cells and non-malignant cells residing in or near the tumor are involved in the process of inflammation-associated malignant progression (Figure 13.3). Chemoattractive molecules, called chemokines, are involved in the recruitment and infiltration of leukocytes, including tumor-associated macrophages (TAMs), into the tumor. The expression of growth factors, cytokines, and chemokines by both TAMs and tumor cells impacts on the cells in that location and promotes proliferation and survival. TNF-α produced by TAMs and a variety of tumor cells is a key player in the inflammatory response, as mentioned earlier in this section, and, when unregulated, can act as a tumor promoter. TNF-α can affect cell motility and tumor metastasis. The enzyme inducible nitric oxide (NO) synthase is one target stimulated by TNF-α. This enzyme is implicated in several stages of carcinogenesis, including

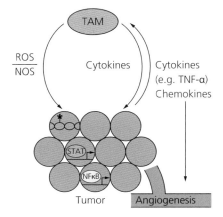

Figure 13.3 Molecular events of the inflammatory response at the site of a tumor. Tumor cells (gray) secrete chemokines that recruit leukocytes, including tumor-associated macrophages (TAMs). TAMs produce cytokines and reactive oxygen/nitrogen species (ROS/NOS, respectively) which can mutate DNA (shown by the asterisk). Inflammatory cytokines activate transcription factors STAT and NF-κB in tumor cells and lead to gene expression. Tumor cells also produce cytokines. Chemokines promote angiogenesis.

cell transformation and growth of transformed cells. The cytokine IL-6, produced by liver macrophages, has been shown to be important in hepatocarcinogenesis (Naugler *et al.*, 2007), and it is the downregulation of IL-6 by macrophages in response to estrogen that is responsible for females being less susceptible to liver cancer.

STAT and NF-κB signaling also induces EMT by inhibiting epithelial differentiation markers.

In addition to their role in recruiting leukocytes to the site of inflammation, chemokines play an important role in the angiogenic switch—pro-inflammatory chemokines promote angiogenesis. Thus, inflammation also leads to metastasis and angiogenesis.

◎ NF-κB is a key player in the inflammatory response

A key mediator of the inflammatory response is the transcription factor NF-κB. It is induced in several cell types, such as macrophages and other target cells of inflammation, and in cancer cells. The gene expression profile induced by NF-κB will depend on the tissue or cell type in which it is induced. Figure 13.4 shows some of the upstream activators of NF-κB in both macrophage and cancer cells (Karin, 2006; Perkins, 2012). NF-κB is activated by specific inflammatory agents, including cytokines (e.g. TNF-α), *H. pylori* Cag A protein, viral proteins (e.g. KSHV), carcinogens

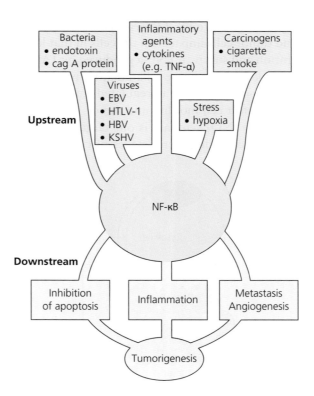

Figure 13.4 Upstream activators and downstream effects of NF-κB.

(cigarette smoke), stress (DNA damage and hypoxia), and chemotherapeutic agents.

In addition to inflammation, NF-κB has other downstream effects that contribute to tumorigenesis such as the inhibition of apoptosis and the promotion of metastasis and angiogenesis (Figure 13.4). Thus, NF-κB provides a molecular link between inflammation and cancer.

Aberrant NF-κB activation in cancer can occur by mutation of components in its regulatory pathway (see later text) or by continuous exposure to cytokine release during chronic inflammation. Oncogenic activation of the *NF-κB* gene has been identified in human tumors, including multiple myeloma, acute lymphocyte leukemia, and prostate and breast cancers.

Let us describe the molecular regulation of the NF-κB pathway (Figure 13.5). NF-κB (shown in light blue) is a dimeric transcription factor made up of hetero- or homodimers of protein members in the NF-κB family. The five NF-κB family members are placed into two groups. The first group consists of p65 (RelA), Rel B, and c-Rel. The second group consists of NF-κB 1 (p50) and NF-κB 2 (p52). The first group of proteins are synthesized as mature products, whereas the second group of proteins must be proteolytically processed to produce the mature p50 and p52 proteins. Only the first group contains transactivation domains, and therefore

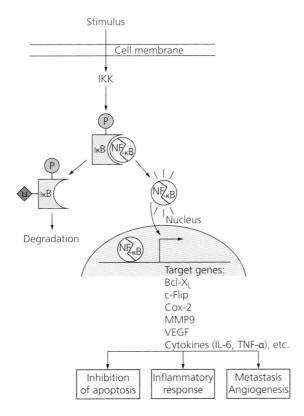

Figure 13.5 Regulation of the NF-κB pathway. See text for details.

PAUSE AND THINK

Can you remember similar types of molecular mechanisms regulating other transcription factors? Hint: one is an important transcription factor involved in angiogenesis, and another is involved in an important developmental program.

proteins of the second group cannot activate transcription on their own. The most predominant NF-κB dimer activated by the classical pathway is p65–p50. (Note: an important finding linking NF-κB to carcinogenesis was the discovery that *c-rel* is the proto-oncogene of the *v-rel* oncogene.) Normally, NF-κB is sequestered in the cytoplasm by an inhibitor of NF-κB (IκB). Upon cell activation, the kinase IκB kinase (IKK) phosphorylates IκB and targets it for degradation via a ubiquitin ligase complex. This causes the release and translocation of NF-κB to the nucleus. Here, NF-κB can transcriptionally regulate its target genes at NF-κB DNA response elements (GGGRNNYYCC). NF-κB regulates more than 200 genes to elicit its varied effects.

One of the most important cellular effects of activation of NF-κB is the inhibition of apoptosis via the induction of anti-apoptotic gene expression (e.g. induction of Bcl-X$_L$, cellular inhibitor of apoptosis (c-IAP), cFLIP). In this way, NF-κB prevents an important tumor-suppressing mechanism and promotes carcinogenesis.

 HOW DO WE KNOW THAT?

Functional knock-out mice

The role of NF-κB in tumorigenesis was investigated in a mouse model system whereby the mouse strain develops hepatitis and is prone to hepatocellular carcinoma (Pikarsky *et al.*, 2004). Transgenic animals containing a hepatocyte-inducible NF-κB inhibitor, IκB, were examined as a means of creating an equivalent of a NF-κB knockout.

Why and how was this hepatocyte-inducible regulation accomplished?

As NF-κB is involved in many crucial cellular responses, a total eradication of activity would lead to embryonic lethality. An experimental system that can be turned on and off would allow for normal development and investigation of the role of NF-κB in one tissue-specific context.

The Iκ*B* gene was linked to a tetracycline-regulated promoter, so that IκB was only expressed in the presence of a tetracycline transactivator. The gene for the transactivator was under the control of a hepatocyte-specific promoter, so that it was only expressed in liver cells. In the presence of the tetracycline derivative doxycycline, the tetracycline-controlled transactivator could not bind to its target DNA sequences (sequences of the tetracycline operon), and transcription of IκB was blocked. Treatment of mice with doxycycline suppressed the transgene (in this case, the inhibitor of NF-κB). So doxycycline-treated transgenic mice contained active NF-κB, while untreated mice contained inactive NF-κB owing to the expression of its inhibitor (see "Pause and think" on p. 315).

The results showed that, when NF-κB is inactive, only a small percentage (10%) of precancerous adenomas progressed to carcinoma, compared with controls. Blocking NF-κB activity induced hepatocyte apoptosis, as detected by antibody staining for activated caspase-3, and also led to a dramatic decrease in tumor progression, as shown by MRI and histological analysis. In addition, they showed that the inflammatory factor TNF-α produced by residing inflammatory cells controlled the activation of NF-κB in the hepatocytes. PCR analysis of cell fractions showed that the source of TNF-α was in the non-hepatocyte fraction of the liver. A block in NF-κB activation was demonstrated by using an anti-TNF-α antibody to block TNF-α function. Thus, NF-κB in the liver is controlled through a paracrine manner via TNF-α produced by inflammatory cells.

This work supported the findings from another laboratory (Greten *et al.*, 2004) that also used knock-out transgenic mice to ablate (knock out) NF-κB activity. An essential activator of NF-κB, called IKKβ, was ablated in intestinal epithelial cells in a mouse model of colon cancer. These mice showed an 80% reduction in tumor incidence relative to control animals. ➡

→ Tumor size was not affected. Analysis of the colon revealed that apoptosis was not inhibited by NF-κB in these altered mice. Thus, apoptosis acts as an important tumor-suppressing mechanism in the absence of NF-κB activation. Ablation of IKKβ in myeloid cells in this mouse model reduces the expression of cytokines that influence tumor growth and results in a decrease in tumor size. Thus, inhibition of the NF-κB pathway in these two cell types affects tumorigenesis in two different ways (Table 13.2).

Table 13.2 Summary of NF-κB pathway inhibition experiments

Deletion of IKKβ[a]	Intestinal epithelial cells	Myeloid cells
Tumor incidence	Decrease	
Tumor size	No effect	Decrease
Production of pro-inflammatory cytokines		Decrease

[a] Functionally equivalent to inactive NF-κB.

Other roles of NF-κB

NF-κB activates the *cyclin D1* gene and thus plays a role in regulating the cell cycle. NF-κB can induce the transcription of the *MDM2* gene and inhibit p53 activity, and thus can affect tumor suppression pathways. In addition, NF-κB activates the expression of pro-inflammatory genes (e.g. the *COX-2* gene and cytokine genes) and metastasis and angiogenic genes (e.g. *MMP9*, chemokine receptors, *VEGF*). COX-2 is an enzyme that is involved in the synthesis of prostaglandin PGE-2, a potent pro-inflammatory molecule (see Section 13.6, "Inhibition of inflammation" on p. 320). A NF-κB DNA-binding element has been identified in the *MMP9* gene. Thus, NF-κB helps maintain the inflammatory response and promotes metastasis. It is important to note that NF-κB also has anti-tumorigenic effects in certain tissues (e.g. skin), and so complexities of cell context exist and must be kept in mind when considering therapeutic strategies. Note STAT3 and AP-1 are two other transcription factors that are activated in tumor and inflammatory cells and orchestrate gene programs involved in inflammation and tumor promotion. STAT3 is required for the maintenance of NF-κB in tumors.

PAUSE AND THINK

See if you can sketch a diagram that shows the transgene (labeling promoter and coding sequences), the transactivator, and the consequences of no treatment versus doxycycline treatment. Check your answer with Figure 13.6.

Inflammation and tissue injury in gastric cancer recruits bone marrow stem cells

In Chapter 8, we discussed that cancer may arise from CSCs and reviewed the data in support of tissue-specific CSCs. An alternative proposal suggests that the stem cells that contribute to cancer may originate from a different tissue. The environment of tissue injury and inflammation has been linked to the recruitment of bone marrow-derived stem cells. Bone marrow-derived stem cells respond to inflammatory mediators and tissue injury and, because of their demonstrated plasticity, may serve as a backup when tissue-specific stem cells are damaged. In a mouse model of gastric cancer induced by chronic infection with *H. pylori*, bone marrow-derived stem cells were shown to be recruited to the stomach and to

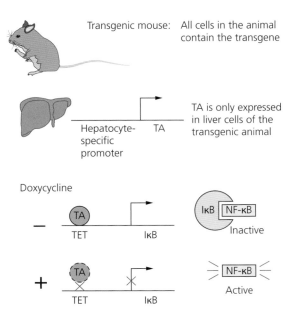

Transgenic mouse: All cells in the animal
                   contain the transgene

TA is only expressed
in liver cells of the
transgenic animal

Hepatocyte-specific promoter   TA

Doxycycline

Figure 13.6 Hepatocyte-inducible regulation of a transgene used to experimentally control NF-κB activity. TA, transactivator; IκB, inhibitor of NF-κB; TET, tetracycline-regulated promoter sequence.

contribute to gastric cancer (Houghton *et al.*, 2004). Acute inflammation and/or injury did not cause recruitment of these stem cells.

Commensal microbiota and inflammation

The human body is colonized by trillions of microorganisms, including bacteria, viruses, parasites, and fungi. Recent evidence has demonstrated that these microbiota are involved in tumor-promoting inflammation. One study showed that obesity induces changes of gut microbiota that result in an increase of a bacterial metabolite known to cause DNA damage (Yoshimoto *et al.*, 2013). The metabolite called deoxycholic acid causes the secretion of inflammatory and tumor-promoting factors and leads to liver cancer after treatment with a carcinogen in mice. Antibiotic treatment prevented the development of liver cancer after treatment with the carcinogen. In addition, inflammation is also triggered in mice when types of bacteria become unbalanced—a condition called dysbiosis. The Human Microbiome Project should help scientists determine if similar events occur in humans.

In summary, as can be seen from the preceding discussions, several mechanisms of infection and inflammation are involved in the process of cancer initiation and promotion. We have seen examples of infectious agents that carry oncogenes or produce products that inhibit tumor suppressors. Many infective agents trigger chronic inflammation, and this plays a part in their mechanism of action. However, chronic inflammation, even in the absence of an infectious agent, plays its own role in carcinogenesis. Key cells of the inflammatory response produce factors that

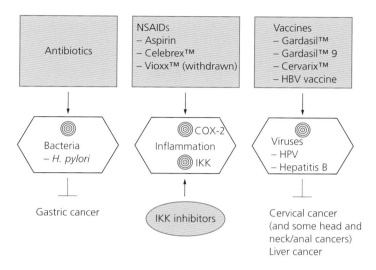

Figure 13.7 Cancer prevention strategies targeted against infection and inflammation.

influence major processes involved in carcinogenesis and, as reported more recently, trigger the migration of stem cells that may themselves contribute to cancer.

◎ Therapeutic strategies

Cancers caused by infections could be prevented if infections could be eradicated. The impact of this can be fully appreciated by looking at some numbers: there are 405,000 deaths per year worldwide from gastric carcinoma, and 35,000 women per year die from cervical cancer in the USA and Europe. Infections are implicated in the occurrence of these diseases. Would it not be extraordinary if we could prevent the suffering and death of these people?

What follows is a discussion of several strategies targeted against infection and inflammation for the prevention of cancer (Figure 13.7). Some of these are used today; others have just been approved or are on our doorstep, and one has been tried but is no longer administered because of severe side effects.

13.3 A national vaccination program against hepatitis B virus in Taiwan

In Taiwan, in 1982, 15–20% of the population were carriers of HBV. In addition, 20% of all cancer deaths were a result of hepatocellular carcinoma, and more than 80% of these cases were caused by chronic hepatitis B infection. As a result of this public health problem, a nationwide hepatitis B vaccination program was initiated in 1984. In a 20-year follow-up study, results demonstrated that vaccination has provided long-term protection and that the incidence of liver cancer was significantly lower among children 6–19 years old in vaccinated, compared with unvaccinated, birth cohorts (Chang *et al.*, 2009; Ni *et al.*, 2012). The trend suggests that Taiwan will nearly eradicate HBV infection in the future and demonstrates how intervention can reverse an endemic.

13.4 Eradication of *H. pylori* and the relationship to prevention of gastric cancer

The effects of *H. pylori* eradication on the prevention of gastric cancer has been studied in several intervention trials. The 14.7-year follow-up of the Shandong Intervention Trial was the first single trial that demonstrated *H. pylori* eradication significantly reduced the incidence of gastric cancer (Ma *et al.*, 2012). The intervention consisted of a 2-week course of treatment with an antibiotic (amoxicillin) and a drug that reduces the production of stomach acid (omeprazole) to subjects who tested positive for *H. pylori*. A statistically significant reduction in the incidence of gastric cancer by 39% was observed. It should be noted that 15 years of follow-up were needed to demonstrate this effect, and time is a constraint for prevention studies. The results of this trial are promising for the prevention of gastric cancer in high-risk geographical regions.

13.5 Cancer vaccines to prevent cervical cancer

The world's first cervical cancer vaccine called Gardasil™ (Merck) was approved in June 2006 by the US FDA. The hope is that, over time, after implementation of vaccination strategies, the number of women diagnosed with cervical cancer in the world (currently about 500,000 cases per year) will decrease dramatically.

The major capsid protein L1 from four HPV types (6, 11, 16, 18) was used to form virus-like particles (VLPs) for the production of this quadrivalent vaccine. When the capsid protein is expressed in eukaryotic cells using recombinant DNA techniques, the L1 capsid protein self-assembles into particles that mimic the virus. These particles, along with an adjuvant, are administered as the vaccine. The vaccine has been demonstrated to prevent cervical cancer caused by HPV16 and HPV18, and also precancerous lesions and warts caused by HPV6, -11, -16, and -18, in clinical trials (Villa *et al.*, 2005). As the ideal time to administer a preventative vaccine is before infection, Gardasil™ has been approved for administration in adolescent girls/young adults, aged 9–26 years, and more recently in males for the prevention of genital warts and anal cancers. The vaccine is likely to help ease the HPV-associated head and neck cancer epidemic (Marur *et al.*,

2010). Cervarix™ (GlaxoSmithKline), a HPV16 and HPV18 bivalent vaccine, also showed beneficial results against cervical cancer in clinical trials (Harper *et al.*, 2004), and was chosen for use in the UK national vaccination program and approved by the FDA (2009). Gardasil™ and Cervarix™ use different adjuvants, and it has been suggested that this is the reason behind the findings that the bivalent vaccine is relatively similar to, or better than, the quadrivalent vaccine with respect to immunogenicity.

A new nonavalent vaccine, Gardasil™ 9, has joined targets HPV6, -11, -16, -18, -31, -33, -45, -52, and -58, and has the potential to prevent 90% of cervical, vulvar, vaginal, and anal cancers. It was approved by the FDA in 2014 for use in females and males after successful clinical trials (Joura *et al.*, 2015).

PAUSE AND THINK

Why were these four HPV types chosen for the production of Gardasil™? We can see from Figure 13.1 that HPV types 16 and 18 account for approximately 70% of cases of cervical cancer. What is not shown on the graph is that HPV6 and -11 cause approximately 90% of genital warts and give rise to lesions that are clinically indistinguishable from precancerous lesions caused by HPV16 and -18. Abnormal lesions caused by HPV6 and -11 lead to false positives and trigger additional investigations. Thus, by eliminating these lesions, abnormal Pap results (see Box "A little lesson about Pap and HPV DNA screening … ") will be easier to interpret.

LIFESTYLE TIP

Although we now have a HPV vaccine to help prevent against cervical cancer, screening programs are still important for detecting cervical cancer caused by HPV strains not covered by the vaccine or for previously infected women.

A little lesson about Pap and HPV DNA screening …

The most common method of screening for cervical cancer is the Papanicolaou cytology technique or Pap smear test. Cells from the cervix are harvested for microscopic analysis and examined for changes in cell morphology that may represent precancerous cell changes called cervical intra-epithelial neoplasia (CIN) or cancer. Limitations of this technique include poor sample collection and/or slide preparation leading to abnormal cell morphology, and errors of interpretation by the laboratory scientist. However, routine Pap screening in some countries has resulted in a substantial reduction in cervical cancer mortality. Significantly, cervical cancer mortality rates in the UK have decreased by over 60% in the last 30 years (http://info.cancerresearchuk.org/cancerstats/types/cervix/mortality/). The addition of testing for HPV DNA may be a great aid to screening procedures in the near future. The value of HPV testing in both primary cervical screening and in the management of abnormal cervical cytology has been stated in recommendations from the IARC and/or in US/European guidelines (Cox and Cuzick, 2006). The Hybrid Capture 2 (hc2) high-risk HPV DNA test (Digene) is a diagnostic test that has been approved by the FDA. In this test, RNA probes are used to detect the genomic DNA of 13 high-risk HPV types. Specific RNA–DNA hybrids are formed in solution and captured by specific antibodies coated at the bottom of a microtiter plate. Additional antibodies that generate a luminescent product in the presence of hybrids are used to detect the captured hybrids, and a specialized instrument called a luminometer analyzes the signal. Note that the majority of HPV infections will be transient (clearance time 6–18 months), and conclusions from these tests need to take this fact into account.

13.6 Inhibition of inflammation

NSAIDs, such as aspirin, have been shown to decrease cancer risk and may be used for the prevention and treatment of cancer. For example, a dose- and duration-dependent protective effect of aspirin on colorectal cancer incidence was demonstrated in an over 30-year ongoing prospective observational study of 121,000 women called the Nurses' Health Study.

One mechanism of action of NSAIDs in inhibiting inflammation is by inhibiting COX activity. There are two COX isoforms: COX-1 is constitutively active, and COX-2 is inducible in response to inflammatory signals, including pro-inflammatory cytokines. These enzymes catalyze the synthesis of prostaglandins from arachidonic acid. Thus, inhibition of COX results in the decrease of prostaglandin synthesis. Prostaglandin synthesis produces mutagenic metabolites, and prostaglandins induce the production of cytokines and stimulate cell proliferation. NSAIDs may also act via inhibition of NF-κB. Several studies support the use of aspirin for chemoprevention of colorectal cancer: one study found that daily aspirin for 5 years reduced mortality due to colon cancer by 34% over 20 years' follow-up; another found that aspirin reduced the risk of colon cancer by 50% in patients with a hereditary syndrome called Lynch syndrome (Chan *et al.*, 2011). Other studies suggest that aspirin can also prevent other cancers, such as breast, lung, and prostate, and can improve survival after diagnosis of colorectal cancer (Wang and DuBois, 2010).

The use of aspirin is not without side effects; it can cause severe stomach irritation and ulcers. As COX-1 contributes to the maintenance of the inner membrane of the stomach and was found to have protective effects on the stomach lining, selective COX-2 inhibitors were developed to eradicate these side effects. However, this avenue of drug discovery is not without problems. Vioxx™ (Merck; http://www.merck.com/), a COX-2 inhibitor, had to be withdrawn from the market in 2004, owing to an increased risk of heart attacks and stroke.

Another COX-2 inhibitor, celecoxib (Celebrex™; Pfizer; http://www.pfizer.com/), has been approved for the disease called familial adenomatous polyposis (see Chapter 8). As you may remember from Chapter 8, patients with this disease carry a germline mutation in the *APC* gene and have almost a 100% risk of colon cancer. One study has demonstrated a 30% reduction in the number of polyps in these patients after treatment with celecoxib (Steinbach *et al.*, 2000). Additional trials of celecoxib for cancer prevention/treatment (Meyerhardt, 2009; Wang and DuBois, 2010) are continuing, although, under high doses, side effects similar to those induced by Vioxx™ have been reported. More recently, the risks of side effects are correlated with drug dose and also previous history of heart disease. Again, tumor profiling may be important for drug testing, as the results of a trial for non-small-cell lung cancer reported that celecoxib benefited patients with increased COX-2 expression. Risk–benefits will need to be examined for specific subpopulations of patients.

Antagonists of TNF-α have demonstrated disease stabilization and some partial responses in early clinical trials. Other cytokine antagonists, for example antagonists of IL-6, are currently being tested in clinical trials.

Inhibiting the NF-κB pathway

Many research groups are focusing on strategies that selectively inhibit IKK activity as an approach to inhibit NF-κB activation. Numerous compounds have been identified by large-scale screening and combinatorial chemistry, and include ATP-competitive and allosteric inhibitors of IKK (Lee and Hung, 2008). One of these compounds, PS-1145, was developed from a β-carboline natural product and was shown to prevent NF-κB activation and inhibit the growth of multiple myeloma cells. Three IKKβ inhibitors have been tested in clinical trials, but none have yet been approved, as toxicity is a common problem. The complexities of NF-κB signaling have made it a difficult molecule to target (DiDonato *et al.*, 2012).

> **PAUSE AND THINK**
>
> What other strategies can be used to inhibit the NF-κB pathway? Think about the molecular mechanisms involved and how NF-κB exerts its effects.
>
> You may have thought of compounds that interfere with NF-κB binding to DNA, block nuclear translocation, block degradation of IκB, or inhibit gene expression of IKK by antisense oligonucleotides or siRNA.

▓ CHAPTER HIGHLIGHTS—REFRESH YOUR MEMORY

- Infectious agents and chronic inflammation account for 15–20% of all cancers.

- Specific DNA and RNA viruses, bacteria, and liver flukes have been classified as carcinogens.

- The immune system plays a dual role in cancer: it can both suppress and promote tumor growth.

- One hundred percent of cervical cancers are associated with HPV. This virus is also a causative agent for several other cancers.

- Common mechanisms of infectious agents and chronic inflammation in carcinogenesis include:

  – Induction of growth factors/signaling programs (autocrine or paracrine)

  – Inactivation of tumor suppressor genes

  – Activation of nuclear transcription factor NF-κB.

- The Cag A protein is implicated in the mechanism of the induction of gastric cancer by the bacterium *H. pylori*.

- Cag A is a phosphoprotein that interferes with kinase signaling and regulation in the cell.

- Chronic inflammation, even in the absence of infection, is characteristic of most tumors and is considered a hallmark of cancer.

- The site of chronic inflammation is characterized by cytokines, chemokines, and reactive oxygen/nitrogen species that can act as a carcinogen.

- The transcription factor NF-κB is an important mediator between inflammation and cancer.

- Bone marrow stem cells migrate to sites of inflammation and injury and may contribute to gastric carcinogenesis.

- Gardasil™ (Merck), the first preventative cervical cancer vaccine, was approved in 2006.

- Current HPV vaccines are not able to protect against all HPV infections. Therefore, screening procedures must be maintained.

- Vaccination programs have been, and will continue to be, important preventative measures for some cancers.

▤ ACTIVITY

1. It is well established that tobacco smoke contains tumor-initiating carcinogens. New evidence suggests that tobacco smoke is also a tumor promoter and that this role depends on inflammation. Look at the 2010 article by Takahashi *et al*. Describe the model system, experimental procedure, and the methods of analysis that provided *in vivo* evidence that tobacco smoke promotes lung tumorigenesis by inducing inflammation. What possible role may anti-inflammatory therapy play in lung cancer?

▤ FURTHER READING

Aggarwal, B.B., Shishodia, S., Sandur, S.K., Pandey, M.K., and Sethi, G. (2006) Inflammation and cancer: how hot is the link? *Biochem. Pharmacol.* **72**: 1605–1621.

Colotta, F., Allavena, P., Ica, A, Garlanda, C., and Mantovani, A. (2009) Cancer-related inflammation, the seventh hallmark of cancer: links to genetic instability. *Carcinogenesis* **30**: 1073–1081.

de Martel, C., Ferlay, J., Franceschi, S., Vignat, J., Bray, F., Forman, D., *et al.* (2012) Global burden of cancers attributable to infections in 2008: a review and synthetic analysis. *Lancet Oncol.* **13**: 607–615.

Elinav, E., Nowarski, R., Thaiss, C.A., Hu, B., Jin, C., and Flavell, R.A. (2013) Inflammation-induced cancer: crosstalk between tumours, immune cells and microorganisms. *Nat. Rev. Cancer* **13**: 759–771.

Escarcega, R.O., Fuentes-Alexandro, S., Garcia-Carrasco, M., Gatica, A., and Zamora, A. (2007) The transcription factor nuclear factor-kappa B and cancer. *Clin. Oncol.* **19**: 154–161.

Grivennikov, S.I., Greten, F.R., and Karin, M. (2010) Immunity, inflammation, and cancer. *Cell* **140**: 883–899.

Kim, K.S., Park, S.A., Ko, K.-N., Yi, S., and Cho, Y.J. (2014) Current status of human papillomavirus vaccines. *Clin. Exp. Vaccine Res.* **3**: 168–175.

Li, Q., Withoff, S., and Verma, I.M. (2005) Inflammation-associated cancer: NF-κB is the lynchpin. *Trends Immunol.* **26**: 318–325.

Lowy, D.R. and Schiller, J.T. (2006) Prophylactic human papillomavirus vaccines. *J. Clin. Invest.* **116**: 1167–1173.

Mantovani, A., Allavena, P., Sica, A., and Balkwill, F. (2008) Cancer-related inflammation. *Nature* **454**: 436–444.

Mesri, E.A., Cesarman, E., and Boshoff, C. (2010) Kaposi's sarcoma and its associated herpesvirus. *Nat. Rev. Cancer* **10**: 707–719.

Moody, C.A. and Laimins, L.A. (2010) Human papillomavirus oncoproteins: pathways to transformation. *Nat. Rev. Cancer* **10**: 550–560.

Polk, D.B. and Peek, R.M.Jr (2010) *Helicobacter pylori*: gastric cancer and beyond. *Nat. Rev. Cancer* **10**: 403–414.

Rayburn, E.R., Ezell, S.J., and Zhang, R. (2009) Anti-inflammatory agents for cancer therapy. *Mol. Cell Pharmacol.* **1**: 29–43.

Tan, T.-T. and Coussens, L.M. (2007) Humoral immunity, inflammation and cancer. *Curr. Opin. Immunol.* **19**: 209–216.

Vogelmann, R. and Amieva, M.R. (2007) The role of bacterial pathogens in cancer. *Curr. Opin. Microbiol.* **10**: 76–81.

Wroblewski, L.E. and Peek Jr, R.M. (2013) *Helicobacter pylori* in gastric carcinogenesis: mechanisms. *Gastroenterol. Clin. N. Am.* **42**: 285–298.

zur Hausen, H. (2002) Papillomaviruses and cancer: from basic studies to clinical application. *Nat. Rev. Cancer* **2**: 342–350.

▦ WEB SITES

Gardasil vaccines and HPV www.gardasil.com and http://www.gardasil9.com/about-hpv/

UK Cervical Cancer Statistics http://info.cancerresearchuk.org/cancerstats/types/cervix/mortality/

▦ SELECTED SPECIAL TOPICS

Boxus, M. and Willems, L. (2009) Mechanisms of HTLV-1 persistence and transformation. *Br. J. Cancer* **101**: 1497–1501.

Chan, A.T, Arber, N., Burn, J., Chia, W.K., Elwood, P., Hull, M.A., *et al.* (2011) Aspirin in the chemoprevention of colorectal neoplasia: an overview. *Cancer Prev. Res.* **5**: 164–178.

Chang, M.H., You, S.-L., Chen, C.-J., Liu, C.-J., Lee, C.-M., Lin, S.-M., *et al.* (2009) Decreased incidence of hepatocellular carcinoma in hepatitis B vaccines: a 20-year follow-up study. *J. Natl. Cancer Inst.* **101**: 1348–1355.

Cox, T. and Cuzick, J. (2006) HPV DNA testing in cervical cancer screening: from evidence to policies. *Gynecol. Oncol.* **103**: 8–11.

Di Donato, J.A., Mercurio, F., and Karin, M. (2012) NF-κB and the link between inflammation and cancer. *Immunol. Rev.* **246**: 379–400.

Greten, F.R., Eckmann, L., Greten, T.F., Park, J.M., Egan, L.J., Kagnoff, M.F., *et al.* (2004) IKKb links inflammation and tumorigenesis in a mouse model of colitis-associated cancer. *Cell* **118**: 285–296.

Guerrieri, F., Belloni, L., Pediconi, N., and Levrero, M. (2013) Molecular mechanisms of HBV-associated hepatocarcinogenesis. *Semin. Liver Dis.* **33**: 147–156.

Harper, D., Franco, E., Wheeler, C., Ferris, D., Jenkins, D., Schuind, A., *et al.* (2004) Efficacy of a bivalent L1 virus-like particle vaccine in prevention of infection with human papillomavirus types 16 and 18 in young women, a randomized controlled trial. *Lancet* **364**: 1757–1765.

Houghton, J., Stoicov, C., Nomura, S., Rogers, A.B., Carlson, J., Li, H., *et al.* (2004) Gastric cancer originating from bone marrow-derived cells. *Science* **306**: 1568–1571.

Hu, Z., Zhu, D., Wang, W., Li, W., Jia, W., Zeng, X., *et al.* (2015) Genome-wide profiling of HPV integration in cervical cancer identifies clustered genomic hot spots and a potential microhomology-mediated integration mechanism. *Nat. Genet.* **47**: 158–163.

International Agency for Research on Cancer (2011). *Monographs on the evaluation of carcinogenic risks to humans, volume 100. A review of carcinogen—Part B: biological agents*. International Agency for Research on Cancer, Lyon.

Joura, E.A., Giuliano, A.R., Iversen, O.E., Bouchard, C., Mao, C., Mehlsen, J., *et al.* (2015) A 9-valent HPV vaccine against infection and intraepithelial neoplasia in women. *N. Engl. J. Med.* **372**: 711–723.

Karin, M. (2006) Nuclear factor-κB in cancer development and progression. *Nature* **441**: 431–436.

Lee, D.-F. and Hung, M.-C. (2008) Advances in targeting IKK and IKK-related kinases for cancer therapy. *Clin. Cancer Res.* **14**: 5656–5662.

Ma, J.-L., Zhang, L., Brown, L.M., Li, J.-Y., Shen, L., Pan, K.-F., *et al.* (2012) Fifteen-year effects of *Helicobacter pylori*, garlic, and vitamin treatments on gastric cancer incidence and mortality. *J. Natl. Cancer Inst.* **104**: 488–492.

Marur, S., D'souza, G., Westra, W.H., and Forastiere, A.A. (2010) HPV-associated head and neck cancer: a virus-related cancer epidemic. *Lancet Oncol.* **11**: 781–789.

Meyerhardt, J.A. (2009) COX-2 inhibitors and colorectal cancer: the end or just a new beginning. *Update Cancer Ther.* **3**: 154–156.

Munoz, N., Bosch, F.X., de Sanjose, S., Herrero, R., Castellsague, X., Shah, K.V., *et al.* (2003) Epidemiologic classification of human papillomavirus types associated with cervical cancer. *N. Engl. J. Med.* **348**: 518–527.

Naugler, W.E., Sakurai, T., Kim, S., Maeda, S., Kim, K.H., Elsharkawy, A.M., *et al.* (2007) Gender disparity in liver cancer due to sex differences in MyD88-dependent IL-6 production. *Science* **317**: 121–124.

Ni, Y.H., Chang, M.H., Wu, J.F., Hsu, H.Y., Chen, H.L., and Chen, D.S. (2012) Minimization of hepatitis B infection by a 25-year universal vaccination program. *J. Hepatol.* **57**: 730–735.

Perkins, N.D. (2012) The diverse and complex roles of NF-κB subunits in cancer. *Nat. Rev. Cancer* **12**: 121–132.

Pikarsky, E., Porat, R.M., Stein, I., Abramovitch, R., Amit, S., Kasem, S., *et al.* (2004) NF-κB functions as a tumour promoter in inflammation-associated cancer. *Nature* **431**: 461–466.

Roithmaier, S., Haydon, A.M., Loi, S., Esmore, D., Griffiths, A., Bergin, P., *et al.* (2007) Incidence of malignancies in heart and/or lung transplant recipients: a single institution experience. *J. Heart Lung Transplant.* **26**: 845–849.

Steinbach, G., Lynch, P.M., Phillips, R.K.S., Wallace, M.H., Hawk, E., Gordon, G.B., *et al.* (2000) The effect of celecoxib, a cyclooxygenase-2 inhibitor, in familial adenomatous polyposis. *N. Engl. J. Med.* **342**: 1946–1952.

Takahashi, H., Ogata, H., Nishigaki, R., Broide, D.H., and Karin, M. (2010) Tobacco smoke promotes lung tumorigenesis by triggering IKKβ- and JNK1-dependent inflammation. *Cancer Cell* **17**: 89–97.

Villa, L.L., Costa, R.L., Petta, C.A., Andrade, R.P., Ault, K.A., Giuliano, A.R., *et al.* (2005) Prophylactic quadrivalent human papillomavirus (types 6, 11, 16, and 18) L1 virus–like particle vaccine in young women: a randomized double-blind placebo-controlled multicentre phase II efficacy trial. *Lancet Oncol.* **6**: 271–278.

Wang, D. and DuBois, R.N. (2010) The role of COX-2 in intestinal inflammation and colorectal cancer. *Oncogene* **29**: 781–788.

Yoshimoto, S., Loo, T.M., Atarashi, K., Kanda, H., Sato, S., Oyadomari, S., *et al.* (2013) Obesity-induced gut microbial metabolite promotes liver cancer through senescence secretome. *Nature* **499**: 97–101.

Chapter 14

Technology and drug and diagnostics development

Introduction

The goal of cancer research is to develop new effective and non-toxic cancer therapies and ways for early detection. In this chapter, advances in research methodology and technology and the process of drug development are reviewed. Advances in technologies drive our ability to gain knowledge and make new diagnostics and therapeutics. Microarrays (followed by next-generation sequencing described in Chapter 4) have made important contributions to our understanding of carcinogenesis and have led to new clinical applications. The use of the CRISPR-Cas9 system for understanding the function of cancer genes is one of the newest additions to our tools for investigating cancer genes and is described in detail. Other technologies, including nanotechnology, are helping to improve the way we detect and target cancer, and have allowed the development of advanced imaging systems.

Developing a new drug uses a logic that is simple and progresses through the following steps: define what is causing the disease; create an experimental model of the disease for drug testing; determine how to stop it and identify or create an inhibitor that can be used as a drug; test and test again. The reality is that each step takes years to achieve and huge sums of money. The development of the drug imatinib (Gleevec™), one of the most successful cancer therapies of recent years, is used to illustrate the process of drug development. The understanding of both the mechanism of action of this drug and the development of drug resistance is being used to produce "second-generation" therapeutics. The chapter ends with a statement about our progress.

14.1 Microarrays and gene expression profiling

Microarrays and their associated technologies enable the expression of tens of thousands of genes to be analyzed at the same time. This technique is a tremendous asset, providing data sets that were not previously available. Previous methods could only examine individual genes or, at most, small sets of genes at a time. Microarrays can identify a group of transcripts associated with a specific outcome or phenotype, called a gene signature or profile. Next-generation sequencing is now superseding microarrays in many applications, but there can be no doubt that microarrays were responsible for a leap in knowledge within the cancer field.

Experimental procedure

Microarrays are grids, usually made on glass slides or silicon chips. They hold DNA representing thousands of genes that act as probes (sequences that are complementary and can hybridize to specific RNAs) for RNA. Analysis of a sample using a microarray will identify RNAs that are present by hybridization to the probes on the grid and indicate genes that are being expressed (transcribed). A typical protocol will be described. Thousands of gene-specific hybridization probes are applied to a glass slide or silicon chip (Figure 14.1a). The DNA probes are usually bound to defined locations on the grid by robotic or laser technology. RNA is isolated from a biological sample, such as a tumor, and copied to incorporate fluorescent nucleotides or a fluorescent tag (Figure 14.1b). The chip is then incubated with labeled RNA or complementary DNA (cDNA) from the tumor sample (Figure 14.1c). Unhybridized RNA is washed off, and the microarray is then scanned under a laser and analyzed by computer (Figure 14.1d). A sample microarray image is shown in Figure 14.1e, in this case probed simultaneously with two samples labeled in different colors (red and green). By analyzing the fluorescent intensities of the RNA or cDNA hybridized to the probes using computerized scanners, gene expression can be quantified. There are two common types of microarray: cDNA microarrays and oligonucleotide microarrays. The difference between the two is due to the nature of the probes. In cDNA microarrays, each probe has its own ideal hybridization temperature (based on factors such as GC content), and thus intensities of a test sample must always be compared with a control sample processed at the same time. In oligonucleotide microarrays, the synthetic probes are designed such that all the probes have identical hybridization temperatures, allowing absolute values of expression to be measured within one sample. Results from microarrays can be visualized in different formats. One type of data display format is a heat map that uses color to represent levels of gene expression (Figure 14.1f). Genes may be arranged in rows, and time points may be arranged in columns. A red box may be used to indicate an increase in

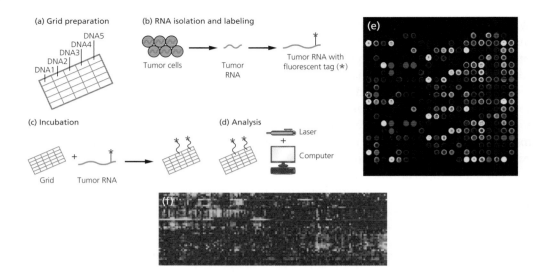

Figure 14.1 (a–d) The basic protocol for microarrays. (e) Sample microarray. (f) Representative heat map.

expression relative to a control; a green box may be used to indicate a decrease in expression relative to a control; and a black box to represent no change. Similarities in the patterns of gene expression can then be depicted in clusters in a cluster analysis diagram. Limitations of this method may be defined by the specificity of the probes. For example, variants of alternative splicing may or may not be detected, depending on the sequences included in the probe. Modifications of the technique, including protein microarrays and antibody microarrays, demonstrate that the microarray design has far-reaching potential.

Application of microarrays

The applications of microarrays in cancer biology are far-reaching. As cancer is a disease of the genome and epigenome at the cellular level, genes that are involved in carcinogenesis can be identified by changes in gene expression taking place as cancers develop.

Different gene expression signatures can be used to identify molecular subtypes of cancers and lead to the development of more precise diagnosis and treatment. Let us look at one example of how a microarray was applied to characterize subclasses of a specific cancer. The application of a lymphochip, a microarray that screens for genes important in cancer, immunology, and lymphoid cells, helped to define two molecularly distinct forms of a particular lymphoma (Alizadeh *et al.*, 2000). Results from the lymphochip showed two distinct patterns of gene expression that were differentiation stage-specific and also corresponded to different clinical outcomes (76% of one subgroup was alive after 5 years, compared with only 16% of the other). The classification of one cancer was refined to two distinct cancers. These findings have laid down an important foundation for further investigations. More recently, whole-genome expression profiling has been used for classification and prognosis of lymphoma (Barrans *et al.*, 2012), and numerous trials are testing different protocols for identifying patients based on their molecular subtype for personalized treatment. DNA arrays have also been used to refine molecular classifications of other cancers, including breast cancer.

Diagnostic, prognostic, and predictive gene signatures have been elucidated, and some are used in clinical applications. The molecular signatures may allow for the prediction of disease outcome and prescription of the most efficient treatment available for a particular tumor type. Several groups have used microarrays to identify a number of marker genes whose expression can predict metastasis and/or prognosis (Shipp *et al.*, 2002; Van't Veer *et al.*, 2002). Two gene expression profiling tests for breast cancer were launched in 2004. Oncotype DX™ (Genomic Health, Redwood City, CA) and Mammaprint™ (Agendia, Amsterdam) are tests designed to predict breast cancer recurrence. Oncotype DX™ uses 16 genes selected from an array of 250, and Mammaprint™ uses 70 genes from an array of 25,000, to indicate a good or bad prognosis. The main purpose of analyzing these selected genes is to identify tumors that are unlikely to metastasize in order to spare patients the trauma of chemotherapy. Only a small proportion of patients whose breast cancer has not spread to the lymph nodes develop metastases (20%) and really require chemotherapy. Most node-negative breast cancer patients are cured by surgery and radiotherapy alone. Scores of less than 17 (low recurrence score) or scores of 31 and higher (high recurrence score) for the Oncotype DX™ test are being used to predict the benefit of chemotherapy for breast cancer patients. (Note that the significance of intermediate values are being investigated; TAILORx Breast Cancer Trial.) Such gene profiling tests may spare patients from receiving unnecessary chemotherapy that is routinely given to all in order to prevent metastasis in the unidentified few for whom it is essential. This is a true mark of progress. A futuristic vision is to be able to develop tailor-made therapies for individual patients based on the genetic profile of their primary tumor.

14.2 Analysis of biomarkers for diagnostics and prognostics

A biomarker is a biochemical or genetic feature that can be used to estimate risk, detect disease, or measure its progression or the effect of treatment. Biomarkers may include the measurement of specific molecules (e.g. human epididymal protein 4 (HE4), an ovarian cancer serum biomarker approved by the FDA, 2009; see Leung *et al.*, 2013), genetic alterations, gene expression profiles, cell-based markers (circulating cells), and single nucleotide polymorphisms (SNPs). The measurement of a biomarker indicates whether it falls within either a "normal" or an "abnormal" range, and this helps to suggest a diagnosis or prognosis.

PAUSE AND THINK

Several biomarkers have been discussed previously in this text. Try to recall biomarkers that: predict the risk of breast and ovarian cancer, indicate response to therapy, or determine prognosis.

Biomarkers can be identified through microarray gene expression profiling (described in Section 14.1) or through mass spectrometry which allows for the simultaneous examination of thousands of proteins in a biological sample. For mass spectrometry analysis, proteins are treated with enzymes to create peptides. These peptides can be fractionated prior to separation and ionization in the spectrometer. Afterwards, the ionized peptides are fragmented, and the mass-to-charge ratios of the ionized products provide amino acid sequence information through the spectrum produced and bioinformatics. The future promises new and improved biomarkers.

Let us examine some progress that has been made in biomarkers for prostate cancer, the second leading cause of cancer-related deaths in men. Elevated levels of prostate-specific antigen (PSA) detected in the blood has been the conventional prostate tumor marker used for screening in the past. However, this became controversial, as improvements are needed to prevent subsequent negative biopsy rates (70–80%). Negative biopsy rates are due to the fact that PSA is not specific for prostate cancer. Elevated levels are also detected in benign prostate conditions such as prostatitis. Diagnostics based on genomics are leading to improvements in prostate cancer detection. The prostate cancer antigen 3 (*PCA3*) gene (also known as *DD3PCA3*) has been identified as the most prostate cancer-specific gene described thus far. It codes for an nc-RNA that is only expressed in prostate tissue and is strongly overexpressed (60–100 times) in more than 95% of prostate tumors. Urinalysis that detects PCA3 RNA, using a nucleotide amplification method called the PROGENSA PCA3 assay, has been designed for the diagnosis of prostate cancer. This assay carried out on urine samples was compared to the standard PSA tests. The results demonstrated that the PCA3 test had better diagnostic capabilities and prognostic value. The development of the PCA3 RNA biomarker holds great promise as a non-invasive diagnostic tool that may reduce the number of unnecessary biopsies (Merola *et al.*, 2015).

The first non-invasive stool-based DNA screening test, Cologuard, was approved by the FDA in 2014. This test detects the presence of DNA mutations (KRAS) and methylation biomarkers (NDRG4 and BMP3) associated with colon carcinogenesis. The results are normalized to a reference gene. The test also detects haemoglobin for the presence of red blood cells. The results are combined to produce a score that indicates a positive or negative result.

Three future potential diagnostic and prognostic indicators that may be detected in a simple blood test include CTCs or tumor DNA, miR-NAs, and exosomes (discussed in Chapter 9). Early studies suggest that overall numbers of CTCs, tumor cells that have intravasated into the bloodstream, provide a prognostic indicator of disease outcome for some carcinomas. The use of miRNAs for identifying tumor origins was discussed previously in Chapter 3. Levels of several miRNAs have been correlated to metastatic outcome. We may see new instruments to detect CTCs and miRNA microarrays in the clinic in the future. Another futuristic idea is that it may become possible to implant a gene chip under the skin to monitor changes in such biomarkers, thus speeding up diagnosis and facilitating early treatment.

14.3 Studying gene function by CRISPR-Cas9

The study of gene function has been instrumental in understanding the role of genomic modifications in cancer cells and guiding the development of new targeted drugs. The use of homologous recombination to alter genes in embryonic stem cells has been the main tool for genetically engineering mouse models of cancer. However, both low efficiency and the investment of long periods of time for handling embryonic stem cells and mouse breeding slowed progress. Expression of cDNA sequences in cells is another way to examine gene function, but levels are difficult to control and are often higher than physiological levels. RNA interference can be used to knock down expression, but the control of the degree of inhibition is not precise. More recently, CRISPR-Cas (clustered regularly interspaced short palindromic repeats-CRISPR-associated system) has been developed to modify genomes for functional studies and is becoming a great boost for the cancer field.

The CRISPR-Cas9 system is derived from the prokaryotic adaptive immune system. Below is a description of how it works experimentally (Figure 14.2). There are two components of the system: a DNA

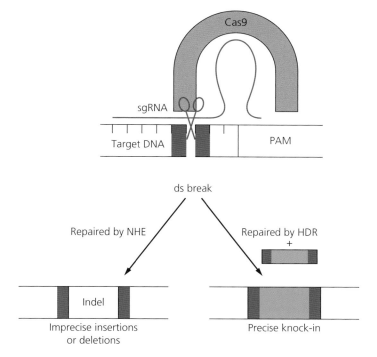

Figure 14.2 The CRISPR-Cas9 system.

endonuclease, Cas9, AND a chimeric single guide RNA (sgRNA) that guides Cas9. The sgRNA contains two parts: a CRISPR RNA part that binds to a 20-nucleotide genomic DNA target site by Watson–Crick base pairing and a trans-activating CRISPR RNA component that binds to the Cas9 endonuclease. The target sequence must be near either an NGG or NAG trinucleotide; these are called protospacer adjacent motifs (PAMs). These molecules generate double-strand breaks 3 bp 5′ of the PAM in the target site that is repaired by either of two endogenous DNA repair pathways: non-homologous end-joining (NHE) or homology-directed repair (HDR). Repair leads to either small insertions or deletions (indels), creating loss-of-function mutations or precise modifications for creating gain-of-function mutations, respectively. (Note: HDR requires exogenous donor DNA templates.) Both Cas9 and sgRNA components can be either transiently or stably transfected into cells via plasmid DNA or viral vectors.

CRISPR can be used to engineer embryonic stem cells rapidly for "traditional" mouse models but has the advantage of being able to target multiple loci at a time, and has been used to engineer tumor-associated chromosomal translocations (Torres *et al.*, 2014). It can also be used to engineer cells *ex vivo* to study leukemias. It has even been used to make somatic mutations *in vivo* in mice to induce tumors of the liver (Xue *et al.*, 2014). The potential for applying this to patients is also exciting and includes *ex vivo* engineering of immune cells for immunotherapy.

14.4 Imaging

Imaging is important for diagnosis, preoperative planning, and follow-up of treatment. In the past, imaging has been restricted mainly to anatomical features. A brief description of each of the main imaging modalities is shown in Table 14.1, and several are described in more detail below.

CT uses X-rays (a type of electromagnetic radiation discussed in Chapter 2). X-rays are absorbed by tissues differently. Differential penetration of the X-rays can be used to generate an image. In modern scanners, the X-ray source and a detector rotate around the patient, as the patient is moved on a table. This generates a volume of data that can be used to construct complex three-dimensional images. It is important to note that there is a risk of radiation-induced cancers with CT scans. One study that used risk models suggested that 29,000 future cancers were related to CT scans given in the USA in 2007 (Berrington de Gonzalez *et al.*, 2009).

MRI is based on the magnetic properties of hydrogen nuclei in tissues. Different tissues have different magnetic properties. In the presence of a strong external magnet, the magnetic fields of the hydrogen atoms align themselves. A pulse of radiofrequency is applied and absorbed by the hydrogen atoms, and this reorientates the magnetic fields. When the pulse

Table 14.1 The main imaging techniques used in cancer

Imaging modality	Mechanism	Comments
Computed tomography (CT)	X-rays	Increases risk of cancer
Positron emission tomography (PET)	Positrons	Requires a glucose analog radiotracer (Figure 11.5)
Optical imaging	Fluorescent/bioluminescent probes	
Magnetic resonance imaging (MRI)	Radiofrequency pulse is applied to tissue in the presence of a static magnetic field	
Ultrasound imaging (US)	High-frequency sound waves	

is stopped, energy is released, as the magnetic field returns to its aligned orientation. This energy is detected from different tissues and used to produce an image. There are no known risks in the current use of MRI (except for patients whose bodies contain metal, e.g. pacemakers).

Ultrasound imaging uses ultrasonic sound waves to analyze anatomical structures. Sound waves are reflected back as an echo upon reaching a structure. The echo is detected and analyzed to create an image. This method is particularly useful for superficial tissues, such as the breast and thyroid, because of the inverse relationship of frequency and depth of penetration. No risks have been reported at the standards of power currently in use.

The future of imaging will reside in molecular and functional imaging (MFI), using several techniques to investigate molecular pathways and tissue function during routine use in the clinic (Glunde *et al.*, 2007). Molecular features that may be examined by MFI include overexpression of receptors, gene expression, or cellular location. Functional aspects may include aspects of angiogenesis (e.g. vascular volume, vascular permeability, hypoxia) and metabolism (e.g. glycolytic activity).

MFI promises to have a major impact on early diagnosis, as well as on the monitoring of disease progression upon treatment. Note, in general, that the detection of early-stage (stage 1) cancers is associated with a 5-year survival rate of more than 90%. Refined imaging may also decrease the number of biopsies taken in the clinic. MFI is another approach that will help match tumor to therapy. Such imaging will also serve as a guide to the administration of future therapeutics such as nano- and microdevices. One of the main things preventing new imaging techniques from reaching the clinic is economics.

14.5 Cancer nanotechnology

A multidisciplinary field that promises to make a huge impact on cancer in the future is cancer nanotechnology. Nanotechnology is the study of devices (or their essential components) that are made by humans and have at least one dimension in the 1–1000 nm range. For scale, the size range is similar to the size of a few atoms to the size of subcellular structures. Nanotechnology has many potential applications in the field of cancer, ranging from diagnostics to imaging agents to drug targeting for therapeutics (for review, see Sanna, Pala, and Sechi, 2014 and Xu *et al.*, 2015). Many different nanomaterials are being explored as drug delivery vehicles and other clinical applications. These include fullerenes which contain a carbon cage (the Nobel Prize was awarded to Robert Curl, Harry Kroto, and Richard Smalley for their discovery of fullerenes in 1996) and gold nanoparticles. Liposomes and polymer conjugates are the two most common types of nanoparticles used for drug delivery in the clinic. Doxil™, a liposome platform that encapsulates doxorubicin, was the first nanomedicine to receive FDA approval in 1996. Several other non-targeted nanomedicines have been approved. Interestingly, nanoparticles can localize to tumors, because those added to the bloodstream can escape through the leaky and abnormal blood vessels of a tumor. This is called the enhanced permeation and retention effect. Other strategies may include stimuli-responsive nanomedicines that release their contents upon exposure to heat or light.

The problem of targeting a cancer drug specifically to a tumor must be better addressed in the future. We are all aware of the harsh side effects observed with conventional therapies that are a result of the exposure of healthy tissue to these agents. Nanostructures that can be filled with anticancer drugs and which also contain targeting moieties on their surface are called nanovectors. Nanovectors hold promise in accomplishing efficient tumor-specific drug delivery, as several are in clinical trials. BIND-014, the first to enter clinical trials, consists of a polymeric core containing docetaxel (a common chemotherapeutic agent) that is coated with molecules that bind to prostate-specific membrane antigen (PSMA). PSMA is overexpressed on the cell surface of prostate cancer cells and the new vasculature of many other tumors.

In addition, nanovectors will be used as imaging contrast agents that greatly amplify signals detected by various imaging techniques. It is easy to envision that nanotechnology will refine microarrays to greater-capacity "nanoarrays."

Lastly, and perhaps most uniquely, nanotechnology could lead to biomolecular sensors that are able to detect many biomarkers simultaneously and will be used for refined diagnosis, prognosis, and treatment monitoring. Two specific designs, the nanocantilever and nanowires, currently show promise. Both can be coated with molecules that bind to biomarkers. Nanocantilevers are deflected upon binding to a biomarker (in a manner similar to piano keys when they are tapped). Lasers are used to detect the deflections. Nanowires undergo a change in conductance upon binding, and this change is detected electronically. Both may change the way and speed at which cancer is monitored.

In summary, nanotechnology may enable specific cancer drug targeting, leading to better therapeutic results and fewer toxic side effects. It promises to enhance imaging and biomarker detection for improved diagnosis. Used as biomolecular sensors, this technology may replace the need for biopsies.

14.6 Strategies of drug development

Drug development follows a series of stages (Figure 14.3), and people with different expertise (biochemists, cell biologists, chemists, clinicians) may carry out the different stages at different facilities. In the preceding chapters, we have seen many examples of important molecular targets and strategies to manipulate them (e.g. kinase inhibitors). Once a potential drug has been identified and prepared as a final product for *in vivo* delivery (drug formulation), subsequent studies can be divided into pre-clinical and clinical studies. **Pre-clinical studies** test a drug in animal models and gather data on safety and efficacy for proof of concept. These studies are required before administration of the drug to humans in **clinical trials** (discussed in Section 1.5, "Clinical trials" on p. 15). The concept of validating a drug target is often used in drug discovery. True cancer drug target validation occurs when a therapeutic agent is shown to act via the molecular target against which it was designed and proved to be clinically effective. However, a more current use of the term "target validation"

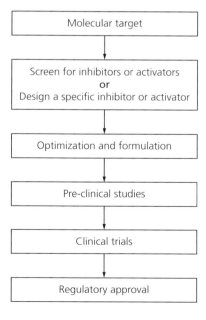

Figure 14.3 Stages of drug development.

refers only to the experimental evaluation of the role of a given gene or protein in cancer and its potential as a therapeutic target; that is, target validation may occur before clinical testing.

In order to develop a new targeted therapy, a three-step approach has been proposed for development of the drug before it goes to the clinic: identify the molecular targets/pathways that drive tumor growth; create a genetically equivalent, high-incidence animal model where tumors of interest develop in their correct anatomical locations and at a developmentally relevant time; and screen for, or design, inhibitors to block the molecular pathway and test their effects in the animal models (Romer and Curran, 2005).

Molecular targets and target validation

There are several different types of molecular target that may be identified and studied for drug development. The most popular involve genetic lesions that have a causal role in cancer. The products of these lesions are oncogenic proteins or mutated tumor suppressor proteins. Targets include the aberrant proteins themselves or components of the pathways they affect. Another type of molecular target may involve tissue-specific characteristics or differentiation pathways. For example, estrogen acts as a mitogen for breast tissue, and inhibitors of estrogen action (e.g. tamoxifen) are effective in the treatment of breast cancer. Similarly, knowledge of the differentiation pathways of the hematopoietic lineage has been applied for the treatment of APL by differentiation therapy (see Section 8.6 "Leukemia and differentiation therapies" on p. 200.). Another type of molecular target affects host processes, rather than tumor biology. For example, molecular regulators of angiogenesis are good therapeutic targets (e.g. VEGF, VEGFR; see Section 10.4, "Anti-angiogenic therapy" on p. 236).

Validating a molecular target may involve several strategies that together provide data for evaluation. Genetic validation of a specific genetic lesion is obtained by investigating the patterns of somatic mutation in a particular tumor (see "Pause and think"). EGFR has proved to be a validated drug target, and its downstream effectors Raf and MEK are also validated targets.

Cell-based systems are also valuable methods used for target validation. Transformation assays that rely on the introduction of putative oncogenes into normal cells (see Chapter 4) are an example and have been instrumental in the field. RNA interference techniques in cells are also important approaches to test potential targets. However, perhaps the most valuable strategy is the use of transgenic animals whereby a target may be analyzed in the context of a tumor *in situ*.

Let us look at an example. Raf, a component of the EGF signal transduction pathway (see Chapter 4), is an important regulator of cell proliferation. Oncogenic mutations resulting in constitutive activation of Raf

PAUSE AND THINK

What pattern of somatic mutation would you expect to observe? One feature to expect is that the genetic lesion is tumor-specific and occurs early in tumor development. Absence in healthy tissue would also be expected.

have been identified in cancers, especially melanomas and thyroid cancers. Evidence was needed to support the hypothesis that Raf is a potential therapeutic target. The results from an experiment utilizing Raf antisense oligonucleotides demonstrated growth inhibition in human tumor xenografts in mice. This provided proof of concept that Raf is a valid target for future drug development. (See "Activity" at the end of the chapter on p. 349 for a discussion of an approved drug that targets Raf.)

Experimental models

There is difficulty in finding a cancer model that can reliably predict the effect of a new drug in human patients. A tumor is similar to an organ, with tumor cells interacting with host cells, the immune system, blood vessels, and the ECM. The obvious difficulty in testing new drugs in cultured cells is that the system is far from replicating a true tumor environment. The next step up from cells in culture is the use of organ cultures and organotypic cultures, as these systems possess a three-dimensional aspect. Organ cultures are made of tissue slices. Organotypic cultures are made of cells grown in a specific matrix to mimic the tissue of interest. The use of induced pluripotent stem cells promises new approaches for modeling cancers. The generation of human gastric organoids *in vitro* through the differentiation of human pluripotent stems cells to study gastric cancer has been reported (McCracken *et al.*, 2014) and may be useful for testing drugs developed to treat *H. pylori* infection and stomach cancer. The most common model systems used in drug discovery are *in vivo* mouse models. There are several approaches for using a mouse model system. The older approach is to use high doses of a single carcinogen, often with little relationship to the etiology and/or molecular defect of the tumor of interest. The most widely used approach is the creation of human tumor xenografts. Xenografts are generated by injecting human cancer cells under the skin of immunodeficient (nude) mice. The use of nude mice is necessary to avoid rejection of human cells by the immune system of the mouse. Disadvantages of this system are that reactions of the immune system cannot be monitored and that the environment, although *in vivo*, is foreign to the tumor. An improvement to this model is to inject human cells into the organ of the mouse from which they were derived (orthotopic). Injecting tumor cells of mice into mice (syngeneic tumor) is another alternative. The approach that offers a better alternative is genetically altered mice, created by transgenic, knock-out, RNA interference, or CRISPR-Cas9 technologies. For example, the multiple intestinal neoplasia (min) mouse carries a germline truncation of the *APC* gene. The min mouse develops multiple adenomas and is used to study colon carcinogenesis. Tissue-specific and inducible promoters can be used in these models to better mimic the etiology/molecular defect of the disease and the anatomical and temporal characteristics of human cancer. Creating mice with defective

DNA repair, telomere dysfunction, or impaired DNA damage checkpoints generates "instability models" of cancer in which DNA damage is more likely to give rise to mutations. The resultant genome instability gives rise to tumors with a level of complexity comparable to human tumors and thus is an important tool for further studies of carcinogenesis (see Maser *et al.*, 2007 and references within). It must be remembered that species differences exist, and results from any of these models may give very different results in the clinic.

Drug screening

High-throughput screening is a common approach to select lead compounds for drug development (Figure 14.4). It permits the testing of millions of compounds in a short period of time. Plates containing hundreds of wells of biological material (e.g. cells) are used to test various chemical compounds for a desired biological effect (e.g. apoptosis). Robotics can be deployed to prepare plates and to analyze up to 100,000 compounds per day. Many screening protocols use synthetic molecules synthesized via combinatorial chemistry—methodologies that rapidly and systematically assemble molecular entities to synthesize a large number of different, but structurally related, compounds. Note that many successful drugs are based on natural compounds that have also been identified using screening procedures. Another approach for selecting lead compounds for development is virtual screening. In this approach, computer analysis is used to select compounds that will bind to a molecular target based on three-dimensional structural information about the target (e.g. crystal structure). As this is an *in silico* approach, laboratory materials are not

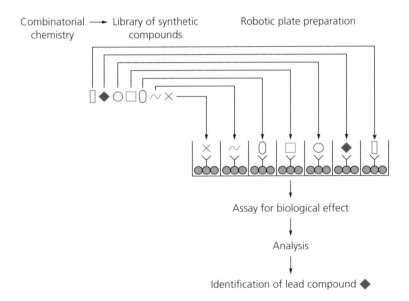

Figure 14.4 High-throughput screening. See text for details.

required, and the compounds examined may not necessarily exist. Studies suggest that high-throughput screening and virtual screening are complementary approaches, each yielding potentially promising results.

14.7 Development of imatinib

Let us look at the development of imatinib (Gleevec™, USA; Glivec™, UK, Europe) in relation to the stages of drug development (Capdeville *et al.*, 2002). It began with the understanding of a genetic lesion: the chromosomal translocation that is characteristic of CML. The identification of the molecular consequence of this translocation pointed to the resulting fusion protein BCR–ABL, as a *molecular target*. As the translocation results in elevated tyrosine kinase activity that is crucial for transformation, the obvious strategy was to develop a specific tyrosine kinase inhibitor. A lead compound (a compound that shows a desired activity, e.g. kinase inhibition) was identified in a chemical screen. In this case, kinase inhibition was demonstrated by the inhibition of protein kinase C. Optimization of the lead compound by the addition of small chemical groups (Figure 14.5) produced a drug that inhibited BCR–ABL tyrosine kinase activity (instead of protein kinase C) and showed good bioavailability. Bioavailability relates to the ability of the drug to reach its site of action after administration. Pre-clinical testing demonstrated fairly selective inhibition of BCR–ABL kinase activity and induction of apoptosis

Figure 14.5 Optimization of a lead compound leading to production of imatinib (Gleevec™). Chemical modifications are shown in red. (a) The addition of a 3′-pyridyl group enhances cellular activity. (b) An amide group conferred BCR–ABL tyrosine kinase inhibition. (c) A methyl group abolished unwanted protein kinase C activity. (d) An N-methyl piperazine moiety enhanced oral bioavailability and solubility.

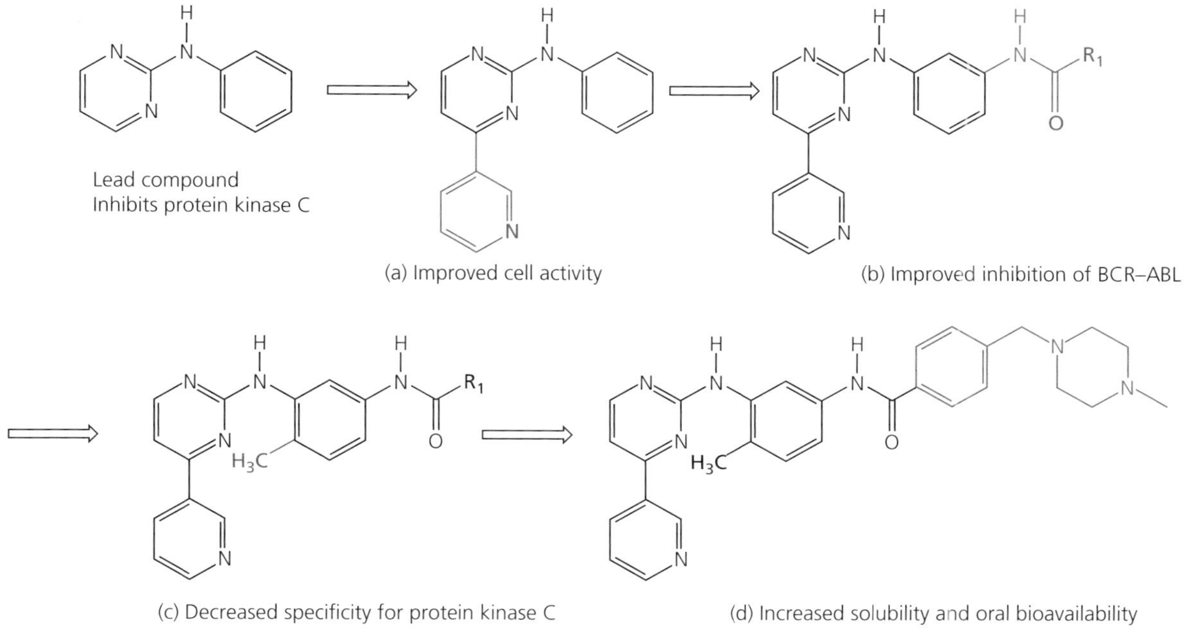

Lead compound
Inhibits protein kinase C

(a) Improved cell activity

(b) Improved inhibition of BCR–ABL

(c) Decreased specificity for protein kinase C

(d) Increased solubility and oral bioavailability

in cell culture experiments and in leukemic cells from patients. Inhibition of tumor growth was also observed in animal models. **Clinical trials** (see Table 1.1 for phases of clinical trials) documented safety and efficacy. Results from a Phase I dose escalation study were extremely positive: 53 out of 54 patients (98%) achieved a complete hematological response with minimal side effects, usually within the first 3 weeks of treatment (Druker *et al.*, 2001).

Positive results continued to be demonstrated for Phase II and Phase III trials. Approval by the US FDA was given on May 2001, and subsequent approval for use in Europe and Japan followed shortly afterwards. The European Medicines Evaluation Agency (EMEA) and the Japanese Ministry of Health and Welfare, together with the FDA, are working towards an internationally harmonized system for drug approval that will facilitate the delivery of drugs to more patients, more quickly.

14.8 Second- and third-generation therapeutics

Initial drug treatment can select for cancer cells that are resistant to the treatment. Resistant cells continue to proliferate. Consequently, new drugs are required when the initial treatment is no longer effective. Although imatinib is a highly successful treatment for the treatment of CML in early-phase disease, patients with later-phase (accelerated and blast stage) disease often develop resistance to imatinib. New drugs called *second-generation inhibitors* have been developed to block the progression of CML in those who become resistant to imatinib. Strategies in the development of these new second-generation inhibitors often rely on structural biology data that analyze drug binding (Figure 14.6).

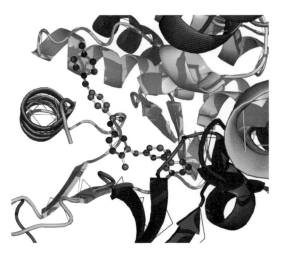

Figure 14.6 Molecular modeling of the binding of the drug imatinib (Gleevec™) to its target Abl.

Point mutations in the *Bcr–Abl* gene that interfere with the binding of imatinib, but maintain a functional tyrosine kinase domain, are the most common mechanism involved in imatinib resistance. Mutations may affect physical bonding/contact points or cause changes in shape that sterically obstruct binding. Alternatively, mutations may destabilize the inactive conformation, the conformation to which imatinib binds.

Although more than 50 different mutations associated with imatinib resistance have been identified, only six account for 60–70% of all mutations. One of these six, a mutation at Thr315 (T315I; a substitution by isoleucine; Gorre *et al.*, 2001), is noteworthy because it has proved to be resistant to many second-generation inhibitors. Dasatinib (Figure 14.7a) is a second-generation inhibitor that, unlike imatinib (Figure 14.7b), binds to the active conformation of the ABL kinase and therefore inhibits many mutant forms of BCR–ABL. The FDA and the EMEA approved dasatinib for the treatment of imatinib-resistant CML. Nilotinib is another second-generation inhibitor that has been approved. Unfortunately, the T315I mutation noted earlier confers both dasatinib and nilotinib resistance.

A potent aurora kinase and JAK2 inhibitor called MK-0457 (VX-680, Merck) was the first agent to show activity in patients with the BCR–ABL T315I mutation. It is thought that this kinase inhibitor shows an effect on BCR–ABL, because it shares the common mechanism of action for protein kinase inhibitors: it competes for binding to the ATP-binding site. Due to cardiac toxicity, further development has been stopped. The drug ponatinib (Iclusig™; ARIAD Pharmaceuticals) has been approved for patients with the T315I mutation.

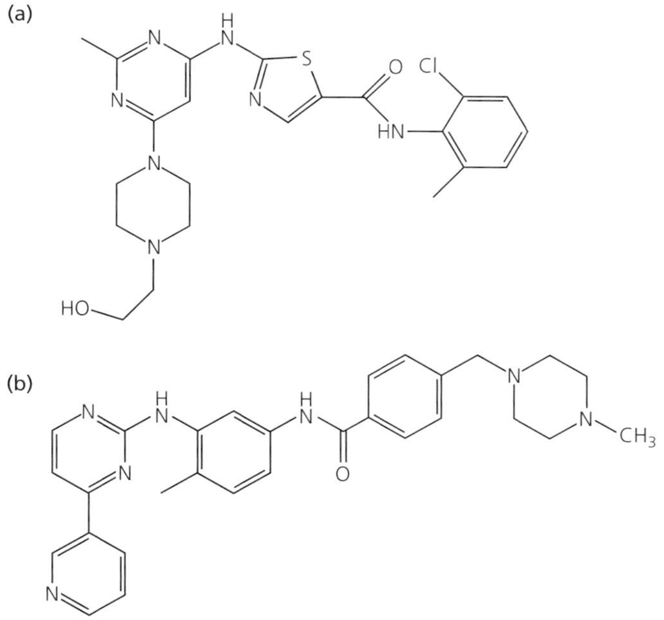

Figure 14.7 An example of a second-generation inhibitor of BCR–ABL: (a) dasatinib. Compare with (b) imatinib.

Another category of agents are being developed for patients who have failed sequential therapy with imatinib and a second-generation tyrosine kinase inhibitor. These are called third-generation tyrosine kinase inhibitors and include allosteric inhibitors that do not compete for ATP binding. Some have shown potent activity against the T315I mutation, and some are in clinical trials, as described below (Greuber *et al.*, 2013; Quintas-Cardama *et al.*, 2010).

A novel allosteric ABL kinase inhibitor, ABL001 (Novartis), is currently in clinical trials. This drug binds to a pocket on the BCR–ABL kinase domain (not the ATP-binding site) that is involved in self-inhibition and normally interacts with the modified N-terminus of ABL1. The interaction of ABL001 with this pocket restores negative regulation of kinase activity. Note that this inhibitor is effective for most mutations, including T315I. We await results from clinical trials. Additional allosteric inhibitors, such as GNF-2, are being developed.

The origin of resistance is still not completely understood. Some data suggest that all mutant *Bcr–Abl* clones may exist *before* imatinib treatment and that clones carrying these may grow in response to selective pressure created by imatinib. When a patient initially responds to a second-generation inhibitor but subsequently develops resistance with a second mutation, the new mutation responsible for the resistance is not found in addition to the first mutation but instead is identified on the original *Bcr–Abl* background. Thus, the clone carrying the new mutation may grow as a result of selective pressure of the second-generation inhibitor. Combinations of inhibitors may be effective in preventing the expansion of drug-resistant clones; thus, such strategies are being examined.

14.9 Improved clinical trial design

There are some flaws in the design of clinical trials performed today that result in an inflated number of unsuccessful drug results. A problem that is difficult to overcome is that many new drugs are tested on older patients who have advanced cancer and who have not responded to a range of conventional therapies. Drugs that may be effective in early-stage cancer may fail to have an effect on late-stage cancer in clinical trials. The development of such drugs would be wrongly discontinued. Ethically acceptable suggestions are needed on how we can move forward. The inclusion of controls is strongly recommended, but surprisingly they have not been included in a majority of Phase II oncology trials in the past, with comparisons being made to information on previous patient outcome. Modifications are being made to the classical Phase I, II, and III trials. The FDA has approved Phase 0 trials. Phase 0 trials allow the testing of small doses of experimental drugs in people for 7 days or less. They aim to gather data on drug targeting, action, and metabolism in the body. Initially, only small batches of a drug are required to be produced for these studies. In addition, Phase IV trials may be carried out after a drug has been approved and licensed. The purpose of Phase IV trials is to further investigate side effects, safety, and/or long-term risks and benefits. They are also useful to help develop new applications for a given drug. Phase IV trials for the HPV vaccine Gardasil™ (discussed in Section 13.5, "Cancer vaccines to prevent cervical cancer" on p. 318) are planned to test different vaccination policies in Nordic countries and to monitor effectiveness and the occurrence of rare adverse events.

It is expected that sample size calculations for a clinical trial should be carried out properly and reported in associated publications. This is important for trial evaluation and the production of statistically meaningful results. Many researchers use computer software programs for this task. However, it may be informative to describe a basic formula (Schulz and Grimes, 2005). For trials with two outcomes (e.g. the patient remains ill or the patient becomes well), four components are required for calculating sample sizes. They are: type I error (α), power, event rate in the control group, and event rate in the treatment group (treatment effect). Type I error is the probability of concluding that two treatments (e.g. treatment versus placebo) differ, when, in fact, they do not, that is the chance of a false-positive result. Power is derived from type II (β) error: $P = 1 - \beta$. Type II error is the opposite of type I: it is the probability of not detecting a statistically significant difference when a difference actually exists. It is the chance of a false-negative result: the treatment is different from the control, but the difference is not detectable. Therefore, power is the probability of detecting a statistically significant difference when a difference really exists. Investigators would usually like less than a 5% chance of making a false-positive error, and thus, by convention, α is usually set to 0.05. Similarly, if investigators want less than a 10% chance of making a false-negative error, they will set β to 0.10. In this case, power will be $1 - 0.10 = 0.90$.

The following equation illustrates a simple formula for calculating sample size for a case with two possible outcomes, assuming $\alpha = 0.05$, power $= 0.90$, and an equal sample size in the two groups:

$$n = \frac{10.51[(R + 1) - p_2 (R^2 + 1)]}{p_2 (1 - R)^2},$$

where n is the sample size, p_1 is the event rate in the treatment group, p_2 is the event rate in the control group, and R is the risk ratio (p_1/p_2).

Defining the target population and clinical endpoints requires special attention in clinical trial design. Important lessons have been gained recently about defining the target population for clinical trials. The results previously discussed that illustrated variable response rates among two different populations to the drug gefitinib (Chapter 4) argue for the benefit of including geographically and genetically diverse populations in clinical trials. Molecular profiling of the tumor prior to, and/or during, clinical trials is also important in order to gain a true measure of drug efficacy. However, obtaining tumor tissue for analysis in some types of cancer, such as lung, may be difficult.

Patient selection (that is choosing patients who have the molecular defect that a drug is designed to target) is important, so that drugs are shown to be efficient when tested on the "right" patients. Otherwise, a variable response will be observed, and the observed efficacy will not be accurate. For example, as we saw earlier with gefitinib (Iressa™), if a drug acts on tumor cells that contain a mutated receptor, the efficacy of the drug should be tested on patients whose tumor contains these mutations. For other drugs, patients may be selected on the basis of gene copy number, rather than mutation. Digital karyotyping is a method that allows determination of DNA copy number on a genomic scale.

When designing drug trials, it is important to define parameters that will indicate the effectiveness of the drug. How will we know if the drug is working? Before the initiation of the trial, clinical endpoints must be defined. Tumor shrinkage is commonly used to indicate a response, but it may not necessarily correlate with survival. In particular, cytostatic drugs inhibit tumor growth and/or metastasis, and thus progression-free survival time is a more suitable endpoint, rather than tumor regression. The classical endpoints used are survival, improved time to progression, plus improvement in symptoms or quality of life.

A novel trial design is the randomized discontinuation trial design. In this design, patients receive one or two cycles of a cytostatic drug, and, upon completion, those who have stable disease are then randomized to placebo or to continue the same treatment. The aim of this design is to enrich the patient population with those with slowly progressive cancer and eliminate those with rapidly progressive cancer.

In addition to clinical endpoints, clinical trials that evaluate new molecular therapeutics warrant defined molecular endpoints. Molecular endpoints involve the evaluation of molecular target inhibition and are important to ensure that the drug is eliciting its effects in the expected manner. For example, testing whether imatinib (Gleevec™) inhibits BCR–ABL tyrosine kinase activity was an important molecular endpoint (data shown in Druker *et al.*, 2001). A phosphorylated substrate of Abl tyrosine kinase was used as a biomarker. The measurement of a biomarker has important implications for the administered dose, as the MTD, often prescribed for conventional chemotherapies, may not be necessary. In the earlier example, the dose of imatinib required to inhibit substrate phosphorylation would be appropriate to test in clinical trials. Also, some studies have shown that expression levels of drug targets can change during progression of the disease, and therefore drug response may be expected to change during the course of the disease.

A new and novel paradigm for a personalized approach to lung cancer clinical trials was demonstrated in The Biomarker-integrated Approaches of Targeted Therapy for Lung Cancer Elimination (BATTLE) trial (Kim *et al.*, 2011). Mandatory fresh tumor biopsies informed future treatment later in the trial. For instance, after an initial period of equal randomization, patients who did not respond to chemotherapy were adaptively randomized to targeted therapies, based on relevant molecular biomarkers analyzed in fresh biopsied material. Overall, the conclusions supported the idea that targeted drugs are superior to chemotherapy for patients with specific activating mutations in targeted molecules. There is a need for a shift from population-based unselected approaches to this type of biomarker-adaptive and hypothesis-testing clinical trial, as we move forwards towards personalized medicine (de Bono and Ashworth, 2010).

 HOW DO WE KNOW THAT?

Clinical trial terms and clinical endpoints

Let us examine some of the characteristics of the Phase II trial of Gardasil™ reported by Villa *et al.* (2005). As the title of the article states, it is a randomized, double-blinded, placebo-controlled, multicenter Phase II study. Each term will be explained.

Randomized: treatment parameters (dose and type: treatment or placebo) were assigned to participants using computer-randomized schedules.

Double-blinded: neither participants, nor hospital staff, nor investigators knew which participants received treatment versus placebo.

Placebo-controlled: the placebo consisted of the same adjuvant used for the vaccine, so that participants in the control group received the exact same treatment as participants in the treatment group, *except* they did not receive the vaccine.

Multicenter: participants were recruited from Brazil, Europe, and the USA. Several criteria were set, including age (16–23 years old) and aspects of previous medical history.

Phase II: a dose escalation study involving several hundreds of participants was used to assess safety, immunogenicity, and efficacy. Methods used to monitor efficacy included gynecological examination, Pap test, PCR analysis to detect HPV, serum samples, and biopsies of any lesions.

The primary endpoint assessed was persistent infection by HPV6, -11, -16, or -18 or disease of the cervix or external genitals. Cervical cancer was not used as an endpoint, because cervical cancer can be prevented by the treatment of precancerous lesions identified by screening. Identifying precancerous lesions and allowing them to progress without treating them would be unethical. Furthermore, long-term studies would be required because of the lag time between infection and cancer.

14.10 Personalized medicine and bioinformatics

Physicians currently make decisions about patient treatment on the basis of population-based statistics, but the future holds hope for a time when physicians will rely on individual differences instead. Your genome is the most personalized "item" you possess. Only cancer patients have a more personalized "item", and that is the genome of their tumors. The study of pharmacogenomics has demonstrated that specific mutations of a tumor may determine a patient's response to a cancer drug, and several mutation screening tests for specific drugs are currently in use (see Section 4.3, "Kinases as drug targets" on p. 97). The time will soon come when hospitals and health services will easily be able to subject an individual tumor to genomic analysis.

Such genomic analysis could take place over the course of the tumor's known history: upon detection, during, and after treatment. This type of information could be pooled from thousands of geographic locations and thousands of clinical trials, providing an extraordinary tool for future treatment and research. The molecular profile of an individual patient will be able to be compared and analyzed in order to select the best known therapy available. Researchers will be able to identify cancer-specific molecular targets for drug design more rapidly. Several new bioinformatics initiatives have been launched. The US National Cancer Institute and the UK National Cancer Research Institute will collaborate in developing tumor information databases and networks (see "Web sites" at the end of the chapter on p. 350). These initiatives will make tissue data, biological contexts or ontologies, and clinical trial information readily available to researchers. In 2015, the President of the USA announced a Precision Medicine Initiative that will aim to collect health information, including genome sequences, from one million or more Americans, with one of the goals being to make an impact on oncology. Computer capabilities and facilities will need to be expanded and enhanced to handle the enormous number of data generated. Some improvements in image compression have already been made, allowing histological images to be analyzed over the Internet.

It is important to remember that the gene profile of a tumor lies against a background of an individual's inherited genome. Genetic variations among individuals can define subsets of a population that respond differently to drugs. A strong determinant of drug response is variation in drug-metabolizing enzymes (Wu, 2011). Genetic variation in the enzyme thiopurine methyltransferase (TPMT) is one example. TPMT-deficient patients accumulate certain chemotherapies, such as 6-mercaptopurine and 6-thioguanine used to treat some leukemias, and this can lead to severe toxicity. Successful treatment of these patients can occur with 10- to 15-fold lower dosage upon identification using pharmacogenic tests. Other patients may clear a drug from the body more efficiently and require higher doses of a particular drug. In the future, the dose of a drug may be adjusted for an individual, based on their genomic information.

The discussed examples and similar observations illustrate the concept that knowing the genetic profile of a tumor, in addition to inherited genomic information, helps doctors to choose the best treatment for an individual.

PAUSE AND THINK

So how close are we to being able to sequence an individual's genome? This was first done in 2007 for one famous person, James Watson the co-discoverer of the structure of DNA. The 2-month project was the result of a collaboration between 454 Life Sciences and the Baylor College of Medicine Human Genome Sequencing Center, and cost under a million dollars (to learn how they did it, see Activity 2 at the end of this chapter on p. 349). But technologies have greatly advanced and are improving all the time, and companies and governments are helping to drive progress. The sequencing of one individual human genome for $1000 is within reach.

14.11 Are we making progress?

Do you think we are making progress? Despite several media articles that raise doubts, the real answer to this question is certainly "yes!" The statistics are available. For example, the overall survival for all stages of prostate cancer combined has increased from 67% to 89% over the past 20 years. The increased survival is attributable to both earlier and better detection, and advances in therapeutics. Furthermore, cancer death rates decreased in both men and women from 1998 through 2011 in the USA (Siegel *et al.*, 2015; see Chapter 1), and similar progress has been reported in other parts of the world. Although our knowledge about cancer has grown enormously, there is still so much more to learn. Perhaps there are some secrets held in the heart—literally. Primary cardiac tumors, of which only one-quarter are malignant, are rare (0.02%). Investigations into why cancer is rare in this particular tissue may lead to knowledge of protective mechanisms that can be applied to other tissues.

You will notice that most of the newly approved therapies, shown in Table 14.2, are directed against molecules that are tyrosine kinases (e.g. EGFR, VEGFR, ABL). There are several tyrosine kinases that are known to play important roles in carcinogenesis (e.g. fibroblast growth factor receptor, FGFR), but inhibitors that target them, although in clinical trials, have yet to be approved. As we saw for imatinib, tumor cells may develop resistance to initial single agent treatment. This suggests that combinations of drugs and drug strategies are important for future treatment regimens. The report of the development of a "two-in-one" antibody that interacts with either HER2 or VEGF suggests that the way we administer treatment combinations may change (Bostrom *et al.*, 2009). Also, a new type of gene therapy is on the horizon. The first in-human Phase I clinical trial has demonstrated that systemic administration of siRNAs via targeted nanoparticles to patients with melanoma can reduce both a specific mRNA and its associated protein (Davis *et al.*, 2010). CRISPR-Cas9 is another new tool at our disposal. These approaches may open the door to the development of gene-specific therapeutics.

As we saw in previous chapters, there are many potential molecular strategies, such as angiogenesis inhibitors, anti-endocrine drugs, apoptotic inducers, cell cycle inhibitors, HDAC inhibitors, and inhibitors of cell renewal signaling pathways, in development. I regret that some strategies have not been discussed such as proteasome inhibitors. The potential for immunotherapies to give long-lasting protection against many types of cancer is most promising. For many of these drugs, the therapeutic index is enhanced, compared with conventional chemotherapies. We await the elongation of the list of newly approved molecular cancer therapeutics, some of which are shown in Table 14.2.

Table 14.2 A selection of targeted cancer therapeutics approved in 2016

Trademark	Drug	Description	Target	Cancer	Company
Avastin™	Bevacizumab	Humanized mAb	VEGF	Colorectal	Genentech
Erbitux™	Cetuximab	Humanized mAb	EGFR	Colorectal	Imclone
Gleevec™ (USA), Glivec™ (UK, Europe)	Imatinib	Small-molecule inhibitor	BCR–ABL, Kit, PDGFR	CML, GIST	Novartis
Herceptin™	Trastuzumab	Humanized mAb	HER2	Breast	Genentech
Iressa™	Gefitinib	Small-molecule inhibitor	EGFR	NSCLC	AstraZeneca
Keytruder™	Pembrolizumab	Humanized mAb	PD-1	Melanoma	Merck (MSD)
Nexavar™	Sorafenib	Multi-kinase inhibitor	Raf, VEGFR, PDGFR, Kit, RET	Renal cell carcinoma	Bayer Pharm
Opdivo™	Nivolumab	Human mAb	PD-1	Melanoma, NSCLC	Bristol-Myers Squibb
Sprycel™	Dasatinib	Small-molecule inhibitor	BCR-ABL, Src family	Imatinib-resistant leukemias	Bristol-Myers Squibb
Sutent™	Sunitinib (SU11248)	Small-molecule inhibitor	PDGFR, VEGFR, Kit	Renal cell carcinoma, GIST	Pfizer
Tarceva™	Erlotinib	Small-molecule inhibitor	EGFR	NSCLC, pancreatic	Genetech, OSI Pharm
Tykerb™	Lapatinib	Small-molecule inhibitor	EGFR, HER2	Breast	GlaxoSmithKline
Vectibix™	Panitumumab	Human mAb	EGFR	Colorectal	Amgen
Velcade™	Bortezomib	Proteasome inhibitor		Myeloma	Millennium Pharm
Xalkori™	Crizontinib	Small-molecule inhibitor	ALK gene fusion, MET	NSCLC with ALK gene fusions	Pfizer
Yervoy™	Ipilimumab	Human mAb	CTLA-4	Melanoma	Bristol-Myers Squibb
Zactima™	Vandetanib (ZD6474)	Small-molecule inhibitor	VEGFR, EGFR, RET	Orphan drug for rare types of thyroid cancer	AstraZeneca
Zelboraf™	Vemurafenib (PLX4032)	Small-molecule inhibitor	BRAF V600 E	Melanoma	Genentech
Zolinza™ (vorinostat)	SAHA (suberoylanilide hydroxamic acid)	Small-molecule inhibitor	HDAC	Non-Hodgkin's lymphoma	Merck & Co.

CML, chronic myelogenous leukemia; GIST, gastrointestinal stromal tumor; mAb, monoclonal antibody; NSCLC, non-small-cell lung cancer.

A career in cancer research?

People are the most important asset in cancer research and drug discovery. A career in cancer research promises to be interesting and rewarding. You are guaranteed to meet and work with intelligent and talented people and to never reach the boundaries of knowledge. Please consider a career in cancer research.

▨ CHAPTER HIGHLIGHTS—REFRESH YOUR MEMORY

- Microarrays analyze the expression of thousands of genes at once.

- Microarrays have several applications, including identifying new oncogenes, helping to refine cancer classifications, and predicting cancer prognosis.

- Gene signature tests can distinguish some breast cancer patients into groups that would not or would benefit from chemotherapy.

- The CRISPR-Cas9 system is a new method for gene editing.

- A biomarker is a biochemical or genetic feature that can be used to measure disease progress or the effect of treatment.

- MFI is imaging that investigates molecular pathways and tissue function.

- Nanovectors are/will be used for tumor-specific drug delivery.

- Biomolecular sensors using nanocantilevers and nanowires are being developed.

- Drug development follows a series of stages from defining a molecular target through to approval.

- Target validation refers to the experimental evaluation of the role of a given gene or protein in cancer and its potential as a therapeutic target.

- Combinatorial chemistry, in conjunction with high-throughput screening, are common methodologies used in drug discovery.

- The development of imatinib (Gleevec™) is a paradigm for drug development of new molecular cancer therapeutics.

- Many patients with late-phase CML often develop resistance to imatinib (Gleevec™).

- Only six mutations in the *Bcr–Abl* gene account for 60–70% of mutations that lead to imatinib (Gleevec™) resistance.

- Second-generation therapeutics are being developed to overcome resistance to imatinib (Gleevec™).

- Phase 0 and IV clinical trials have been added to the three conventional phases of clinical trials, Phases I, II, and III.

- Defining the target population and clinical endpoints are two important aspects requiring careful consideration during the design of clinical trials.

- We *are* making progress in the field of molecular cancer therapeutics.

▨ ACTIVITY

1. Read the case history of the discovery and development of sorafenib (*Nat. Rev. Drug Discov.* **5**: 835–844). Compare the process with that of imatinib. Pay particular attention to the target population used in the early clinical trials. Note the timescale from discovery to approval.

2. If you are interested in learning more about DNA sequencing and reading the paper that reported the sequencing of James Watson's genome see: Metzker, M.L. (2010) Sequencing technologies: the next generation. *Nat Rev. Genet.* **11**: 31–46 and Wheeler, D.A., *et al*. (2008) The complete genome of an individual by massively parallel DNA sequencing. *Nature* **452**: 872–877.

▨ FURTHER READING

Arteaga, C.L. and Baselga, J. (2003) Clinical trial design and end point for epidermal growth factor receptor-targeted therapies: implications for drug development and practice. *Clin. Cancer Res*. **9**: 1579–1589.

Benson, J.D., Chen, Y.-N.P., Vornell-Kennon, S.A., Dorsch, M., Kim, S., Leszczyniecka, M.,*et al*. (2006) Validating cancer drug targets. *Nature* **441**: 451–456.

Cogbill, T.H. and Ziegelbein, K.J. (2011) Computed tomography, magnetic resonance, and ultrasound imaging: basic principles, glossary of terms, and patient safety. *Surg. Clin. N. Am*. **91**: 1–14.

Druker, B.J. (2002) STI571 (Gleevec™) as a paradigm for cancer therapy. *Trends Mol. Med.* **8**: S14–20

Henry, N.L. and Hayes, D.F. (2012) Cancer biomarkers. *Mol. Oncol.* **6**: 140–146.

Klebe, G. (2006) Virtual ligand screening: strategies, perspectives, and limitations. *Drug Discov. Today* **11**: 580–594.

Sanchez-Rivera, F.J. and Jacks, T. (2015) Applications of the CRISPR-Cas-9 system in cancer biology. *Nat. Rev. Cancer* **15**: 387–395.

Sanoudou, D., Mountzios, G., Arvanitis, D.A., and Pectasides, D. (2012) Array-based pharmacogenomics of molecular-targeted therapies in oncology. *Pharmacogenomics J.* **12**: 185–196.

Schiller, J.H. (2004) Clinical trial design issues in the era of targeted therapies. *Clin. Cancer Res.* **10**: 4281S–4282S.

Strausberg, R.L., Simpson, A.J.G., Old, L.J., and Riggins, G.J. (2004) Oncogenomics and the development of new cancer therapies. *Nature* **429**: 469–474.

Weisberg, E., Manley, P.W., Cowan-Jacob, S.W., Hochhaus, A., and Griffin, J.D. (2007) Second generation inhibitors of BCR-ABL for the treatment of imatinib-resistant chronic myeloid leukemia. *Nat. Rev. Cancer* **7**: 345–356.

WEB SITES

Bioinformatics initiatives. National Cancer Informatics Program (NCIP), USA https://cbiit.nci.nih.gov/ncip and National Cancer Research Institute Informatics Initiative, UK http://www.nesc.ac.uk/talks/745/AbiAjose-Adeogun.pdf

Clinical trials: National Cancer Institute http://www.cancer.gov/

Hematology/Oncology Approvals and Safety Notifications www.fda.gov/Drugs/InformationOnDrugs/ApprovedDrugs/ucm279174.htm

SELECTED SPECIAL TOPICS

Alizadeh, A.A., Eisen, M.B., Davis, R.E., Ma, C., Lossos, I.S., Rosenwald, A., *et al.* (2000) Distinct types of diffuse large B-cell lymphoma identified by gene expression profiling. *Nature* **403**: 503–511.

Barrans, S.L., Crouch, S., Care, M.A., Worrillow, L., Smith, A., Patmore, R., *et al.* (2012) Whole genome expression profiling based on paraffin embedded tissue can be used to classify diffuse large B-cell lymphoma and predict clinical outcome. *Br. J. Haematol.* **159**: 441–453.

Berrington de González, A., Mahesh, M., Kim, K.P., Bhargavan, M., Lewis, R., Mettler, F., *et al.* (2009) Projected cancer risks from computed tomographic scans performed in the United States in 2007. *Arch. Intern. Med.* **169**: 2071–2077.

Bostrom, J., Yu, S.F., Kan, D., Appleton, B.A., Lee, C.V., Billeci, K., *et al.* (2009) Variants of the antibody herceptin that interact with HER2 and VEGF at the antigen binding site. *Science* **323**: 1610–1614.

Capdeville, R., Buchdunger, E., Zimmermann, J., and Matter, A. (2002) Glivec (STI571, Imatinib), a rationally developed, targeted anticancer drug. *Nat. Rev. Drug Discov.* **1**: 493–502.

Davis, M.E., Zuckerman, J.E., Choi, C.H., Seligson, D., Tolcher, A., Alabi, C.A., *et al.* (2010) Evidence of RNAi in humans from systemically administered siRNA via targeted nanoparticles. *Nature* **464**: 1067–1070.

De Bono, J.S. and Ashworth, A. (2010) Translating cancer research into targeted therapeutics. *Nature* **467**: 543–549.

Druker, B.J., Talpaz, M., Resta, D.J., Peng, B., Buchdunger, E., Ford, J.M., *et al.* (2001) Efficiency and safety of a specific inhibitor of the BCR-ABL tyrosine kinase in chronic myeloid leukemia. *N. Engl. J. Med.* **344**: 1031–1037.

Glunde, K., Pathak, A.P., and Bhujwalla, Z.M. (2007) Molecular-functional imaging of cancer: to image and imagine. *Trends Mol. Med.* **13**: 287–297.

Gorre, M.E., Mohammed, M., Ellwood, K., Hsu, N., Paquette, R., Rao, P.N., *et al.* (2001) Clinical resistance to STI-571 cancer therapy caused by Bcr-Abl gene mutation or amplification. *Science* **293**: 876–880.

Greuber, E.K., Smith-Pearson, P., Wang, J., and Pendergast, A.M. (2013) Role of ABL family kinases in cancer: from leukaemia to solid tumours. *Nat. Rev. Cancer* **13**: 559–571.

Kim, E.S., Herbst, R.S., Wistuba, I.I., Lee, J.J., Blumenschein Jr, G.R., Tsao, A., *et al.* (2011) The BATTLE Trial: Personalizing Therapy for Lung Cancer. *Cancer Discov.* **1**: 44–x2013;53; CD–100010.

Leung, F., Musrap, N., Diamandis, E.P., and Kulasingam, V. (2013) Advances in mass spectrometry-based technologies to direct personalized medicine in ovarian cancer. *Transl. Proteomics* **1**: 74–86.

Lynch, T.J., Bell, D.W., Sordella, R., Gurubhagavatula, S., Okimoto, R.A., Brannigan, B.W., *et al.* (2004) Activating mutations in the epidermal growth factor receptor underlying responsiveness of non-small cell lung cancer to gefitinib. *N. Engl. J. Med.* **350**: 2129–2139.

Maser, R.S., Choudhury, B., Campbell, P.J., Feng, B., Wong, K.-K., Protopopov, A., *et al.* (2007) Chromosomally unstable mouse tumours have genomic alterations similar to diverse human cancers. *Nature* **447**: 966–971.

McCracken, K.W., Cata, E.M., Crawford, C.M., Sinogoga, K.L., Schumacher, M., Rockich, B.E., *et al.* (2014) Modelling human development and disease in pluripotent stem-cell-derived gastric organoids. *Nature* **516**: 400–404.

Merola, R., Tomao, L., Antenucci, A., Sperduti, I., Sentinelli, S., Masi, S., *et al.* (2015) PCA3 in prostate cancer and tumor aggressiveness detection on 407 high-risk patients: a National Cancer Institute experience. *J. Exp. Clin. Cancer Res.* **34**: 15–20.

Paez, J.G., Janne, P.A., Lee, J.C., Tracy, S., Greulich, H., Gabriel, S., *et al.* (2004) EGFR mutations in lung cancer: correlation with clinical response to Gefitinib therapy. *Science* **304**: 1497–1500.

Quintas-Cardama, A., Kantarjian, H., and Cortes, J. (2010) Third-generation tyrosine kinase inhibitors and beyond. *Semin. Hematol.* **47**: 371–380.

Romer, J. and Curran, T. (2005) Targeting medulloblastoma: small-molecule inhibitors of the sonic hedgehog pathway as potential cancer therapeutics. *Cancer Res.* **65**: 4975–4978.

Sanna, V., Pala, N., and Sechi, M. (2014) Targeted therapy using nanotechnology: focus on cancer. *Int. J. Nanomedicine* **9**: 467–483.

Schulz, K.F. and Grimes, D.A. (2005) Sample size calculations in randomized trials: mandatory and mystical. *Lancet* **365**: 1348–1353.

Shipp, M.A., Ross, K.N., Tamayo, P., Weng, A.P., Kutok, J.L., Aguiar, R.C., *et al.* (2002) Diffuse large B-cell lymphoma outcome prediction by gene-expression profiling and supervised machine learning. *Nat. Med.* **8**: 68–74.

Siegel, R., Miller, K., and Jemal, A. (2015) Cancer statistics, 2015. *CA Cancer J. Clin.* **65**: 5–29.

Torres, R., Martin, M.C., Garcia, A., Cigudosa, J.C., Ramirez, J.C., and Rodriguez-Perales, S. (2014) Engineering human tumor-associated chromosomal translocations with the RNA-guided CRISPR-Cas9 system. *Nat. Commun.* **5**: 3964.

Van't Veer, L.J., Dai, H., van de Vijver, M.J., He, Y.D., Hart, A.A.M., Mao, M., *et al.* (2002) Gene expression profiling predicts clinical outcome of breast cancer. *Nature* **415**: 530–535.

Villa, L.L., Costa, R.L., Petta, C.A., Andrade, R.P., Ault, K.A., Giuliano, A.R., *et al.* (2005) Prophylactic quadrivalent human papillomavirus (types 6, 11, 16, and 18) L1 virus-like particle vaccine in young women: a randomised double-blind placebo-controlled multicentre phase II efficacy trial. *Lancet Oncol.* **6**: 271–278.

Wu, A.H.B. (2011) Drug metabolizing enzyme activities versus genetic variances for drug of clinical pharmacogenomic relevance. *Clin. Proteomics* **8**: 12.

Xu, X., Ho, W., Zhang, X., Bertrand, N., and Farokhzad, O. (2015) Cancer nanomedicine: from targeted delivery to combination therapy. *Trends Mol. Med.* **21**: 223–232.

Xue, W., Chen, S., Yin, H., Tammela, T., Papagiannakopoulos, T., Joshi, N.S., *et al.* (2014) CRISPR-mediated direct mutation of cancer genes in the mouse liver. *Nature* **514**: 380–384.

APPENDIX 1: CELL CYCLE REGULATION

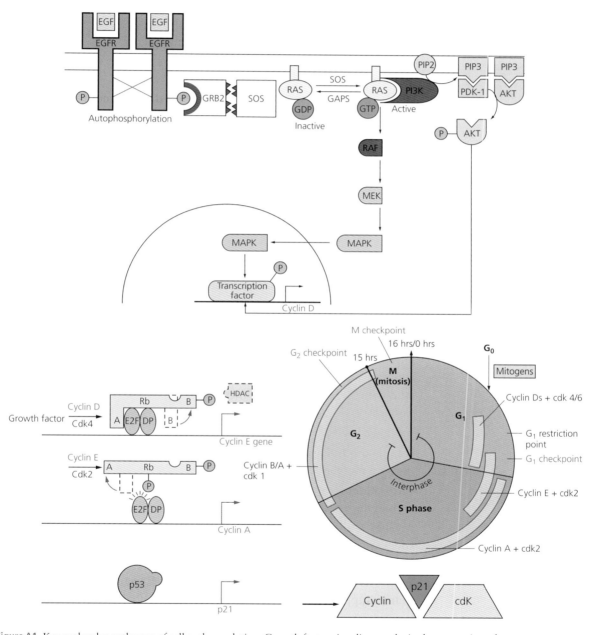

Figure A1 Key molecular pathways of cell cycle regulation. Growth factor signaling results in the expression of target genes, including cyclin genes. Cyclin proteins are key for progression of the cell cycle and are involved in the regulation of the tumor suppressor protein RB. The protein product of the *p21* gene acts as an inhibitor of cyclin–cdk (cyclin-dependent kinase) complexes. These pathways are discussed in detail in the text.

GLOSSARY

Adenocarcinomas malignant tumors of a gland.

Adjuvant a vaccine additive that enhances the immune response to an antigen.

Aflatoxin a carcinogenic compound produced by the mold *Aspergillus flavus* that contaminates some food products such as peanuts.

Alkylating agents chemicals that introduce an alkyl group onto DNA; they act as carcinogens but are also used in chemotherapy.

Alleles alternative forms of a gene at the same locus or relative position in a chromosomal pair. One allele may be dominant over the other.

Allografts transplants of tissue from one individual to another (e.g. a heart transplant).

Aneuploidy the state of having an abnormal number of chromosomes in a cell such as having 45 or 47 when 46 is the normal number expected.

Angiogenesis the process of forming new blood vessels from pre-existing ones by the growth and migration of endothelial cells in a process called "sprouting." The induction of angiogenesis is a hallmark of cancer.

Anoikis apoptosis triggered in response to a lack of extracellular matrix ligand binding.

Antibody a protein produced by lymphocytes in response to an antigen, and which can specifically bind the antigen as part of an immune response.

Antigen a molecule capable of generating an immune response.

Antimetabolites agents that resemble an endogenous metabolite and block a metabolic pathway.

Antioxidants compounds that significantly inhibit or delay the damaging action of reactive oxygen species, often by being oxidized themselves.

Antisense oligonucleotides synthetic nucleotide fragments that hybridize to complementary DNA or RNA in order to inhibit gene expression.

Apoptosis a process of "neat" programmed cell death. It plays a role in tumor suppression; inhibition of apoptosis is a hallmark of cancer.

Attenuated reduced virulence (infectivity) of a pathogenic microorganism.

Autoimmunity a condition in which an individual's immune system starts reacting against their own tissues, causing disease.

Autophagy a process whereby proteins and organelle components that are no longer required are targeted to the lysosomes for degradation. Excessive autophagy leads to a specific type of non-apoptotic cell death program.

Basement membranes acellular supports of endothelial, epithelial, and some mesenchymal cells made up of a complex mix of extracellular matrix proteins, including laminins, collagens, and proteoglycans. They act as passive barriers that separate tissue compartments.

Benign characteristic of a tumor that does not invade surrounding tissues or metastasize.

Bioinformatics the use of computers and information technology to store and analyze nucleotide and amino acid sequences and related information.

Biomarker a biochemical or genetic feature that can be used to measure the progression of disease or the effect of treatment.

Cachexia a metabolic defect often associated with cancer that is characterized by progressive weight loss caused by the depletion of adipose tissue and skeletal muscle.

Cancer stem cells cells within a tumor that have the ability to self-renew and to give rise to phenotypically diverse cancer cells.

Carcinogens chemicals or forms of energy that cause cancer.

Carcinogenesis the process of inducing cancer.

Carcinomas malignant tumors of the epithelium.

Caspases specific aspartate proteases (that cleave target proteins at aspartate residues) involved in apoptosis.

cDNA the DNA sequence that is complementary (c) to a messenger (m)RNA.

Cell cycle the sequence of stages through which a cell passes between one cell division and the next. The cell cycle can be divided into four main stages: the M phase, when nuclear and cytoplasmic division occurs; the G_1 phase; the S phase in which DNA replication occurs; and the G_2 phase.

Cell-mediated immunity an immune response that involves the activation of antigen-specific T cells or phagocytes via T-cell-released cytokines, and results in the destruction of antigen-bearing cells.

Chemoprevention the use of naturally occurring or synthetic agents to prevent, inhibit, or reverse the process of carcinogenesis in pre-malignant cells.

Chimeric antigen receptors proteins that have been engineered to allow T cells to recognize a specific antigen, usually created by fusing a single-chain variable fragment from a monoclonal antibody to a T-cell co-receptor protein sequence.

Chromatin fibers made up of DNA, RNA, and protein that form chromosomes.

Chromosome a structure composed of a DNA molecule and associated RNA and protein. Humans have 46 chromosomes in the nucleus of their somatic cells.

Chromothripsis a single catastrophic event in a cell that leads to multiple mutations.

Chronic refers to a long-lasting condition; opposite of acute.

Clinical trials involve the testing of a new drug in humans under medical supervision to test for drug safety and efficacy. Clinical trials proceed in sequential and defined phases: Phases I, II, and III.

Clonal originating from one cell.

Coding region the nucleotide sequences of a gene that are transcribed into mRNA, and those coding for exons are translated into protein.

Coley's toxin a bacterial vaccine made from a mixture of heat-inactivated *Streptococcus pyogenes* and *Serratia marcescens* bacteria, first used as an immunotherapy by William Coley in the 1880s to treat sarcoma patients.

Combinatorial chemistry methodologies that rapidly and systematically assemble molecular entities to synthesize a large number of different, but structurally related, compounds.

CpG islands regions of DNA that contain clusters of CG dinucleotides. They are often located in the promoter regions of genes and are not normally methylated. CpG islands of tumor suppressor genes may be found methylated in cancer cells, resulting in epigenetic gene silencing.

Cytokines small secreted proteins which are the major cell signaling proteins of both B and T cells (e.g. interferon-γ, interleukins, chemokines).

Cytostatic drug a drug that stops cell growth.

Cytotoxic drug a drug that kills cells.

Differentiation the functional specialization of a cell as a result of the expression of a specific set of genes.

Disseminated tumor cells tumor cells found in a distant metastatic organ.

DNA response elements short sequences of DNA that act as binding sites for transcription factors in gene promoters.

Dominant negative a mutation that produces a protein that interacts and/or interferes with the function of a wild-type protein.

Downstream refers to DNA sequences that are nearer 3′ as a point of reference. Note that, by convention, a DNA sequence is read from the 5′ end to the 3′ end.

Dysplastic abnormal growth or development of cells, tissues, or organs.

Electromagnetic radiation a naturally occurring energy that moves as waves resulting from the acceleration of electric charge and the associated electric and magnetic fields. The characteristics of the radiation depend on its wavelength.

Electromagnetic spectrum the range of wavelengths over which electromagnetic radiation extends. The longest waves (wavelength 10^5–10^{-3} m) are radio waves, and the shortest are gamma rays (wavelength 10^{-11}–10^{-14} m).

Electrophilic molecules that are electron-deficient and are therefore attracted to compounds with a net negative charge.

Embryonic stem cells cells derived from the inner cell mass of an early embryo. When transferred into another early embryo, they combine with the inner cell mass cells of the host and contribute to embryo formation.

Epigenetic refers to inheritable information that is encoded by modifications of the genome and chromatin components and affects gene expression. It does not include changes in the base sequence of DNA.

Epithelium–mesenchymal transition (EMT) involves cells leaving an epithelial layer and becoming a loose mass of mesenchymal cells which can migrate individually. EMT is crucial to gastrulation and early development and may also play a role in metastasis.

Estrogens steroid hormones secreted by the ovary, but also produced by adipose cells, that act to maintain female characteristics and as a mitogen for breast cells.

Extravasation the process whereby a cancer cell exits a blood vessel or lymphatic vessel.

First-pass organ the first organ *en route* via the bloodstream that lies downstream from the primary tumor site.

Gene a region of DNA that occupies a specific position on a chromosome and includes the regulatory region and coding region for a protein.

Gene amplification the multiple replication of a section of DNA that results in the production of many copies of the genes involved.

Gene expression the process by which the information encoded by a gene is converted for the making of a protein. In terms of molecular biology, this usually refers to transcription.

Genome-wide association study (GWAS) a study that investigates the entire genome for common genetic variants in different individuals to see if there are any associations between SNPs and a trait/disease such as cancer.

Genomics the study of all the genes contained in a set of chromosomes.

Genotoxic the ability of a substance to damage DNA.

Genotype the genetic characteristics of a cell or organism. Also the combination of the alleles at a particular locus.

Germline mutation a mutation in either egg or sperm cell DNA (as opposed to a somatic mutation). Mutations in germ cells only can be passed on to the next generation.

Haploinsufficiency occurs with the loss of one (of two) wild-type alleles, and the single functional copy of the gene is insufficient to maintain normal function.

Hematopoietic refers to tissue that can give rise to blood cells in the process of hematopoiesis.

Heterodimer a functional protein that is made up of two different subunits.

Heterozygous having different alleles at a given locus on homologous chromosomes.

Histones basic proteins within chromatin that bind DNA at regular intervals.

Homodimers functional proteins that are made up of two identical subunits.

Homozygous having the same two alleles at a given locus on homologous chromosomes.

Horizontal transfer the movement of genetic material and protein from one cell to another that occurs without cell division, but rather by means such as through exosomes.

Hypoxia a state of low levels of oxygen.

Immunoediting the concept that tumor cells modulate the host anti-tumor immune response and the host immune response shapes tumor immunogenicity and clonal selection.

Immunosurveillance a monitoring function of the immune system that recognizes cancer cells as foreign agents and eliminates them.

Immunotherapy treatment that uses components of the immune system to fight cancer.

Incidence the number of new cases of cancer (or other disease) in a defined population over a defined period of time.

Indoleamine-2,3-dioxygenase an enzyme that catalyzes the first step in tryptophan catabolism that plays a role in anti-tumor defense and immunoregulation. It is overexpressed in a variety of cancers and causes immunosuppression.

Interferon-γ a soluble cytokine that is critical for stimulating innate and adaptive immunity.

Intravasation the process whereby a cancer cell enters a blood vessel or lymphatic vessel.

Invasion spread of tumor cells into surrounding tissue.

Kataegis a pattern of localized hypermutation identified in some cancer genomes and hypothesized to result from cytosine deaminations catalyzed by APOBEC proteins.

Kinases enzymes that transfer phosphate groups to a protein at serine, threonine, or tyrosine amino acids.

Knock-out mice mice in which both alleles of a gene have been inactivated experimentally. These mice are often used to study gene function.

Lead compounds compounds identified during the development of a drug that show a desired activity, for example, kinase inhibition.

Leucine zipper a protein domain that mediates dimer formation and is normally adjacent to a basic DNA-binding domain. It is characterized by a pattern of five leucine residues, each separated by six residues.

Leukemias a type of cancer characterized by the overproduction of white blood cells or their precursors in the blood or bone marrow.

Ligand an agent that binds to a receptor. A specific hormone is a ligand for its corresponding hormone receptor.

Linear energy transfer (LET) rate of energy loss to the surrounding medium in a radiation track (unit: keV/m).

Loss of heterozygosity loss of the second allele of a gene.

Lymphoma a solid tumor of T or B lymphocytes in the lymph nodes, thymus, or spleen.

M phase the phase of the cell cycle whereby the cell divides to produce two daughter cells and includes mitosis and cytokinesis.

Malignant characteristic of a tumor that is capable of invading surrounding tissue and of metastasizing to secondary locations.

MAP kinase mitogen-activated enzyme that phosphorylates serine and threonine residues on proteins. Also known as extracellular signal-related kinase (ERK).

Metastasis the process of cancer cells spreading from a primary site to secondary sites in the body.

Metastasis suppressor genes genes that inhibit metastasis but do not affect the growth of the primary tumor.

Microarrays (DNA) grids of known DNA samples attached to a solid support and probed with cDNA or genomic DNA. It can be used to monitor gene expression of thousands of genes simultaneously.

MicroRNAs (miRNAs) small non-coding RNAs of 18–25 nucleotides in length that can post-transcriptionally regulate sets of genes. Specificity of miRNAs is based on Watson–Crick complementarities with the 3′ untranslated region of target mRNAs.

Mimetics molecules that mimic, or imitate, the action (such as protein binding) of another molecule.

Missense mutations a type of mutation that converts one codon to another, specifying a different amino acid.

Mitogens substances that can cause cells to divide (i.e. undergo mitosis).

Mitosis the division of the nucleus that occurs in somatic cells. The process maintains a complete set of chromosomes ($2n$) for each of the two daughter cells.

Morphology the study of form and structure of organisms.

Mutagens chemicals or forms of energy that can cause a mutation.

Mutations heritable changes in the bases of DNA, which may include transitions, transversions, deletions, insertions, or translocations.

Nanotechnology the study of devices (or their essential components) that are made by humans and have at least one dimension in the 1–1000 nm range. For scale, the size range is similar to the size of a few atoms to the size of subcellular structures.

Necrosis a type of cell death characterized by membrane disruption and the release of lytic enzymes. This "sloppy" way of dying contrasts with cell death by apoptosis.

Next-generation sequencing (NGS) high-throughput DNA sequencing using a number of new technologies (such as Illumina) that sequences millions of small fragments of DNA in parallel and links the information using bioinformatics. An entire human genome can be sequenced in a day using NGS.

Non-coding RNAs RNA molecules that do not code for a protein. Many have newly discovered roles in gene regulation.

Non-genotoxic carcinogens substances that cause cancer without damaging DNA.

Nonsense mutations a type of mutation that converts a codon that specifies an amino acid to one of the "stop" codons, thus signaling termination of translation and the formation of an incomplete polypeptide.

Nude mice immunodeficient mice (usually hairless) that have no cell-mediated immunity because of the absence of the thymus gland. They can be used experimentally to grow human tumors.

Nutrigenetics the study of the effects of genetic variation on responses to dietary components.

Nutrigenomics the study of the effects of nutrients on gene expression.

Oncogenes genes whose products are capable of transforming a normal cell into a cancer cell. Oncogenes result from the mutation of normal genes (proto-oncogenes).

Oncogene addiction is the dependence of a cancer cell on a specific oncogene for its maintenance.

Oncolytic viruses natural or engineered viruses that can selectively infect and kill tumor cells by cell lysis.

Oncomirs microRNAs (miRNAs) that can function as oncogenes. Amplification or overexpression of miRNAs that downregulate tumor suppressor mRNAs are oncogenic.

Ontogeny the development of an individual.

Organotropism the affinity of specific cancers to metastasize to specific organs or tissues.

Phagocytosis the process whereby particles or cells are engulfed by cells such as macrophages. Cells that undergo apoptosis are consumed by phagocytosis.

Pharmacogenomics the study of the influence of the genome on an individual's response to a drug. Gene variability may lead to differences in drug response among individuals.

Phenotypes the observable characteristics of a cell or organism.

Phosphorylation the addition of a phosphate group PO_4^{3-} to a biomolecule. Phosphorylation may cause conformational changes in proteins or activate particular enzymes.

Polymorphisms the occurrence of two or more alleles for a given locus in a population where at least two alleles appear with frequencies of more than 1%. In simple terms: too common to be due merely to a new mutation.

Polyps tumors that project from an epithelial surface (e.g. polyps of the colon).

Pre-clinical studies studies to test a drug or medical treatment in animals and to gather data regarding safety and efficacy for proof of concept. Pre-clinical studies are required before clinical trials.

Pre-metastatic niche is the site of future metastasis. Signals from the primary tumor direct the migration of bone marrow cells to these sites where they are involved in altering the local microenvironment before arrival of the tumor cells.

Prognosis a forecast or future outlook for a disease.

Promoter the regulatory region of a gene that initiates transcription; usually DNA sequences located 5′ to the coding sequences but which may be located in other regions such as introns and 3′ sequences.

Prophylactic a treatment used to prevent disease.

Proteases an enzyme that degrades proteins.

Proteasomes complexes of proteases in the cytoplasm that degrade proteins marked by covalent modification with ubiquitin.

Proteolysis enzymatic protein degradation involving cleavage of peptide bonds.

Proto-oncogenes normal cellular counterparts of mutated genes that can cause tumors.

Pseudogene a copy of a related gene that has lost its ability to code for a protein, often due to the accumulation of many mutations.

Purine the nitrogenous bases, adenine and guanine, found in DNA.

Pyrimidine the nitrogenous bases, cytosine, thymine, and uracil, found in DNA or RNA.

Radiolysis the use of ionizing radiation to produce chemical reactions.

Reactive oxygen species (ROS) in this book, used to classify reactive intermediates of oxygen (e.g. hydroxyl radicals, hydrogen peroxide, and superoxide radical), although broader definitions exist.

Receptor a transmembrane, cytoplasmic, or nuclear molecule that binds to a specific factor such as a growth factor or hormone.

Recessive an allele that is expressed only when present in the homozygous or hemizygous state (i.e. two such alleles must be present).

Relapse reappearance of a disease.

Remission reduction in the severity of cancer as a result of treatment.

Response element a short sequence of DNA within a gene promoter that is recognized by a specific protein and contributes to the regulation of the gene.

Retinoblastoma cancer of the retinal cells of the eye. A germline mutation in the retinoblastoma (*Rb*) gene is found in familial cases.

S phase the phase of the cell cycle in which DNA synthesis occurs.

Sarcomas malignant tumors of the mesenchyme, for example, bone cancer.

Self-renewal the process whereby a stem cell (or progenitor cell) gives rise to a daughter cell with equivalent developmental potential. For example, a stem cell divides to give rise to two daughter cells: another stem cell and perhaps another more differentiated cell.

Senescence irreversible cell cycle arrest.

Signal transduction the transfer of information along a pathway of a cell that converts a signal received from the outside of the cell to the inside, to generate a cell response.

Single nucleotide polymorphisms (SNPs) single base changes in DNA that differ from the usual bases at those positions. Some cause disease, and others are normal variations in the DNA sequence.

Somatic cells all cells other than egg or sperm cells. Mutations in somatic cells cannot be passed on to the next generation.

Sporadic cancer a non-hereditary cancer. It arises in the absence of germline mutations that predispose individuals to an increased risk of developing a specific cancer.

Stem cells cells that can self-renew and give rise to more differentiated cell types.

Super-enhancers large clusters of transcriptional enhancers that drive the expression of genes that define cell identity and may be highly transcribed.

Supplements extra sources of dietary components taken in addition to food.

Telomerase an enzyme that extends telomere length. Elevated levels are observed in many cancer cells.

Telomeres repeated DNA sequences and associated proteins that are located at the ends of chromosomes. The structures shorten upon each round of cell replication.

Therapeutic index the difference between the minimum effective dose and the maximum tolerated dose of a drug. The larger the value, the safer the drug.

Transcription the process of transferring the information encoded by DNA into RNA; also refers to the process that occurs when a gene is expressed.

Transfection the transfer of exogenous DNA into cells by experimental procedures such as microinjection or electroporation.

Transformation the changes that occur as a normal cell converts into a cancer cell.

Transgenic mice mice that carry foreign DNA experimentally introduced in every cell of their bodies.

Transitions DNA mutations whereby a purine (A or G) is exchanged for another purine (G or A) or a pyrimidine (C or T) is exchanged for another pyrimidine (T or C).

Translation the process of transferring the information encoded by RNA into protein using the genetic code.

Translocations DNA mutations whereby the part of one chromosome is transferred to, or exchanged for, another part of a different chromosome.

Transversions mutations whereby a purine is exchanged for a pyrimidine, or vice versa.

Tumors abnormal growths of cells that can be either benign or malignant.

Tumor suppressor genes genes whose products perform functions that inhibit tumor formation, and therefore loss or mutation of (usually both copies of) these genes leads to tumor formation; also, genes in which a germline mutation predisposes individuals to cancer.

Ubiquitin a small polypeptide (76 amino acids) that is covalently attached to proteins at lysine residues as a marker for protein degradation (proteolysis) by proteasomes.

Upstream refers to DNA sequences that are nearer 5′ as a point of reference. Note that, by convention, a DNA sequence is read from the 5′ end to the 3′ end.

Warburg effect an observation originally made by Otto Warburg that tumor cells utilize glycolysis for glucose metabolism, even in the presence of oxygen (aerobic glycolysis).

Wavelength a characteristic of a wave. It is the distance in meters between successive points of equal phase in a wave, for example, the distance between successive peaks.

Xenobiotics substances foreign to living systems.

Xenografts the transfer of tissue from one species to another. A common xenograft model used in cancer research is the transfer of human tumor cells into immunodeficient mice.

■ INDEX